# Digital Signal Processing and Telecommunications

# Digital Signal Processing and Telecommunications

Edited by **Edmond Thor**

**C**WILLFORD PRESS

New York

Published by Willford Press,
118-35 Queens Blvd., Suite 400,
Forest Hills, NY 11375, USA
www.willfordpress.com

**Digital Signal Processing and Telecommunications**
Edited by Edmond Thor

International Standard Book Number: 978-1-68285-048-0 (Hardback)

Printed in the United States of America.

# Contents

# Preface

I am honored to present to you this unique book which encompasses the most up-to-date data in the field. I was extremely pleased to get this opportunity of editing the work of experts from across the globe. I have also written papers in this field and researched the various aspects revolving around the progress of the discipline. I have tried to unify my knowledge along with that of stalwarts from every corner of the world, to produce a text which not only benefits the readers but also facilitates the growth of the field.

Digital signal processing has applications across all sectors such as image processing, audio and speech signal processing, seismic data processing, etc. It has also revolutionized telecommunication systems. This book is a valuable compilation of topics such as local area networks and wide area networks, control systems, etc. that will help the readers to understand the tools and techniques of digital signal processing. It will benefit students and professionals alike.

Finally, I would like to thank all the contributing authors for their valuable time and contributions. This book would not have been possible without their efforts. I would also like to thank my friends and family for their constant support.

**Editor**

# Modified design of bootlace lens for multiple beam forming

**Ravi Pratap Singh Kushwah\* and P. K. Singhal**

Department of Electronics, Madhav Institute of Technology and Science, Gwalior -474 005, India.

As of now, utility of microwave bootlace lens is well established. For broadband and wide scanning network, it is required to optimize and use features of this lens for better communication and scanning device. This paper presents the design of a compact bootlace lenses for multiple beam forming. Equations to design the lens have been given. The designed lenses has been fabricated and tested. The measured results are in close agreement with the designed values.

**Key words:** Multiple beam forming, antenna design, bootlace lens, wide area scanning, rotman lens.

## INTRODUCTION

Microwave bootlace lens forms an important class of multiple beam forming networks. Ruze (1950) suggested a lens for wide angle scanning. Rotman and Turner (1963) and Leonakis (1986) suggested modification in Ruze's lens to improve the scanning capabilities. Four approaches have been reported for the design of bootlace lens (2, 4, 5 and 6). Rotman and Turner (1963) described the design of bootlace lens for parallel plate configuration, however, the same design concept for the design of bootlace lens in microstrip configuration can be used by dividing the lens region by square root of the of the relative dielectric constant of the substrate of the microstrip. In the design approach proposed by Rotman and Turner, the off axis focal points $F_1$ and $F_2$ were located on angle $\alpha$ and $-\alpha$ respectively. Katagi et al. (1984) suggested an improved method to design boot- lace lens, the suggested approach reduces the phase error for large array length. Gagnon (1989) modified the design approach proposed in (Rotman and Turner, 1963) by locating the off axis focal points at angles $\beta$ and $-\beta$, where $\beta$ is determined according to the Snell's law that is, $Sin\beta = \sqrt{\varepsilon_r} Sin \alpha$ where $\varepsilon_r$ is the relative dielectric constant. In the approach proposed by Singhal et al. (2003), the off axis focal points were located at angles $\beta$ and $-\beta$, where the value of $\beta$ is calculated to equalize the height of feed and array contours. In the present work bootlace lenses have been designed by all the four reported design

approaches and analysed by contour integral approach.

All the reported design approaches are based on the phase shift comparison. Lot of work has been reported on the shape and phase error of the bootlace lens (Singhal et al., 2003; Shrama et al., 1992; Peterson and Rausch, 1999; Hansen, 1992; Sbarra et al., 2007). The reported work describes the shape and phase error of the bootlace lens in term of on axis and off axis focal length along with the other design parameters.

In the present work, effects of radius and center of focal arc on the shape and phase error of the lens have been investigated. Bootlace lenses have been designed at UHF band with different number of input and output ports for an angular coverage of $\pm 35°$.

## LENS DESIGN

Figure 1 shows the cross section of a trifocal bootlace lens. One focal point $F_0$ is located on the central axis and two others $F_1$ and $F_2$ are symmetrically located on the either side of a circular focal arc of which (0,-K) is the center and R is the radius. Outer contour $I_2$ is a straight line and defines the position of the radiating elements. $I_1$ is the inner contour of the lens (also called the array contour). The inner and outer contours are connected by TEM mode transmission lines W (N). Two off axis focal points $F_1$ and $F_2$ are located on the focal arc at angles $+\beta$ and $-\beta$. Angle $\beta$ can be suitably selected to equalize the height of feed and array contour, it is required that the lens be designed in such a way that the out going beams make angles $-\alpha$, 0 and $+\alpha$ with the x - axis when feeds is placed at $F_1$, $F_0$ and $F_2$ respectively. A ray originating from $F_1$ may reach the wave- front through a general point P(X, Y) on the inner contour $I_1$, transmission line W(N) and point Q(N) on the outer contour and then trace a straight line at an angle $-\alpha$ and terminate perpendicular

\*Corresponding author. E-mail: kushwah.ravipratapsingh@ rediffmail.com.

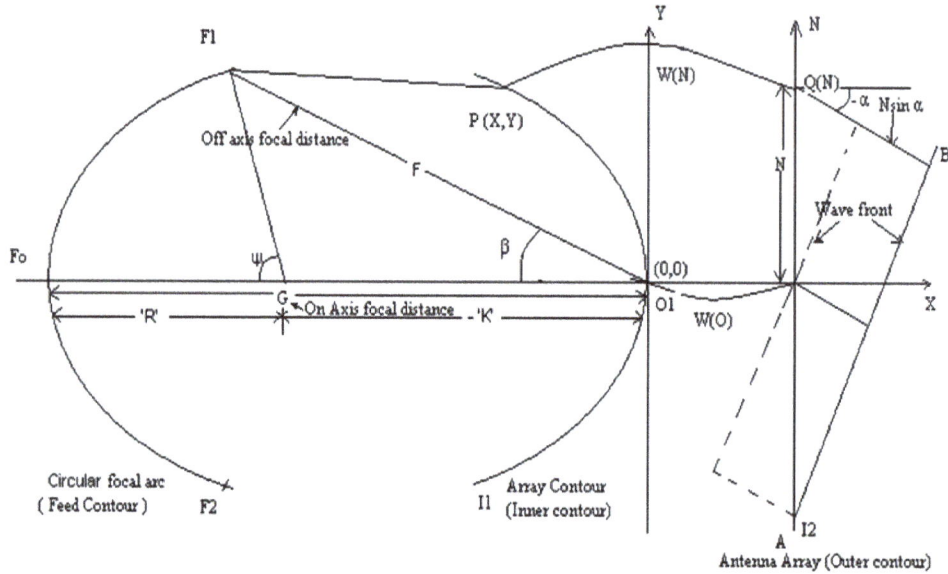

**Figure 1.** Cross-section of the bootlace lens geometry.

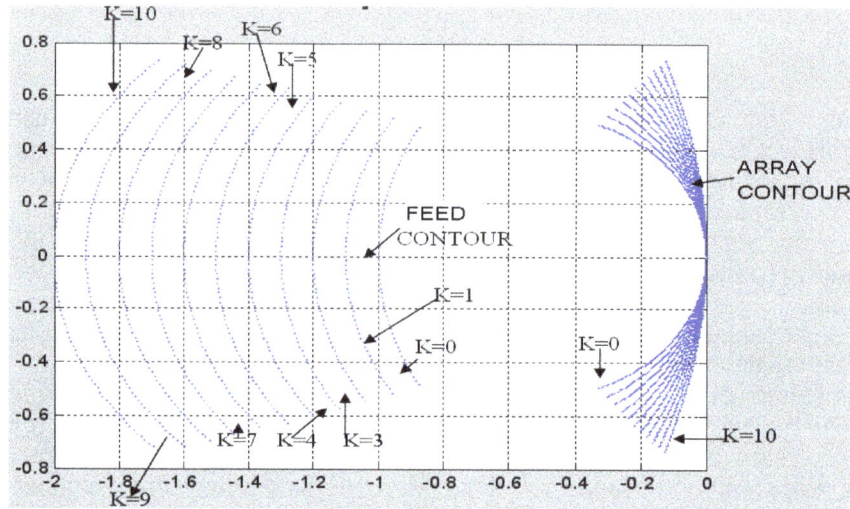

**Figure 2**. Effect of the center the feed contour on the shape of the lens.

to the wave-front. Also, the ray from $F_1$ may reach the wave-front from and $F_1$ to pint $O_1$ and then through transmission line $W(0)$ to the wave-front. Similarly, rays from other feed points may reach their respective wave-front.

Inner contour and transmission lines are designed from the design equations which are derived using the fact that at the wave-front, all of these rays must be in phase independent of the path they travel. This requires that the total phase shift in traversing the path to reach the wave-front in each case be equal. Using this concept, the following design equations can be written

$$\sqrt{\varepsilon_r}\,(F_1P) + \sqrt{\varepsilon_{re}}\,W(N) + N\mathrm{Sin}\,\alpha = \sqrt{\varepsilon_r}\,F + \sqrt{\varepsilon_{re}}\,W(O) \qquad (1)$$
$$\sqrt{\varepsilon_r}\,(F_2P) + \sqrt{\varepsilon_{re}}\,W(N) - N\mathrm{Sin}\,\alpha = \sqrt{\varepsilon_r}\,F + \sqrt{\varepsilon_{re}}\,W(O) \qquad (2)$$
$$\sqrt{\varepsilon_r}\,(F_0P) + \sqrt{\varepsilon_{re}}\,W(N) = \sqrt{\varepsilon_r}\,G + \sqrt{\varepsilon_{re}}\,W(O) \qquad (3)$$

Where

$$(F_1P)^2 = (X + F\,\mathrm{Cos}\,\beta)^2 + (Y - F\,\mathrm{Sin}\,\beta)^2 \qquad (4)$$

$$(F_2P)^2 = (X + F\,\mathrm{Cos}\,\beta)^2 + (Y + F\,\mathrm{Sin}\,\beta)^2 \qquad (5)$$
$$(F_0P)^2 = (X + G)^2 + (Y)^2 \qquad (6)$$

N - Indicate the position of the radiating elements called the lens aperture;
$\varepsilon_r$ - Substrate dielectric constant;
$\varepsilon_{re}$ - Effective dielectric constant of the transmission line.
The other parameters involved in the design equations are shown in Figure 1.

Using the above design equations and the approach suggested in (Singhal et al., 2003), the bootlace lens can be designed.

**Effects of the radius and the center of the feed contour on the shape of the lens**

Figure 2 shows the shape of the feed and array contours for different position at the center (0, -K) of the feed contour, keeping

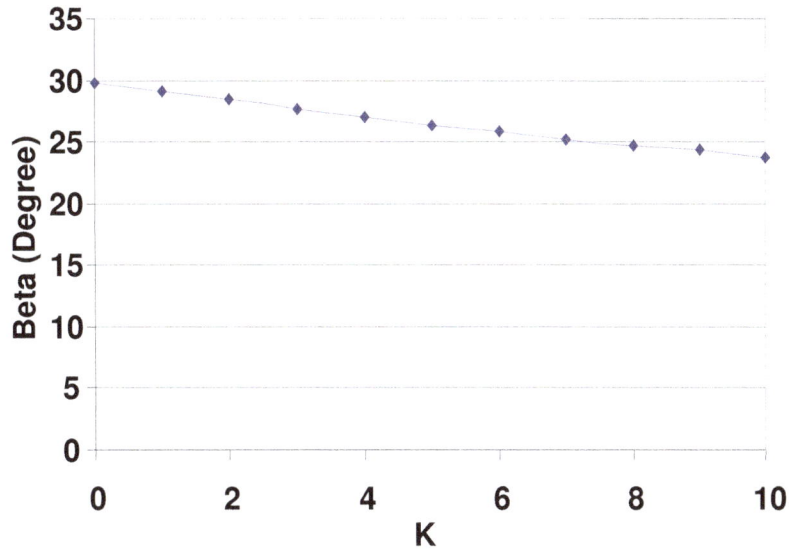

**Figure 3.** Variation of β with Center K.

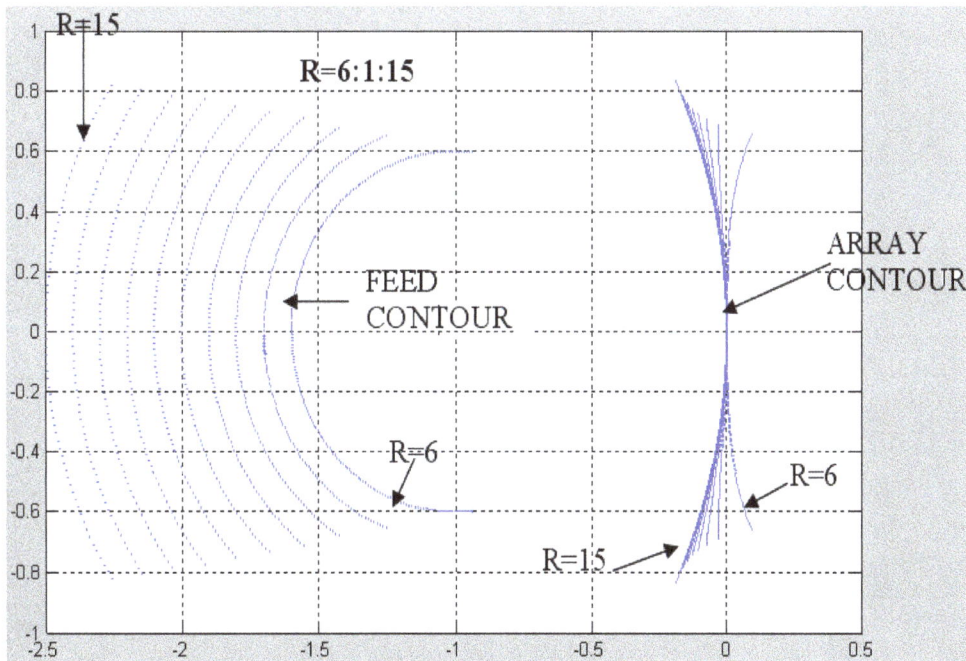

**Figure 4.** Effect of the radius of the feed contour on the shape of the lens.

the other parameters constant As the feed contour moves away from array contour (value of K increases) array contour opens.

Angle β is determined in such a way that the height of feed and array contours remain equal. Figure 3 shows the variation of β with different values of K, as K increases β decreases.

Figure 4 shows the variation of the shape of the feed and array contours with the radius R of the feed contours. As radius decreases, curvature of feed contour increases and array contour opens. There is limit of the radius when the shape of the array contour changes from concave to convex.

As shown in Figure 5 at R = 6.65, the array contour is almost a straight line, with further decrease in R, the array contour will

become convex. Figure 6 shows the variation of angle β with the radius, as the radius increases, β decreases.

**Effects of the radius and the center of the feed contour on the phase error of the lens**

Figure 7(a) - (d) shows the variation of phase along the lens aperture for different values of K (center for the focal arc) with R (radius of the focal arc) as the parameters. It can be observed that for certain values of R and K the phase error is minimum. Therefore, it is required to select the value of R and K carefully; the

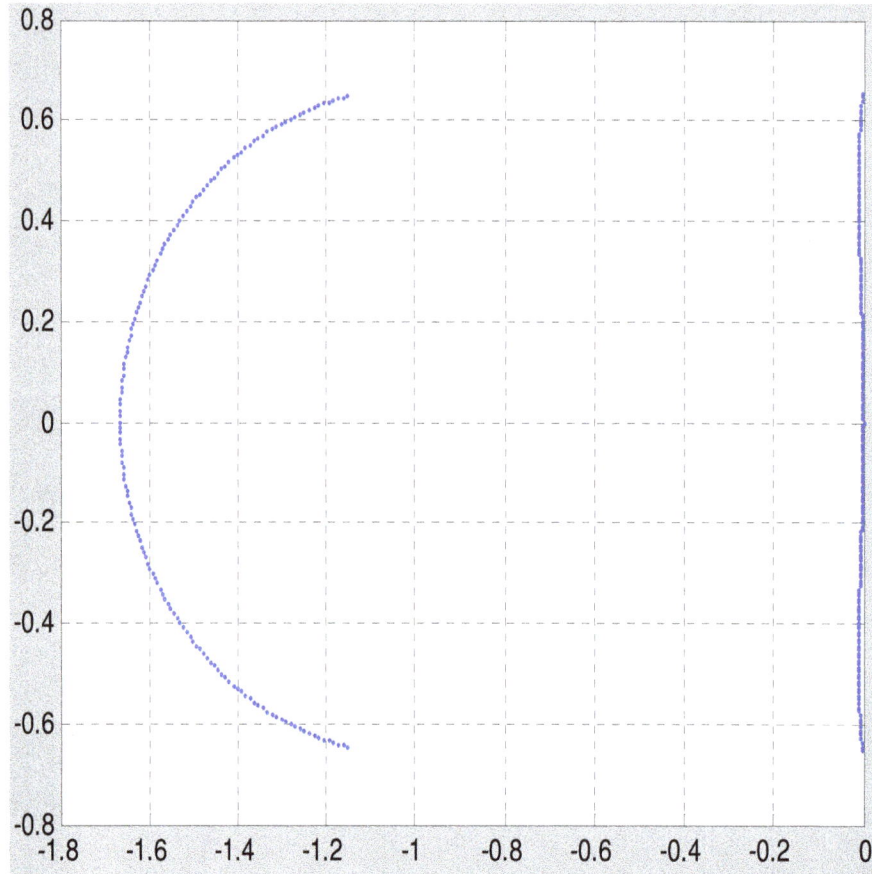

**Figure 5**. Feed and array contour at R = 6.65.

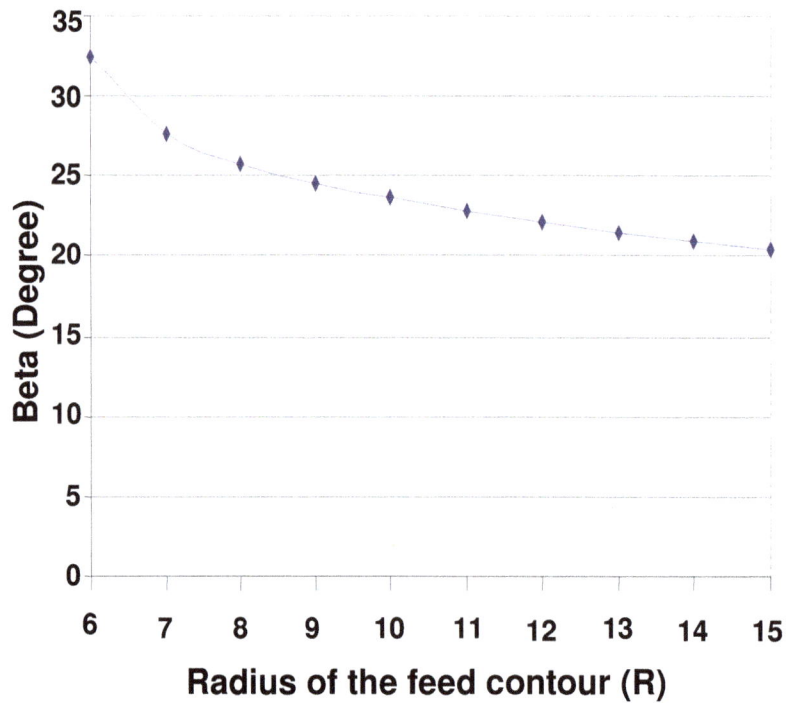

**Figure 6.** Variation of β with the radius of the feed contour.

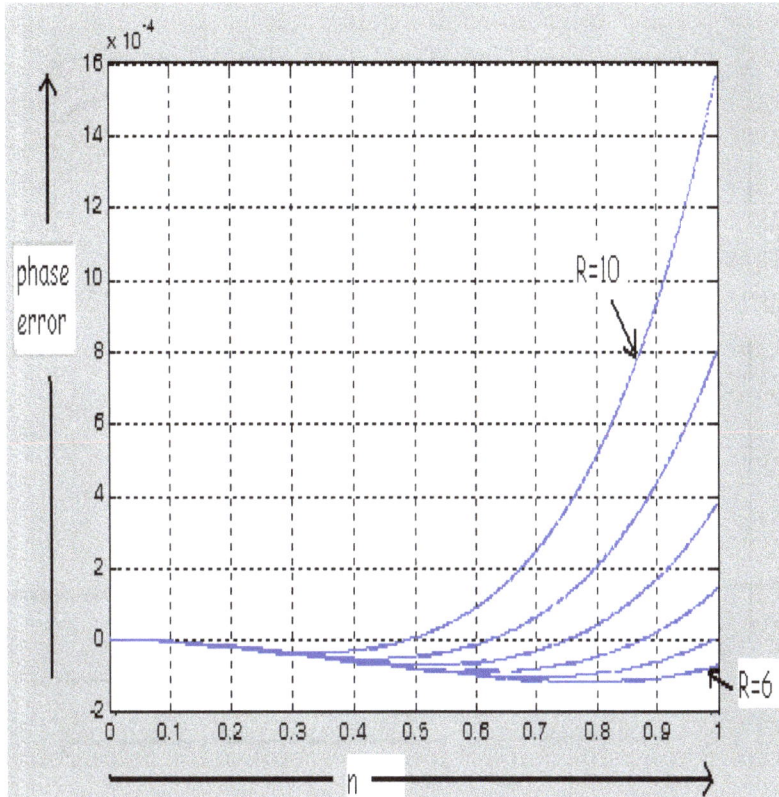

**Figure7(a).** Variation of phase error along the lens aperture for K = 2.

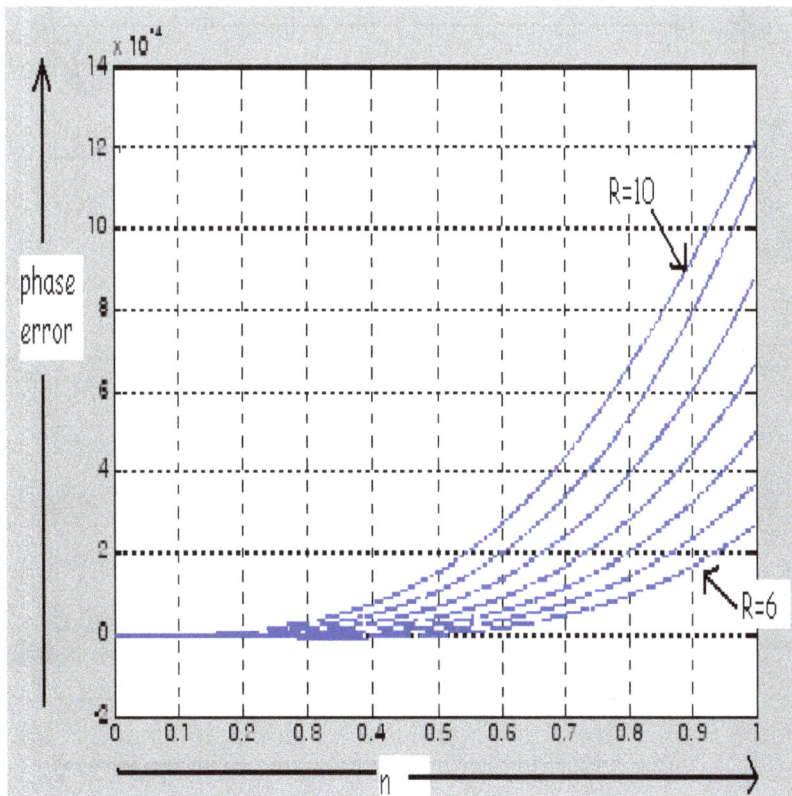

**Figure7(b).** Variation of phase error along the lens aperture for K = 4.

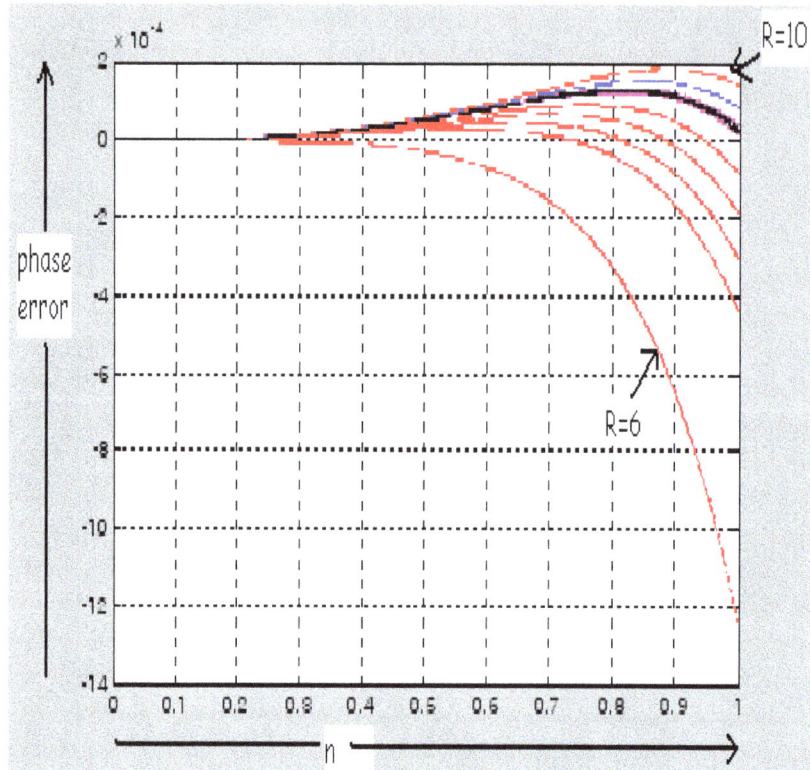

**Figure7(c).** Variation of phase error along the lens aperture for K = 6.

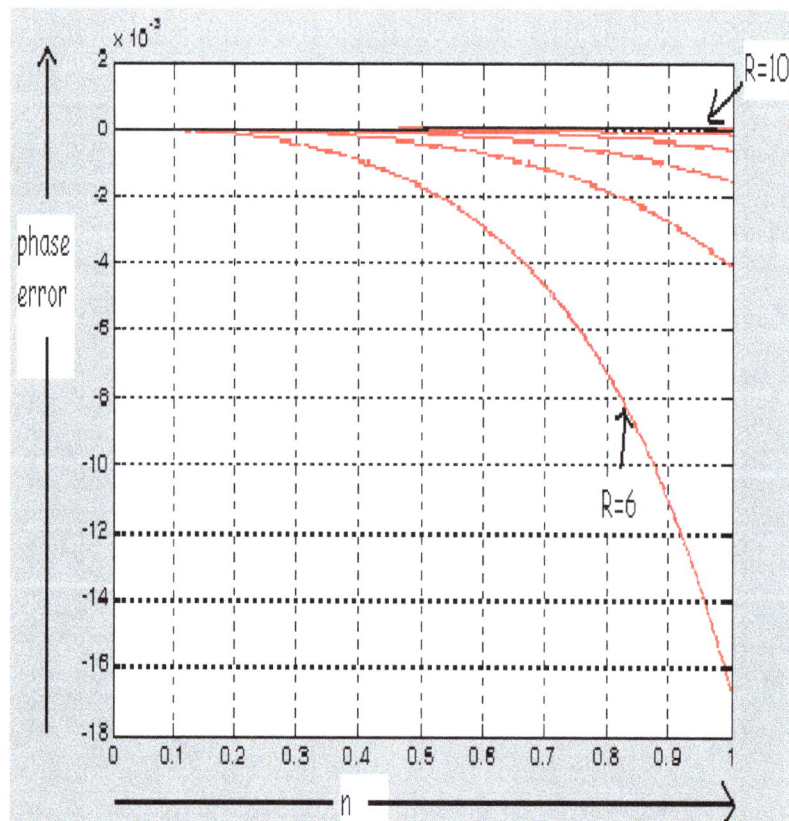

**Figure 7(d).** Variation of phase error along the lens aperture for K = 10.

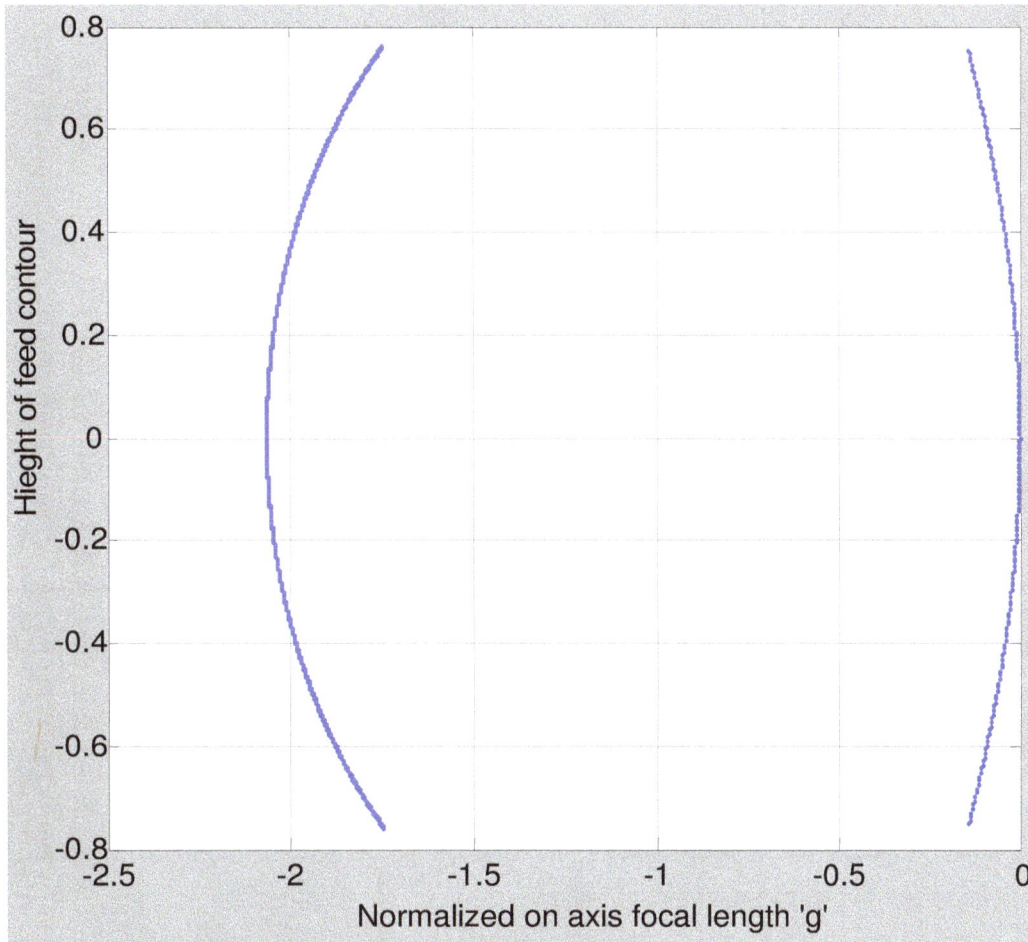

**Figure 8**. Designed feed and array contours for Bootlace Lens 1.

shape of the lens also depends upon R and K.

**Design examples**

Two bootlace lenses have been designed for operation at 0.8 GHz. Lens 1: The first lens was designed for the following requirements.
Angular coverage = ± 35°.
Number of antenna elements = 10.
Number of input beams = 09.
Central Frequency = 0.8 GHz.
Spacing between antenna elements = 3.0 cm.

The complete structure was fabricated in microstrip version on substrate of thickness 1.6 mm and dielectric constant 4.7 and loss tangent is 0.02. Figure 8 shows the designed feed and array contours. Designed parameters have been suitable optimized to equalize the height of feed and array contours. Figure 9 shows the variation of the phase error along the lens aperture for the designed lens. Figure 10 shows the top view of the geometry of the designed bootlace in microstrip configuration. It consists of 39 ports. Port No. 1 to 9 are input ports. Port No. 20 to 29 are the output ports. Port No. 10 to 19 and 30 to 39 are the dummy ports. Function of dummy ports is to cover the gap between feed contour and array contour. These dummy ports are terminated in 50 ohms dummy load. The input ports are connected to source and the output ports are

connected to radiating elements. Lens 2: The second lens was designed for the following requirements:

Angular coverage = ± 35°.
Number of antenna elements = 07.
Number of input beams = 07.
Central Frequency = 0.8 GHz.
Spacing between antenna elements = 4.0 cm. Second lens was fabricated on the same material as lens 1. Figure 11 shows the designed feed and array contours. Figure 12 shows the variation of the phase error along the lens aperture for the designed lens.

Figure 13 shows the top view of the geometry of the designed bootlace lens in microstrip configuration. It consists of total 22 ports; seven input ports, seven output ports and eight dummy ports.

**RESULTS AND DISCUSSION**

Table 1 shows the direction of the outgoing beam for input at different feed ports for bootlace lens 1. The bootlace lens was designed for an angular coverage of ±35° and it is covering an angular area from +36 to -33° and the outgoing beams are also uniformly spaced. The outcome of the designed lens is matching with the design

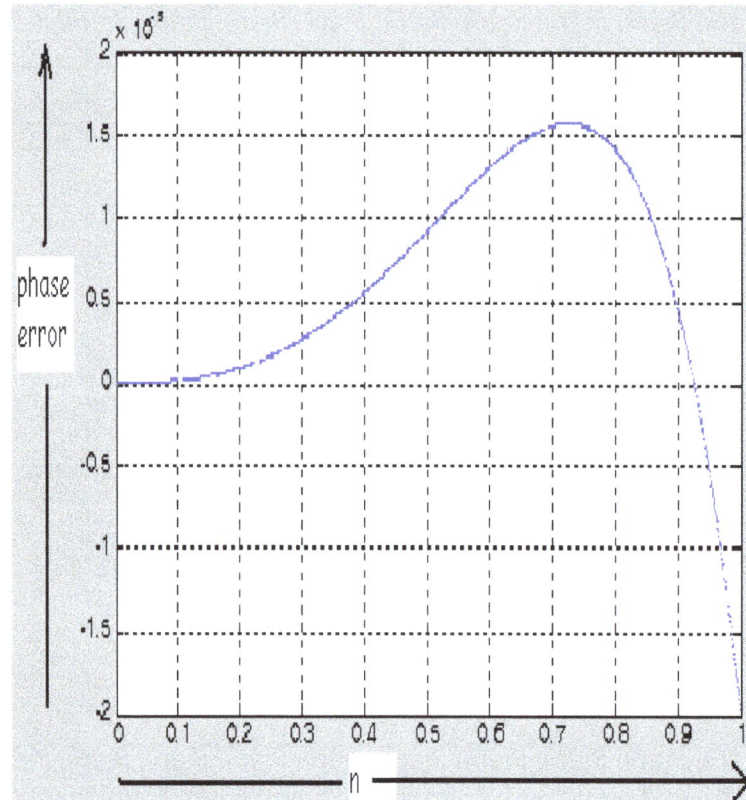

**Figure 9.** Normalized phase error for lens 1.

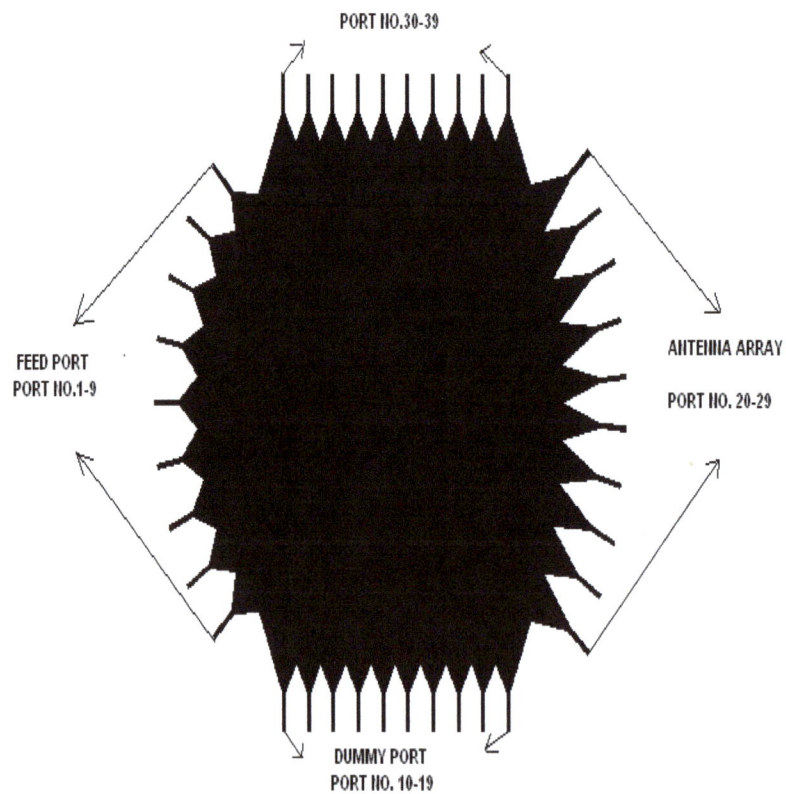

**Figure 10.** To view the designed bootlace lens 1 in microstrip configuration.

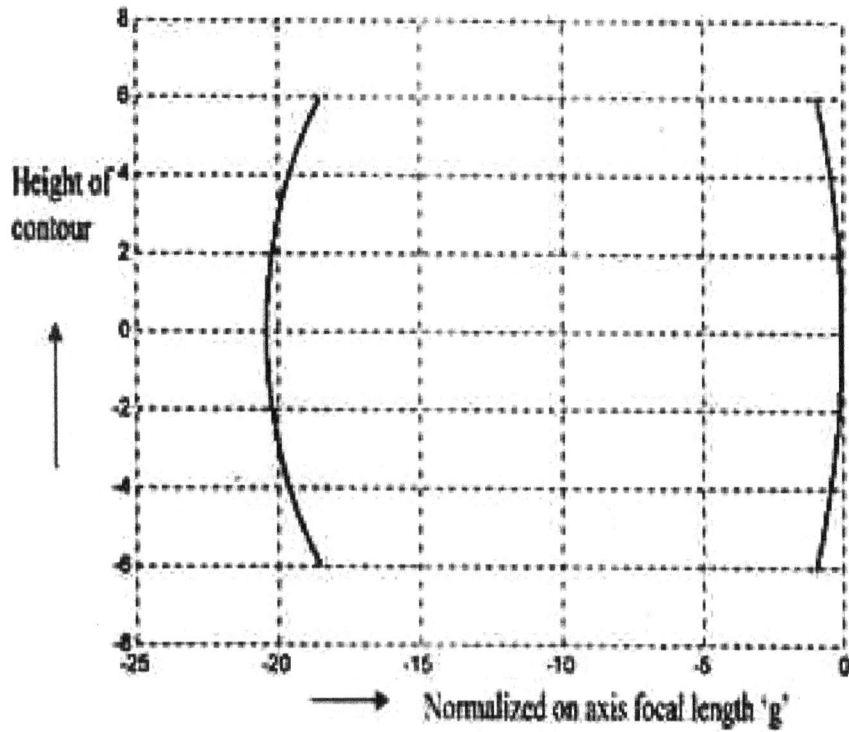

**Figure 11.** Designed feed and array contour for bootlace Lens 2.

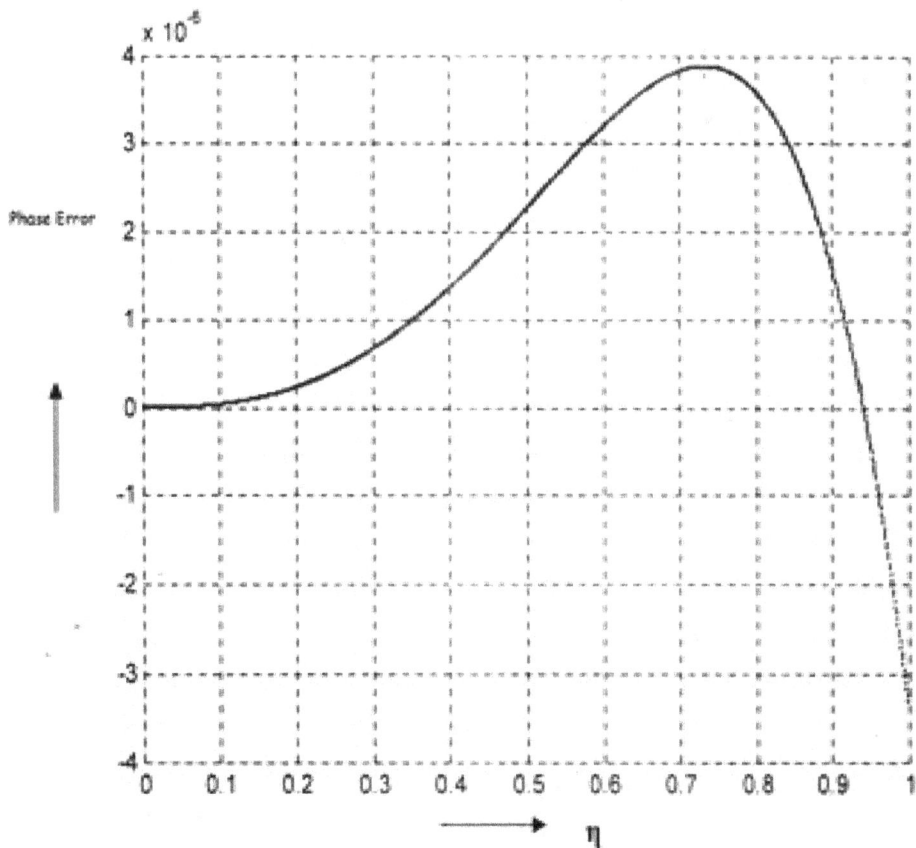

**Figure 12.** Normalized phase error for bootlace lens 2.

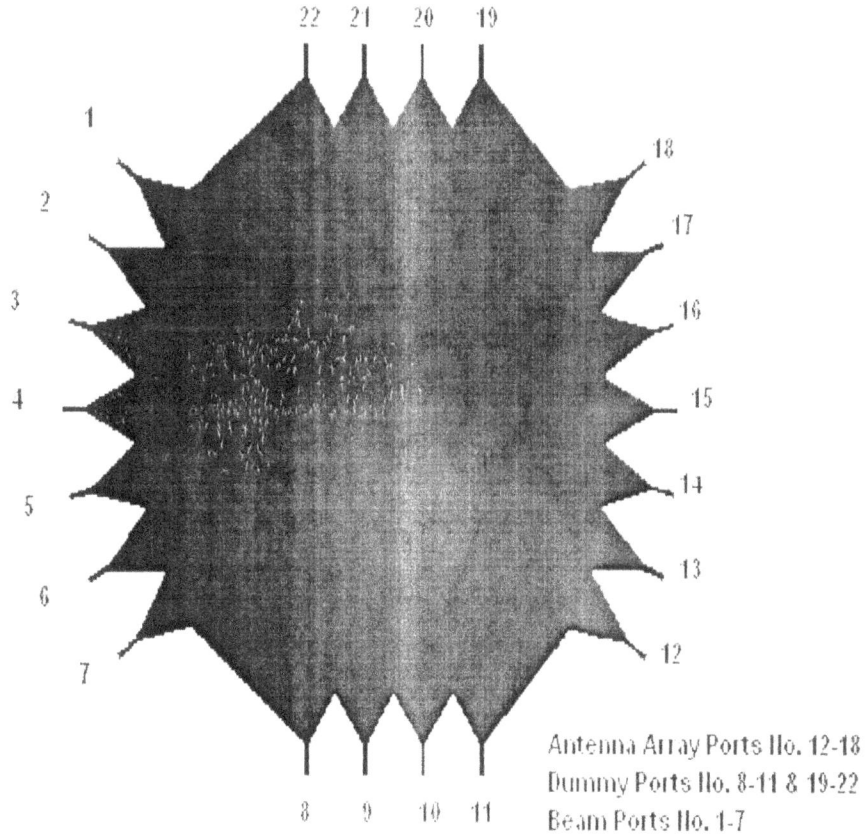

**Figure 13.** To view of the designed bootlace lens 2 in microstrip configuration.

**Table 1.** Direction of outgoing beam for input at different feed ports for lens 1.

| Input port no. | Outgoing beam angle in degrees |
|---|---|
| 1 | 36 |
| 2 | 27 |
| 3 | 17 |
| 4 | 10 |
| 5 | 0 |
| 6 | -9 |
| 7 | -18 |
| 8 | -28 |
| 9 | -33 |

values. Table 2 shows the direction of outgoing beam for input at different feed ports for lens 2. This is covering and angular are from + 38 to -36°.

**Concluding remarks**

Effects of the center position and radius of the feed contour on the shape and phase error of the bootlace lens have been investigated; these investigations are useful in selecting the design parameters. Two bootlace lenses have been designed for different requirements. The measured results are in close agreements with the design values.

**ACKNOWLEDGEMENTS**

Authors would like to acknowledg e  the financial support

**Table 2.** Direction of outgoing beam for input at different feed ports for lens 2.

| Input port no. | Outgoing beam angle in degrees |
|:---:|:---:|
| 1 | 38 |
| 2 | 25 |
| 3 | 11 |
| 4 | 0 |
| 5 | -11 |
| 6 | -25 |
| 7 | -36 |

of Department of Science and Technology, Government of India and the authorities of Madhav Institute of Technology and Science, Gwalior, India.

## REFERENCES

Gagnon DR (1989). Procedure for correct refocussing of the Rotman.

Hansen RC (1992). Design trades for Rotman Lens. IEEE Trans. On Antenna and Propagation 39(4): 464-472.

Katagi Mano T, Shin-Ichi Sato S (1984). "An improved design method of Rotman lens antennas", IEEE Trans. On Antenna and Porpagation, AP-32: 524-527.

Larry Leonakis G (1986). Correction to wide angle microwave lens for line source applications, IEEE Trans Antennas Propagat 36: 1067.

Lens according to Snell's Law, IEEE Trans. On Antenna and Propagation 37(3): 390-392.

Rotman W, Turner RF (1963). Wide angle microwave lens for line source applications", IEEE Trans. on Antennas Propagat AP-11: 623-632.

Ruze J (1950). Wide angle metal plate optics., Proc. IRE 38: 53-59.

Sbarra E, Marcaccioli L, Gatti RV, Sorrentino R (2007). A novel Rotman lens in SIW technology, European Microwave Conference pp. 1515-1518.

Sharma PC, Gupta KC, Tsai CM, Brice JD, Presnell R (1992). Two dimensional Field analysis for CAD of Rotman type beam forming lenses. Int. J. Microwave Millimeter-Wave Computer Aided Engg. 2(2): 82-89.

Peterson AF, Rausch EO (1999). Scattering matrix integral equation analysis of the design of a waveguide Rotman lens. IEEE Trans. On Antenna and Propagation 47(5): 870-878.

Singhal PK, Sharma PC, Gupta RD (2003). "Rotman lens with equal height of array and feed contours", IEEE Trans. On Antenna and Propagation 51(8): 2048-2056.

# 2

# A new scheme of image watermarking based on fuzzy clustering theory

**Sameh Oueslati[1,2]\*, Adnane Cherif[1] and Bassel Solaiman[2]**

[1]Department of Physics, Laboratory of Signal Processing, Faculty of Sciences, University Tunis, El Manar 1060, Tunis.
[2]Department: Image and Information Processing, Higher National School of Telecommunication of Bretagne, Technopole of Brest Iroise, 29285 Brest – France.

Digital watermarking technology has the advantages of easy implementation and capability of providing wide security services such as copyright protection, authentication and secret communication. In this paper, a novel image watermarking approach based on embedding watermarks in different domains, without any distortion of the watermarked image is used. In the spatial domain, the processing method is based on study of segmentation by fuzzy c-means clustering method (FCM) that outputs the zones of watermark embedding and respectively the associated appropriate embedding gain factors. However, in the DCT, we embedded the watermark into the coefficients in mid-band frequency of the selected blocks with different embedding strength. Experimental results show that the proposed scheme has good imperceptibility and high robustness to common image processing operators.

**Key words:** Image watermarking, insertion force, Image segmentation, fuzzy C-mean (FCM), statistical features.

## INTRODUCTION

With the rapid development of Internet and multimedia technology, more and more digital media including images, texts, and video are transmitted over the Internet (Barni et al., 2009). However, as far as we know, transmitting information on computer networks is not safe and the valuable data is easy to be copied, thus, providing copyright protection for digital media is becoming increasingly important (Davoine et al., 2004). A solution to alleviate this problem is to insert watermarks into the media particularly in images, so it can be detected and used as evidence of copyright.

There are three important requirements that are mostly needed for a well-designed watermarking scheme described as follows (Patrick et al., 2007; Anne, 2001; Azza, 2009).

1) Imperceptibility: The host image or original image should not be visibly degraded by the watermark. In other words, we must ensure that an unauthorized user do not perceive the existence of the watermark. Imperceptibility ensures the excellent perceptual quality of the protected image.
2) Robustness: The hidden watermark must survive image processing or operations such as clipping, filtering, and enhancement. The watermark should be retrieved after been compressed by lossy compression techniques such as JPEG. It also must be against the malicious attack that denotes the manipulation of destroying or removing the watermark.
3) The third important factor that allows a watermarking method to be commercially interesting is the watermarking capacity that permits a bigger amount of data embedding in the image.

In this paper, we propose to exploit the robustness of respectively the spatial and frequency domain in the same time (Shih et al., 2003). A set of watermarks is embedded in the DCT frequency domain (Barni et al., 1998; Cox et al., 1997) in different selected blocks coefficients with respect to the JPEG quantization values table. The choice of these coefficients is based on a strategy to minimize the vulnerability of the embedding scheme by the redundancy of the different embedded

*Corresponding author. E-mail: sameh.oueslati@telecom-bretagne.eu.

watermarks (Patrick et al., 2002). In the same time, a second set of watermarks is embedded in the spatial domain.

A study of segmentation by fuzzy c-means method is carried out in order to determine the best embedding locations and the highest usable gain factor with respect to the image zones characteristics. This embedding approach proved that the watermarked image become more robust mutually to the JPEG compression (Wallace, 1992) and wide kinds of synchronous and asynchronous attacks. In addition, because of the recurrence resulting from the multiple embedded watermarks in these two domains, at least all or some of these inserted watermarks survived in each of the applied attacks.

## EMBEDDING PROCESS IN THE FREQUENCY DOMAIN

The first step of this approach is to insert several identical marks in the field of DCT. The watermarks are presented as different binary images, containing data about the author's data about the patient as: The name, medical diagnose...etc. With $P \times P$ size described as the following:

$$M_L = \{M_L(i,j), 0 \le i, j \le P\} \, M \in \{0,1\} \, L \in .\{1,2......,L_{max}\} \quad (1)$$

Among the work of watermarking, the algorithm proposed by Zhao (1995) is coded on a pair of frequency values (0, 1). The use of DCT frequency domain can fulfill not only the invisible through the study of optimizing the insertion gain used, but also security by providing a blind algorithm which use the original image which is not essential and the extraction of the mark made by a secret key (Patrick, 2005; Wolfgang, 1997). In order to invisibly embed the watermark that can survive lossy data compressions, a reasonable

$$C_1(i_1, \; j_1) - C_2(i_2, j_2) \ge \alpha \quad (2)$$

$C_1, C_2$ are the DCT coefficients, $(i_1, \; j_1)$, $(i_2, j_2)$ are respectively the positions of the two selected coefficients with same quantization values and $\alpha$ is the gain factor resultant from this equation.

The redundancy introduced by the insertion of multiple watermarks encoded in the described coefficients, proved through experimental results, that it introduces a higher robustness after different attacks. An explanation of the robustness increase is that some or at least one of the embedded watermarks survives in each time the attack is applied. By applying an inverse DCT transform, we obtain a spatial representation of a watermarked image $I_{DCT^{-1}}$. Trade-off is to embed the watermark into the middle-frequency range of the image. In this paper, the DCT coefficients where the watermark bits will be encoded are chosen from the medium frequency band in order to provide additional resistance to lossy compression while avoiding significant modifications or distortions to the cover image (Bruyndonckx et al., 1995). Instead of choosing arbitrarily the coefficients locations, we can increase the robustness to compression by basing our choice on the recommended JPEG table. In fact, if two locations are chosen as they present identical quantization values, any scaling of the first coefficient will scale the second by the same factor preserving their relative size (Davoine et al., 2004). Furthermore, to augment the survival chances of the embedded watermarks against a large set of attacks and to reduce the probability of detection errors, an additional gain factor noted $\alpha$ is used in the watermark embedding process. Some criteria are presented for the choice of $\alpha$ as shown in the Equation (2), in order to respect the imperceptibility threshold shown by the image distortions. It is found that the computed gain factor value is approximately equal to that given by this equation (Davoine et al., 2004).

## STATISTICAL SEGMENTATION FOR ZONES DETERMINATION

Spatial insertion procedure is preceded by a step of determining zones of insertion. Indeed, a heterogeneous image is composed by different zones (homogeneous textures, low intensity...) (Khaled et al., 2009). This diversity implies that the insertion in these different zones may not be identical. Hence the classification stage of the image used in different zones according to the characteristics of each is indispensable (Lotfi et al., 2009). To enhance the robustness against various image distortions that can be subjected, we use a force insertion α of watermark. This force must be below a predefined threshold of visual perceptibility which depends on the characteristics of the zones of insertion. This force is not always uniform on all components of the inserted watermark, but depends on the characteristics of the zones of insertion (textured, uniform ...), because the eye is less sensitive to noise in regions of the image where the brightness is very high or very low and the human eye is less sensitive to regions of the image with strong texture, more specifically, to zones near the edges. These aspects have been implemented in this article; so the key point to embed a watermark is to determine where the watermark can be embedded and how much strength can be added.

### Fuzzy C-means clustering based on the image statistical features

The most widely used clustering method is probably the fuzzy C-means (Ruspini, 1969; Dunn, 1974; Bezdek, 1981; Mukherjee et al., 1996), called FCM algorithm, which is a "fuzzy relative" to the simple C-means technique (Haralick et al., 1973). FCM is an unsupervised clustering technique which has been utilized in a wide variety of image processing applications such as medical imaging (Hsiang et al., 1999; Bezdek et al., 1993) and remote sensing (Rignot et al., 1992; Chumsamrong et al., 2000; Yang et al., 2005). Its advantages include a straightforward implementation, fairly robust behavior, applicability to multichannel data, and the ability to model uncertainty within the data. A major disadvantage of its use in imaging applications, however, is that FCM does not incorporate information about spatial context, causing it to be sensitive to noise and other imaging artifacts. In fact, an image can be represented in terms of pixels, which are associated with a location and a gray level value. It can also be represented by its derivatives, e.g., regions with statistical features like average grayscale value (Ag), standard deviation (Sd), variance (Va), entropy(E), skewness (Sk), kurtosis(Ku) given in Table 1.

Therefore, a proposed segmentation approach combining pixel characterization by a set of statistical features and fuzzy clustering approach, FCM, is discussed. The proposed approach can be divided into two principal steps. The first consists to characterize each image pixel by a feature vector. Features can be extracted from regions masked by $(n \times n)$ window. The second step is a clustering procedure of the feature vector, initially extracted, using FCM clustering algorithm. By applying FCM, a partition of the feature vectors into new regions can be found. As depicted in Figure 1, the system scans the image using a sliding window and extracts a feature vector for each $(n \times n)$ block. The c-means algorithm is used to cluster the feature vectors into several classes with every class corresponding to one region in the segmented image (Bezdek et al., 1981). An alternative to the block-wise

**Table 1.** A set of statistical features.

$$A_g = \frac{1}{MN} \sum_{i=1}^{M} \sum_{j=1}^{N} g(i,j) \qquad (3)$$

$$S_d = \frac{1}{MN} \sum_{i=1}^{M} \sum_{j=1}^{N} (g(i,j) - M_e)^{1/2} \qquad (4)$$

$$V_a = \frac{1}{MN} \sum_{i=1}^{M} \sum_{j=1}^{N} (g(i,j) - M_e)^2 \qquad (5)$$

$$E = -\sum_{i=1}^{M} \sum_{j=1}^{N} g(i,j) \log(g(i,j)) \qquad (6)$$

$$S_K = \frac{1}{MN} \sum_{i=1}^{M} \sum_{j=1}^{N} (g(i,j) - M_e)^3 \qquad (7)$$

$$K_u = \frac{1}{MN} \sum_{i=1}^{M} \sum_{j=1}^{N} (g(i,j) - M_e)^4 \qquad (8)$$

Note $g(i,j)$ is the grey level of pixel $(i,j)$.

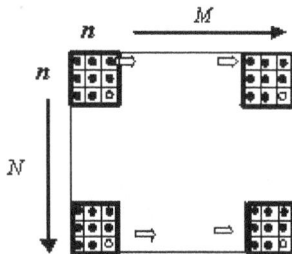

**Figure 1.** Scan an image ($M \times M$) by a window ($n \times n$) with n

segmentation is a pixel wise segmentation by forming a window centered around every pixel. A feature vector for a pixel is then extracted from the windowed block. The spatial scanning order of an image ($M \times M$) is performed, as shown in Figure 1, from left to right and top to bottom, pixel by pixel.

## Spatial embedding procedure

Before embedding in the spatial domain, a study of the image characteristics is preliminary carried out in order to find out different images zones, where each one corresponds to specific image characteristics. In each of these zones, an automatic computation of the gain factor is used in the embedding equation in order to maintain watermark under the imperceptibility limits, this study fuzzy c-means clustering that takes into account the image statistics parameters. Using these parameters, we can identify the zones limits corresponding to the different images characteristics and computes the matching gains. This proposed method is found to be a fine tool of image segmentation with the characteristic been flexible and having the possibility to detect different programmed

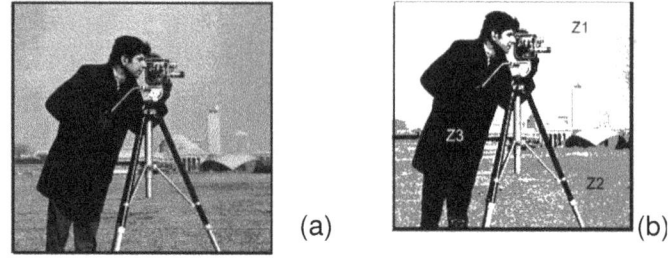

**Figure 2.** (a) Original image "cameraman", (b) Three classified zones.

zones with acceptable accuracy. Using this method, the proposed technique does not allow a wrong classification output. A 2×2 pixel block is used to browse the DCT watermarked image to identify and mark the different existing zones. All the image matrix lines are consecutively processed and browsed by this block with an overlap of one line and one column pixel. The original is automatically classified and marked with different colors as shown in Figure 2. Three different zones are sorted with regards to their respective specificities where Z.1 represents the first zone marked in white that corresponds to a homogeneous zone in the host image (sky), Z.2, the second zone marked in black that corresponds to a textured zone (green), and the third zone called Z.3 corresponds to a dark zone in the host image (cameraman). For every zone, the appropriate gain factor called $\alpha$, is automatically computed as detailed in Equations (10) basing on the Weber's law to keep the watermark unperceivable as the following:

$$\frac{\Delta I(i,j)}{|I(i,j)|} = cte \qquad (9)$$

Where $cte$ denotes a constant, and $\Delta I(i,j)$ is the pixel difference between the watermarked and the original image. The Weber law can be written as the following:

$$(I_W(i,j) - I(i,j))/|I(i,j)| = cte \qquad (10)$$

Once the gains computed, this imperceptibility limit is also protected as shown in Equation (11) by the use of a security factor that forbids the possibility to visualize some details of the embedded watermark even though the image size is zoomed many times.

$$\alpha = \alpha_c \times S_F \qquad (11)$$

Where $\alpha$ the used gain factor, $\alpha_c$ is the computed and adjusted gain factor, $S_F$ is the security factor. This factor used equals 0.75.

The general shape of the insertion procedure takes into account that the image was previously marked in the DCT frequency domain by a set of labels introduced as the following equation:

$$I_{MML}(i,j) = I_{M,L,n}(i,j)[1 + \alpha M_{L,n}(i,j)] \qquad (12)$$

Where $L$ denotes the watermark index, $N$ the number of the segmented zones; $n$ varies with the indexed zones and respectively $L \in \{1,2,...,L_{max}\}$ and $n \in \{1,2,....N\}$. In this equation $I_{MM}$ denotes the double watermarked image in the spatial and frequency domain, $I_{MML}$ is the watermarked image by $L$ watermarks in the two domains, $I_{M,L,n}$ is the frequency

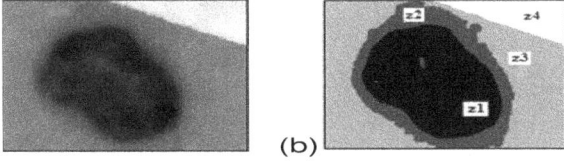

**Figure 3.** (a) Original image: (b) Four classified zones.

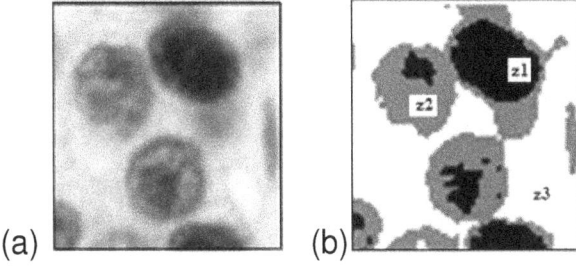

**Figure 4.** (a)Original image; (b) Three classified zones.

watermarked image going to be watermarked in the second time by the watermark number $L$ in the $n^{th}$ spatial classified zone, $M_{L,n}$ is the watermark number $L$ going to be embedded in the $n^{th}$ Zone and $\alpha$ is the variable gain factor determined by the segmentation process. The total embedded watermarks in the spatial and frequency domain are then considered equal to eleven.

Different other images are used in the carried experiments in all the watermarking process, with different classified zones numbers as shown in Figures 3 and 4. These figures show medical images with textures. Figure 3 shows an image of malignant melanoma, a skin cancer seen in France as the leading cause of death among women between 25-29 years 6,000 new cases each year, partitioned into four zones where the watermark can be embedded with several different gain factors. Figure 4 is divided into three different zones. In each image, the position where the watermark is to be embedded changes with the different zone.

## EXPERIMENTAL RESULTS

The watermarking algorithm was tested using a 256 × 256 gray scale images. The peak signal to noise ratio (PSNR) is used to measure the imperceptibility of the watermarked image. By using the proposed scheme, the watermark is almost imperceptibility to the human eyes, as shown in Table 2.

The PSNR is defined as follows:

$$PSNR = 10\log_{10}(\frac{X_{max}^2}{MSE}) = PSNR = 10\log_{10}(\frac{255^2}{MSE}) \quad (13)$$

$X_{max}$ : The maximum luminance.
The MSE is defined as follows:

$$MSE = \frac{\sum_{i=1}^{n}\sum_{j=1}^{m}(I_{ij} - I_{ij}^*)^2}{n \times m} \quad (14)$$

$I$ and $I^*$ are respectively the original image and the image watermarked size $m \times n$ where $I_{ij}$ and $I_{ij}^*$ are their components.

This error is mainly due to the addition of the mark. In the bibliography, quality of the watermarked image is good if the PSNR is equal to or higher than 30 dB. We note from Table 2 that for a given image and following the insertion of the first mark will be worth significantly more than 30 dB PSNR. We note from Table 2 that the greater the number of signatures, PSNR decreases. In fact, the insertion of signatures in the image returns to introduce new information and therefore a degradation of image quality, and every time we increase the number of the mark image, quality decreases.

## Robustness against attacks

Furthermore, several experiments were performed in order to demonstrate the robustness of the algorithm under various attacks, including lossy JPEG compression, Noises attacks, as well as filtering attacks.

## JPEG compression attack

Lossy compression algorithms such as JPEG are commonly used for efficient storage and transmission of images over the Internet. It is therefore crucial to examine whether the proposed watermarking scheme can survive JPEG compression attacks. In order to perform this experiment, the watermarked image shown in Figure 5(c) was compressed using different quality factors. As shown in Table 3, in each experiment, the correlations values are gathered corresponding to the embedded watermarks in the two domains. High correlations are obtained using this method which proves that all or some of the embedded watermarks have survived to the applied attacks. The watermarked image indicates a good perceptual quality and the extracted watermark is similar to original one. As shown in Figure 6, the extracted watermark is still visually acceptable after the watermarked image had undergone several common attacks such as JPEG lossy compression and destructive signal processing; some results are shown in Table 3.

## Fidelity against noise attacks

It is quite relevant to evaluate the robustness of the suggested method against noise. In fact, we have tested our new approach using 10 different noise generations and by modifying variances at each time. From Figure 7, we can observe values of PSNR that are always higher than 30 *dB*. This makes it obvious that the image quality is good and these new watermarked images algorithm is powerful to keep image fidelity even after noise attack.

The watermark detector response when the watermarked image is introduced to additive Gaussian noise with different variance values is shown in Figure 8(a).

**Table 2.** PSNR of watermarking images.

| Number of watermark | Watermark 1 | Watermark 2 | Watermark 3 | Watermark 4 |
|---|---|---|---|---|
| PSNR (dB) | 49.08 | 41.17 | 38.15 | 34.40 |

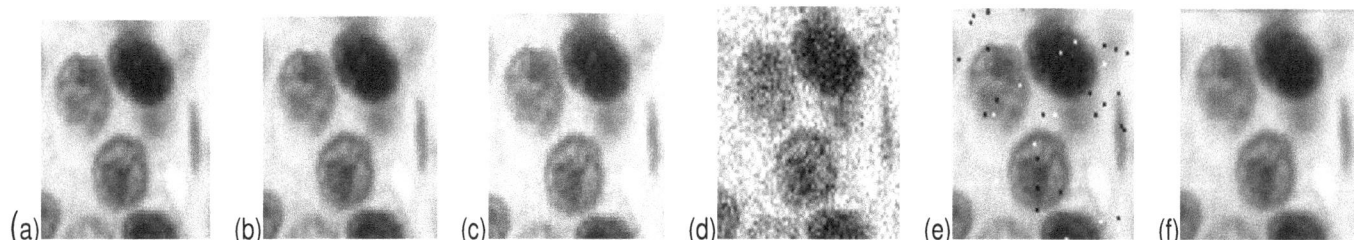

**Figure 5.** The watermarked image introduced to various attacks. (a) Original image (b) Watermarked image, (c) JPEG compression, (d) additive Gaussian noise, (e) Salt and paper noise, (f) Gaussian filter.

**Table 3.** Presentation of the correlation values after attack of JPEG compression 60 between the watermarks obtained in the spatial domain and frequency and the original.

| Number of watermarks | Spatial domain | Frequency domain |
|---|---|---|
| Watermark 1 | 0.9080 | 0.9798 |
| Watermark 2 | 0.9862 | 0.9953 |
| Watermark 3 | 0.8523 | 0.9745 |

**Watermarks detection**

In this section we present experimental results carried out on a database of 800 marks in which the extracted marks are tested. The test technique is made by correlation between the extracted marks and the dictionary. Figures 8 demonstrate that the marks are correctly detected from our simulation results from Watermarked images after Noises attacks, JPEG Compression and Filtering attacks. Figure 8(b) shows that the detection results are satisfactory, even if the watermarked image has deteriorated considerably during the compression of a quality factor as low as 20. We have tested the robustness of our proposed method face to Gaussian filter Figure 8(c) displays the watermark detector response when the watermarked image is attacked by Gaussian filter.

**Rotation attack**

This attack can be seen as an innocent attack or malicious attack. In fact, the analog-digital conversion (scanning) sometimes involves the need for a relationship on the image to better exploit. This same attack can be considered malicious if the attacker makes a slight turn invisible on the image (of the order of

several degrees or radians). This rotation, imperceptible to the naked eye will cause desync total image when looking for the signature and a change in the coefficients related to the image especially in the areas processed (frequency, multi-resolution ...). Regarding the attack by rotating the image, the angles chosen are 2, 5, 10 and 12°. The large angles of rotation have degraded the quality of digital image. However, these values do not pose problems for the extraction of the mark. The marks obtained after this attack showed a similarity with the original image showing that the mark inserted was not damaged by the attack carried.

**Conclusion**

In this paper, a novel image watermarking approach based on a multiple domain watermarking with several watermarks embedding in the spatial and frequency domains. The watermarking process is divided in two separated steps: Embedding in the DCT frequency domains and the spatial domain. In the DCT frequency domain, a strategy to choose the DCT coefficients where the watermarks are embedded in order to minimize the image distortion and obtain the maximum robustness of the different inserted watermarks is carried out. Whereas in the spatial domain, a method fuzzy c-means clustering based on the image statistical features leading to the image classification into different zones. In these zones, an automatic computing of the gain factor used to embed the watermark based on the Weber's law is also considered. The simulation results proved that the proposed technique is robust against different synchronous and asynchronous attacks such as JPEG compression, different filtering and geometrical transformations. In the watermark detection process we proved that between the embedded watermarks, a

(a) Identity:07263015
Manuscript Number:
JEEER-10-030
University: FST

(b) Identity:07263015
Manuscript Number:
JEEER-10 030
University: FST

(c) Identity:07263015
Manuscript Number:
JEEER-10-030
University: FST

(d) Identity:07263015
Manuscript Number:
JEEER-10-030
University: FST

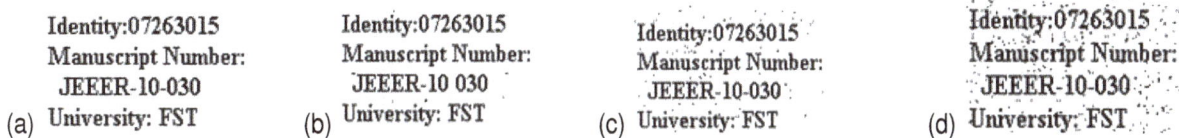

**Figure 6.** Original and extracted watermarks: (a): original (b): JPEG 90% (c): JPEG 50% (d): JPEG 20%.

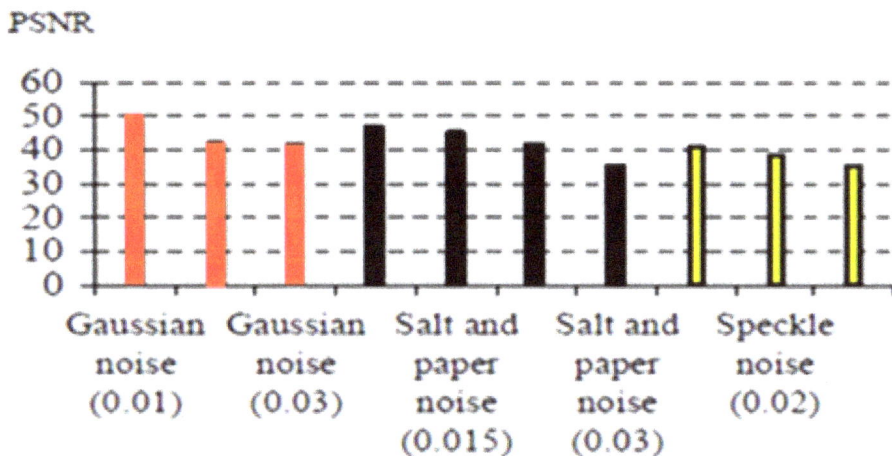

PSNR

**Figure 7.** Mean values of PSNR images Watermarked and attacked by different types of noises.

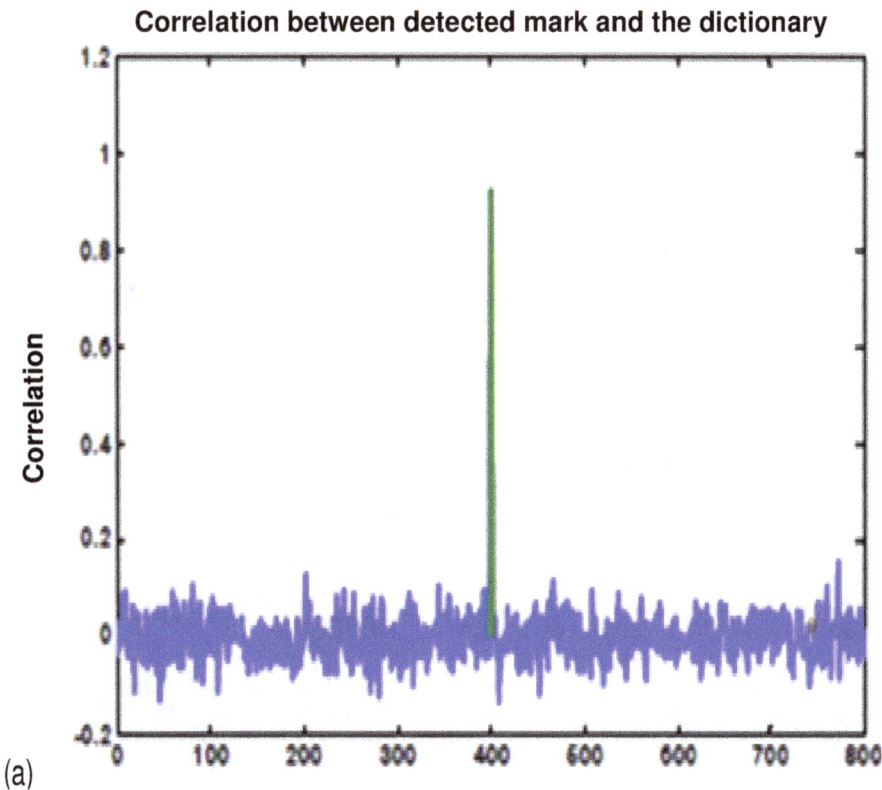

**Correlation between detected mark and the dictionary**

(a)

**Correlation between detected mark and the dictionary**

(b)

**Correlation between detected mark and the dictionary**

(c)                                                              **marks**

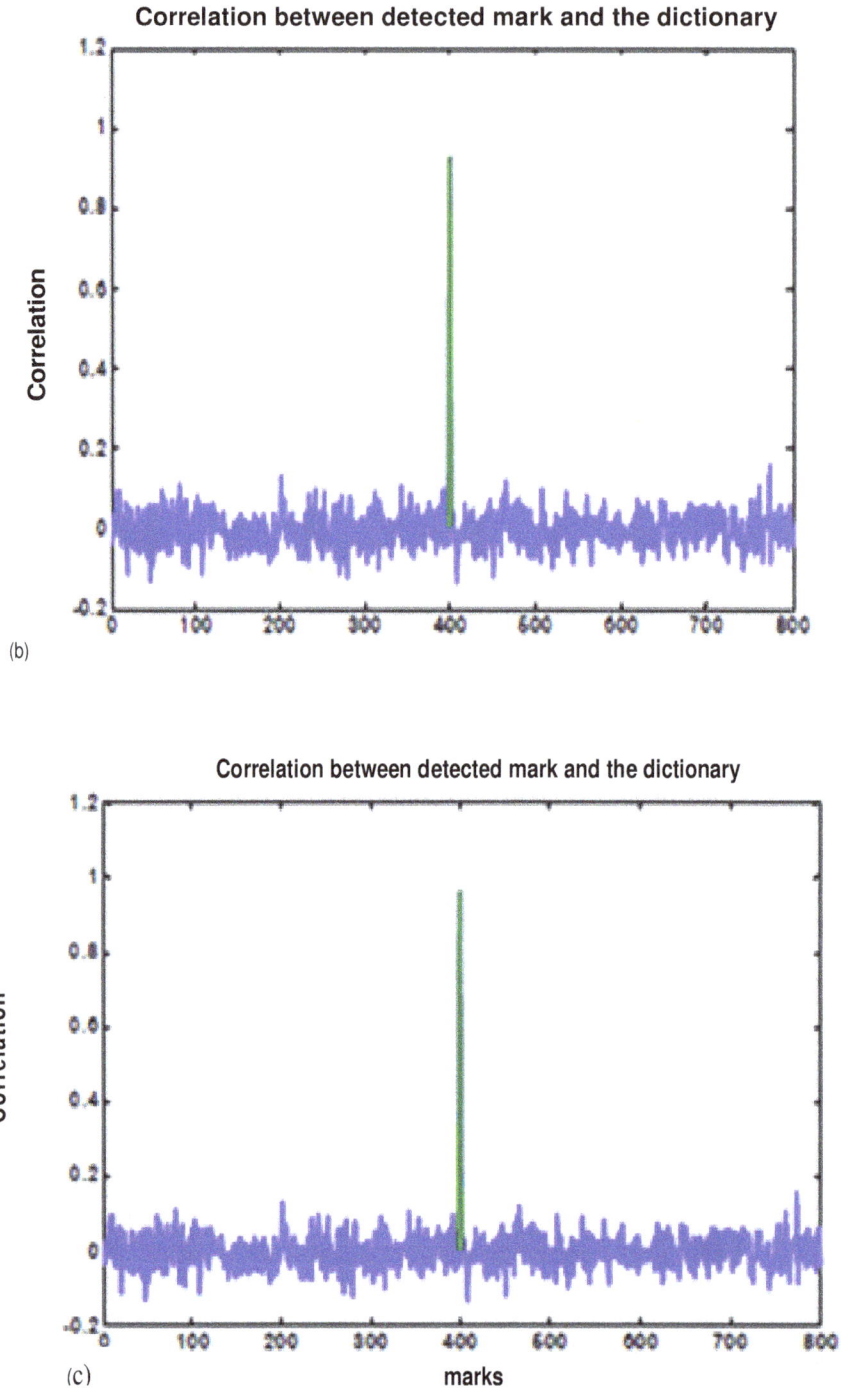

**Figure 8.** (a) Watermark detector response of attacked by Gaussian noise (0.05), (b) Watermark detector response of attacked by JPEG compression of quality 50, (c) Watermark detector response of attacked by Gaussian filter.

different watermark had survived to a large set of the applied attacks kinds. In addition, the redundancy caused by the multiple insertions has not altered our algorithm robustness. High correlations values after the attacked watermarked image are found in all the applied attacks kinds.

**REFERENCES**

Anne MAN (2001). Tatouage d'images numériques par parquets d'ondelettes, thesis, Nantes.

Azza OZ (2009). Compression et tatouage d'images à des fins d'archivage et de transmission : application aux images médicales Habilitation University, Tunis El Manar.

Barni M, Bartolini F, Cappellini V, Piva A (1998). A DCT domain system for robust image watermarking, Signal Process., 66(3): 357-372.

Bezdek JC, (1981). Pattern recognition with fuzzy objective function algorithms, Pleunum, New York.

Bezdek JC, Hall LO, Clarke LP (1993). Review of MR image segmentation techniques using pattern recognition. Med. Phys., 20, 1033–1048.

Bruyndonckx O, Quisquater JJ, Macq B (1995). Spatial method for copyright labeling of digital images", IEEE Workshop Non-linear Signal Image Process., Thessaloniki, Greece, pp. 456 - 459.

Chumsamrong W, Thitimajshima P, Rangsanseri Y (2000). Syntetic aperture radar (SAR) image segmentation using a new modified fuzzy c-means algorithm. Proc. Geosci. Remote Sens. Symp., 2: 624–626.

Cox IJ, Kiliani J, Leighton T, Shamoon T (1997). Secure spread spectrum watermarking for multimedia. IEEE Trans. Image Process., pp. 1673-1687.

Davoine F , Pateux S (2004). Tatouage de documents audiovisuels numériques, Edition Hermes science, Lavoisier.

Dunn JC (1974). A fuzzy relative of the ISODATA process and its use in detecting compact well separated clusters. J. Cybernet., Vol. 3 : 32-57.

Haralick RM, Shanmugam K, Dinstein (1973). Textural Features for Image Classification, IEEE Trans. Syst., Man Cybernet., 3(6), pp. 610-621.

Hsiang K, Ming-Jang C, Chung-Chih L (1999). Model-Free Functional MRI Analysis Using Kohonen Clustering Neural Networks and Fuzzy C-Means, IEEE Trans. Med. Imaging, 18(12), pp. 1025-1036.

Khaled SA, Nizar AB, Ahmed BH (2009). Segmentation statistique non supervisée des images Code 2D par modèle de Markov Caché, 5th International Conférence of Sciences of Electronic, Technologies of Information and Telecommunications.

Lotfi TL (2009). A Fuzzy Segmentation Approach for Images Application, 5th International Conference of Sciences of Electronic, Technologies of Information and Telecommunications.

Mukherjee DP, Pal P, Das J (1996). Sonar image segmentation using fuzzy c-means. Signal Process., 54(3): 295–302.

Patrick BA (2005). Méthodes de tatouage d'images Fondées sur le contenu. Thesis, Institut National Polytechnique, Grenoble.

Patrick BA, Chassery JM, Macq B (2002). Geometrically invariant watermarking using feature points, IEEE Trans. Image Process., 11(9): 1014-1028.

Patrick BA, Furon T, Cayre F, Doërr G (2007). Practical security analysis of dirty paper trellis watermarking. In Information Hiding: 9th international workshop, Saint-Malo, vol. 4567 of Lecture Notes in Computer Science, Springer Verlag.

Rignot E, Chellappa R, Dubois P (1992). Unsupervised segmentation of polarimetric SAR data using the covariance matrix. IEEE Trans. Geosci. Remote Sens., 30(4): 697–705.

Ruspini EH (1969). A new approach to clustering, Inf. Control, 15(1): 22–32.

Shih FH, Wu FYT (2003). Combinational image watermarking in the spatial and frequency domain, Patt. Recognit., 36(4): 969-975.

Wallace GK (1992). The JPEG still picture compression standard. IEEE Trans. Consumer Electronics, 38(1): 18-34.

Wolfgang R, Delp E (1997). A Watermarking Technique for digital imagery: Further studies, International Conference on Imaging Science, Systems and Technology, Los Vegas, Nevada.

Yang Y, Zheng CH, Lin P (2005). Fuzzy C-means clustering algorithm with a novel penalty term for image segmentation. Opto- Electron. Rev., 13(4).

Ying LI (2003). Texture segmentation based on features in wavelet domain for image retrieval, Visual Communications and Image Processing, Lugano, Switzerland.

Zhao J, Koch E (1995). Towards robust and hidden image copyright labelling. IEEE Workshop on Nonlinear Signal Image Processing.

# 3

# Locational marginal pricing approach to minimize congestion in restructured power market

## P. Ramachandran[1]* and R. Senthil[2]

[1]Department of Electrical Engineering, Dr. MGR University, Chennai, India.
[2]University College of Engineering, Villupuram Anna University, Chennai, India.

**The privatization and deregulation of electricity markets has a very large impact on almost all the power systems around the world. Competitive electricity markets are complex systems with many participants who buy and sell electricity. Much of the complexity arises from the limitations of the underlying transmission systems and the fact that supply and demand must be in balance at all times. When the producers and consumers of electric energy desire to produce and consume in amounts that would cause the transmission system to operate at or beyond one or more transfer limits, the system is said to be congested. In this paper, Locational Marginal Pricing approach is adopted to locate the spots of congestion in the Indian utility system under various critical conditions of the system, such as transmission line outage, increase in loads and generation failure and the results are found efficient in minimizing the congestion.**

**Keywords:** Electricity, markets, transmission, congestion.

## INTRODUCTION

Electricity Supply Industry (ESI), throughout the world, is undergoing restructuring for better utilization of the resources and for providing quality service and choice to the consumer at competitive prices. Restructuring of the power industry aims at abolishing the monopoly in the generation and trading sectors, thereby, introducing competition at various levels wherever it is possible. Electricity sector restructuring, also popularly known as deregulation is expected to draw private investment, increase efficiency, promote technical growth and improve customer satisfaction as different parties compete with each other to win their market share and remain in business. Electricity markets throughout the world continue to be opened to competitive forces. The underlying objective of introducing competition into these markets is to make them more efficient. In competitive environment, the price is determined by stochastic supply and demand functions. As a consequence of increased volatility, a market participant could make trading contracts

with other parties to hedge possible risks and get better returns. Congestion occurs when transmission lines or transformers are overloaded and this prevents the system operators from dispatching additional power from a specific generator.

Locational marginal pricing (LMP) is a market-pricing approach used to manage the efficient use of the transmission system when congestion occurs on the bulk power grid. Congestion arises when one or more restrictions on the transmission system prevent the economic, or least expensive, supply of energy from serving the demand. For example, transmission lines may not have enough capacity to carry all the electricity to meet the demand in a certain location. This is called a "transmission constraint." LMP includes the cost of supplying the more expensive electricity in those locations, thus providing a precise, market based method for pricing energy that includes the "cost of congestion." LMP provides market participants a clear and accurate signal of the price of electricity at every location on the grid. Amarasinghe et al (Amarasinghe, 2008) describes the basics of LMP and also when LMPs are used for settlement of transactions, consumers are charged more

---

*Corresponding author. E-mail: rama_p50@yahoo.com.

than the average cost of production of electricity due to the nonlinear nature of the power flow and the constraints imposed by the Optimal Power Flow (OPF). This difference which is accumulated with the Independent System Operator (ISO) is referred to as network rental. It is made up of two components known as loss rental and constraint rental. Loss rental is due to the difference in average losses and marginal losses, caused by the nonlinear nature of losses. This paper develops a method to calculate these different rental components paid by each consumer, by combining the power flow tracing and optimality conditions.

The general formulation of the LMP with necessary components is described elaborately (Eugene et al., 2004; Tina and George, 2007). It gives some insights regarding the evaluation of the LMP components, in general, and the distinct characteristics, including the limitations, of the various proposed decomposition approaches, in particular. It deals with the salient features of the formulation that is the role of the generators with the ability to vary their output, as well as the impact of the network congestion on the price setting, are explicitly recognized. The formulation's comprehensiveness brings numerous insights into the various decompositions, provides a platform for their comparative analysis, and allows us to understand the direct implications and the role of the policy specified. Moreover, the formulation reveals the limitations of any decomposition into the components due to the underlying structural interdependencies among them. Paper (Silpa, 2007) describes the advantages and disadvantages of deregulation of which congestion is the main factor. It gives a brief description about the various congestion management schemes available and also tabulates the methods practiced in various power markets. It also provides two options for congestion management that is load shedding and using VAR support. The given options have been implemented in the Standard IEEE 24 bus system and results have been obtained. It also states that the VAR support is more advantageous than the load curtailment.

The impact of reactive power with regulating devices is described in Srivastava and Verma (2000) and Daniel and Christopher (2007). The effectiveness of locational marginal pricing (Kim, 2006; Xie et al., 2006; Keshi et al., 2006; Hamoud and Bradley, 2001; Fangxing et al., 2004) as a market signal for reserve supply is discussed on the basis of network constrained integrated energy and reserve market arrangement. Revenue recovery based on LMP is explored as an assessment tool in the presence of serious network outage. The model is solved by using DC-OPF. In particular, marginal loss factor is incorporated into energy and reserve integrated market to overcome the absence of network loss price component in DC-based optimal power flow. In these papers, locational marginal pricing as a possible market signal tool of reserve supply for system security is discussed in

the presence of network outage. In particular, revenue recovery of reserve supply on the basis of the locational marginal pricing environments is explored in the platform of energy and reserve integrated market. Paper (Fangxing and Rui, 2007) deals about the different energy prices resulted due to transmission constraints. It suggests a computer program to calculate, for a given period of time, transmission congestion cost (TCC) in dollars per unit time and locational marginal pricing (LMP) in dollars per megawatt-hour (MWh) at any selected bus in the transmission system. In addition, the information provided by the program output on congested transmission elements is used to identify buses in the network whose LMPs are representative of the entire network. The computed LMPs at these buses are used to define zones in the network where each zone has its LMP. The proposed methodology can be used to carry out sensitivity studies to determine the impact of changes in system parameters and operating conditions on the LMPs. The proposed method is illustrated using the IEEE Reliability Test System (RTS). Reference papers (Enzo and Shmuel, 2007; Kwok, 2004; Commission for Energy Regulation (CER), http://www.pjm.com, http://www.iso-ne.com, Fu and Zuyi, 2006; Chen et al, 2002; Shariati et al, 2008; Goncalves et al., 2003; Scott and William, 2000; Jeffrey et al, 1999) describe the overview of Locational Marginal Pricing scheme and how it will be used in the new market structure. It also defines and stresses the need for LMP by discussing the zonal congestion management and how it is alleviated by practicing LMP. It is explained through an illustrative example and also the tariff definition of LMP.

A detailed comparison of Nodal and Zonal Congestion management methods ay analyzing their advantages and limitations through a number of illustrative examples and a supporting theoretical analysis are elaborated in Jeffrey et al. (1999) This paper examines the assertion that administrative aggregation of many nodes into larger zones would ensure competition across a wider area and constrain this power of the monopolist. Also it emphasizes the point of zonal pricing always subsidizes the dominant local generator and increases monopoly profits above those that would occur under nodal pricing. It also states the argument that market power dictates a need for zonal aggregation motivates the detailed demonstration that this is both wrong and creates a set of new problems that could be avoided with nodal pricing or splitting congested zones.

## PROBLEM FORMULATION

In a competitive electricity market, the settlement between the independent system operator (ISO) and the participants is based on locational marginal prices (LMPs). LMP at a given node of a power system is the sensitivity of operational cost to the change in load at that node, and it is calculated using an optimal power flow (OPF) program. When LMPs are used for settlement of transactions, consumers are charged more than the average cost of production of electricity due to the nonlinear nature of the power flow and the constraints imposed by the OPF.

The objective function may be represented as the minimization of total cost of generation (Eugene et al., 2004[2]):

$$\min G(P) = \sum_{k=1}^{N} S_k P_{gk} \qquad (1)$$

If bus $k$ is a pure load bus, then $P_{gk} = s_k = 0$.

The power balance equation considering the losses is:

$$\sum_{k=1}^{N} P_{gk} - P_{dk} = P_{loss} \qquad (2)$$

Let us assume that the line constraint sensitivity matrix is $T$ with elements $t_{jk}$ that give the change in flow on circuit $j$ to a change in real power injection at bus $k$, under a specified slack distribution. Thus:

$$t_{jk} = \frac{\Delta F_j}{\Delta P_k} \qquad (3)$$

In the matrix form, Equation (3) becomes:

$$\underline{F} = \underline{TP} \qquad (4)$$

It is important to note that the sensitivity factors of (3) are computed under a so-called "slack-bus" assumption which indicates how the change $\Delta Pk$ is assumed to be compensated. It could be compensated from another certain bus, or from several other buses, or from all other buses, and the elements of $T$ will change depending on which of these is assumed. It is generally considered best to employ a so-called distributed slack bus assumption here where the compensation is assumed to come from all other generator buses. We know that the "normal" flow constraints on every circuit is:

$$-\underline{F_{max}} \le \underline{F} \le \underline{F_{max}} \qquad (5)$$

Substitution of Equation (4) in (5) results in:

$$-\underline{F_{max}} \le \underline{TP} \le \underline{F_{max}} \qquad (6)$$

We assume at this point that high flows in our network are unidirectional, that is we need not be concerned with high flows in both directions. This does not prevent bidirectional flows; it merely enables us to be concerned with reaching the upper bound in only one direction. Therefore, we may ignore the lower bound in Equation (6) so that our circuit flow constraint is:

$$\underline{TP} \le \underline{F_{max}} \qquad (7)$$

In scalar form, Equation (7) is:

$$\sum_{k=1}^{N} t_{jk} P_k \le F_{j\,max}, \qquad j = 1,......., M \qquad (8)$$

Replacing injection with difference between generation and load, we obtain:

$$\sum_{k=1}^{N} t_{jk} \left( P_{gk} - P_{dk} \right) \le F_{j\,max}, \qquad j = 1,......., M \qquad (9)$$

The OPF problem determines the optimal generator dispatch subject to a set of constraints which represents the operational and physical limits of the power system. A competitive market environment is considered, where generators make offers to sell electricity as price-quantity pairs. For the purpose of simplicity, no demand side bidding is considered and hence, loads are known constants for the dispatch. Therefore, the OPF can be written as a problem of minimizing the total cost of generation subjected to real and reactive power are balanced, real power generation is within the limits specified by the offer quantity, reactive power generation is within the limits, line flows are within the thermal limits, and voltages are within specified limits, respectively. Therefore, The Lagrangian function for linear optimized power flow (LOPF) becomes:

$$L(P_g, \lambda, \mu) = \sum_{k=1}^{N} S_k P_{gk} - \lambda [\sum_{k=1}^{N} P_{gk} - P_{dk} - P_{loss}]$$

$$-\sum_{j=1}^{M} \mu_j [\sum_{k=1}^{N} t_{jk}(P_{gk} - P_{dk}) - F_{j\,max}] \qquad (10)$$

The first order conditions for finding the optimum to LOPF include:

$$k \in gen: \frac{\partial L}{\partial P_{gk}} = S_k - \lambda(1 - \frac{\partial P_{loss}}{\partial P_{gk}}) - \sum_{j=1}^{M} \mu_j t_{jk} = 0$$

(11)

But we are more interested in the load buses. Consider:

$$k \in load: \frac{\partial L}{\partial P_{dk}} = \lambda(1 + \frac{\partial P_{loss}}{\partial P_{dk}}) + \sum_{j=1}^{M} \mu_j t_{jk}$$

(12)

Note that $P_{dk}$ is not a decision variable, and therefore we do not set it as 0. Equation (12) gives the change in the optimal value of the objective function due to a small change in the parameter $P_{dk}$.

We call $\frac{\partial L}{\partial P_{dk}}$ the LMP for bus $k$, that is,

$$k \in load: LMP_k = \lambda(1 + \frac{\partial P_{loss}}{\partial P_{dk}}) + \sum_{j=1}^{M} \mu_j t_{jk}$$

Written slightly different,

$$k \in load: LMP_k = \lambda + \lambda \frac{\partial P_{loss}}{\partial P_{dk}} + \sum_{j=1}^{M} \mu_j t_{jk}$$

(13)

Equation (13) consists of three components.

$k \in load:$

$LMP_k = \lambda$ ..... Energy component

$+ \lambda \frac{\partial P_{loss}}{\partial P_{dk}}$ .... Loss component

$+ \sum_{j=1}^{M} \mu_j t_{jk}$ .... Congestion component

Delivered energy component

Where LMP$_k$ is the Locational Marginal Price at bus 'k', $\lambda$ is the Lagrange Multiplier associated with the Power balance equation, $\frac{\partial P_{loss}}{\partial P_{dk}}$ is the real power loss sensitivity factor at bus 'k', $\mu_j$ is the vector of Lagrange multipliers associated to the network constraints on line 'j' and t$_{jk}$ is the sensitivity factor of the network at bus 'k' due to network constraints on line 'j'. The Lagrange multipliers determined from the solution of the optimum power flow provide important economic 'information' regarding the power system. A Lagrange multiplier can be interpreted as the derivative of the objective function with respect to enforcing the respective constraint. Therefore, the Lagrange multipliers associated with enforcing the power flow Equations of the OPF can be interpreted as the marginal cost of providing addition energy ($/MWh) to that bus in the power system. This marginal cost is known as LMP and sometimes is called the shadow price of the power injection at the node. The LMP is decomposed into three components which are the cost of energy, cost of marginal losses and cost of marginal congestion.

The main aim of decomposition is to reflect the cost of system marginal cost, loss compensation and congestion management as well as voltage support. These components are all important cost terms in the deregulated electricity market and can be forwarded to the generators and consumers as control signals to regulate the level of their generations and consumptions. Many methods have been followed to minimize the transmission congestion (Silpa, 2007) viz.

1. Adding a transmission line across the congestion path
2. Increase the capacity of power system components
3. Generation Re-dispatch
- Modification of generating schedules
4. Load Re-dispatch
- Shedding - reduce specific loads
- Encouraging some specific load serving
5. Using VAR Support

Increasing the capacity of the components is much complicated as the components have to be completely disconnected from the power system. Hence this method does not hold good. Also, in the competitive market, it becomes a serious issue to re-dispatch certain generators or loads as in some cases the generating companies fail to accept the modifications of their schedules and therefore the re-dispatching methods are also not preferred. Hence the best ways to minimize congestion were found to be the addition of transmission line and the usage of VAR support.

**STEP – BY – STEP ALGORITHM**

The procedure to manage the congestion in the system is given as the following Step-by-Step algorithm. The procedure for finding the location for adding a transmission line and for installing regulating devices to reduce congestion is given as thus explained in the following steps:

**Step 1**: Obtain the Generator data, line data, bus data, generator cost data and other power flow constraints of the utility test system.

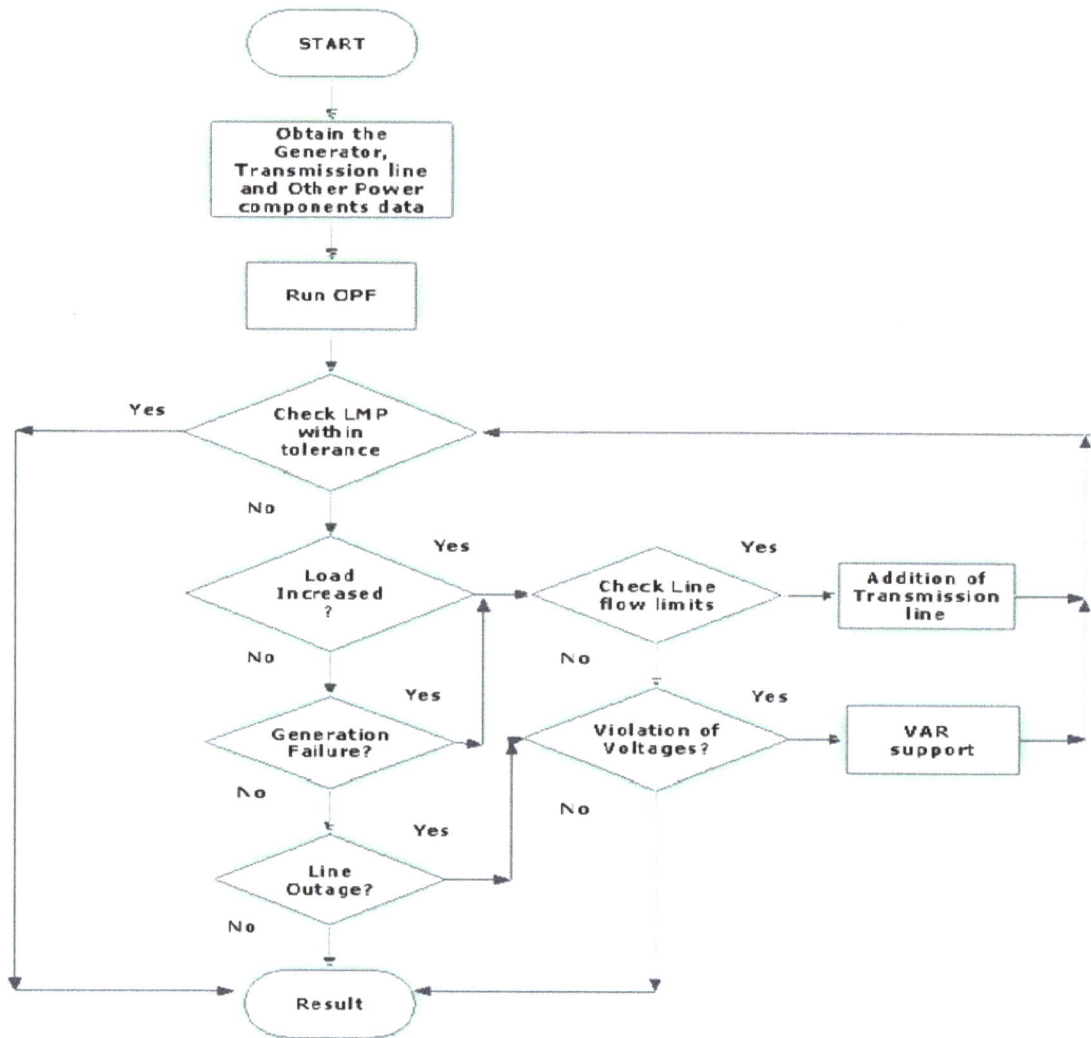

**Figure 1.** Flow diagram – Computation of locational marginal price.

**Step 2:** Run optimal power flow (OPF) analysis and obtain the LMP values from it and check out the tolerance limit.

**Step 3:** If the LMP is not within limits, then check whether it is an abnormal condition of load increased.

**Step 4:** If the load is found to be increased then check for line flow limits. If the limits exceeds then add a new transmission line in the required buses.

**Step 5:** If the line flow limits are not seen to be violating then the voltage profile has to be checked. If the voltages are seen to be violating then reactive power is injected in voltage defective buses.

**Step 6:** If the load is not seen to be increased then the generation is checked. If the generator outage is to be occurred, then repeat the steps 4 and step 5.

**Step 7:** If the congestion is not due to generation failure, then the transmission line limits are checked. If the line outage is found to occur then voltage profile is checked

and VAR is injected in voltage deficient buses.

**Step 8:** The congestion relief method is carried out and OPF is to be executed again and LMP values are computed again and the system is checked for congestion.

**Step 9:** Stop.

The flow diagram of the aforementioned procedure is shown in Figure 1.

**SIMULATION RESULTS AND DISCUSSION**

The locational marginal price is computed by executing the optimal power flow program in MATPOWER software. It is computed in base load, increase in load and vulnerable conditions. If there is any increase in LMP value in certain buses, the remedy action is carried out to

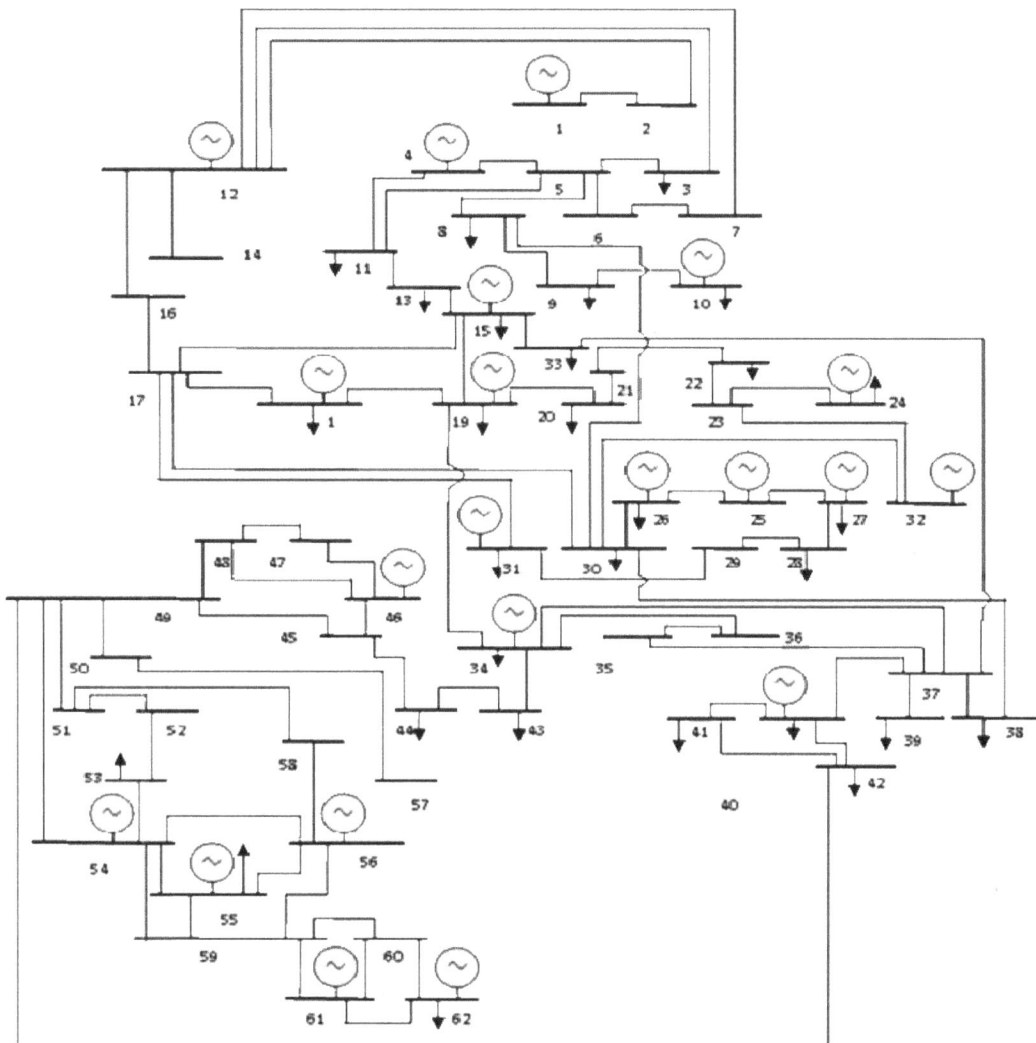

**Figure 2.** Oneline diagram of indian utility system.

make the LMP almost same in all the buses. The aforementioned cases studies were demonstrated on the practical Indian utility bus system. The Indian utility system consists of 25 generators, 89 (220KV) transmission lines and 11 tap changing transformers and 57 loads. Its one line diagram is shown in Figure 2. The total demand in the system is 2909 MW. The optimal power flow of the above test system was carried out and the corresponding LMP values in the different buses were obtained. Using the values of LMP, congested spots are identified and the congestion relief methods are adopted to relieve congestion.

## Increased in load condition

The loads at all nodes of the Indian utility system were increased by 25% and the optimal power flow was carried out. The LMPs of the corresponding buses were

determined using the above execution. During this condition, it was observed that there was a drastic increase in LMP at the bus 13, 20, 21, 33, 35, 36 and also notable increase in some buses. Five transmission lines are added in the system to minimize the congestion. The transmission lines are added in the place where the LMP is found to be increased drastically. In this system the line is added from bus 11 to 13, 9 to 10, 22 to 23, 26 to 30 and 53 to 54. After adding the transmission lines, OPF program was run again and the results were given in Table 1. In case if immediate relief was needed and there was violation of voltage limits, then injection of reactive power in voltage affected buses served good in neutralizing the congestion. In this case, the reactive power was injected in buses 13, 20, 22, 33, 37, 38, and 53 and again OPF was executed and results were given in Table 1. Figure 3 shows the comparison of LMP values for abnormal condition of all the loads increased by 25% and the reduced values of LMPs resulted from the

**Table 1.** Increased in load condition.

| Bus No. | Increase in loads (25%) | | Line added | | Injection of reactive power | |
|---|---|---|---|---|---|---|
| | Voltage (p.u) | LMP ($/MWh) | Voltage (p.u) | LMP ($/MWh) | Voltage (p.u) | LMP ($/MWh) |
| 13 | 0.989 | 67.036 | 0.991 | 56.651 | 1.004 | 51.142 |
| 20 | 0.980 | 65.884 | 0.991 | 56.260 | 1.001 | 48.314 |
| 21 | 0.978 | 64.393 | 0.996 | 54.902 | 1.023 | 42.319 |
| 33 | 1.011 | 64.870 | 1.016 | 55.697 | 1.045 | 49.300 |
| 35 | 1.025 | 63.477 | 1.029 | 54.602 | 1.046 | 48.890 |
| 36 | 1.028 | 63.226 | 1.032 | 54.379 | 1.049 | 48.658 |

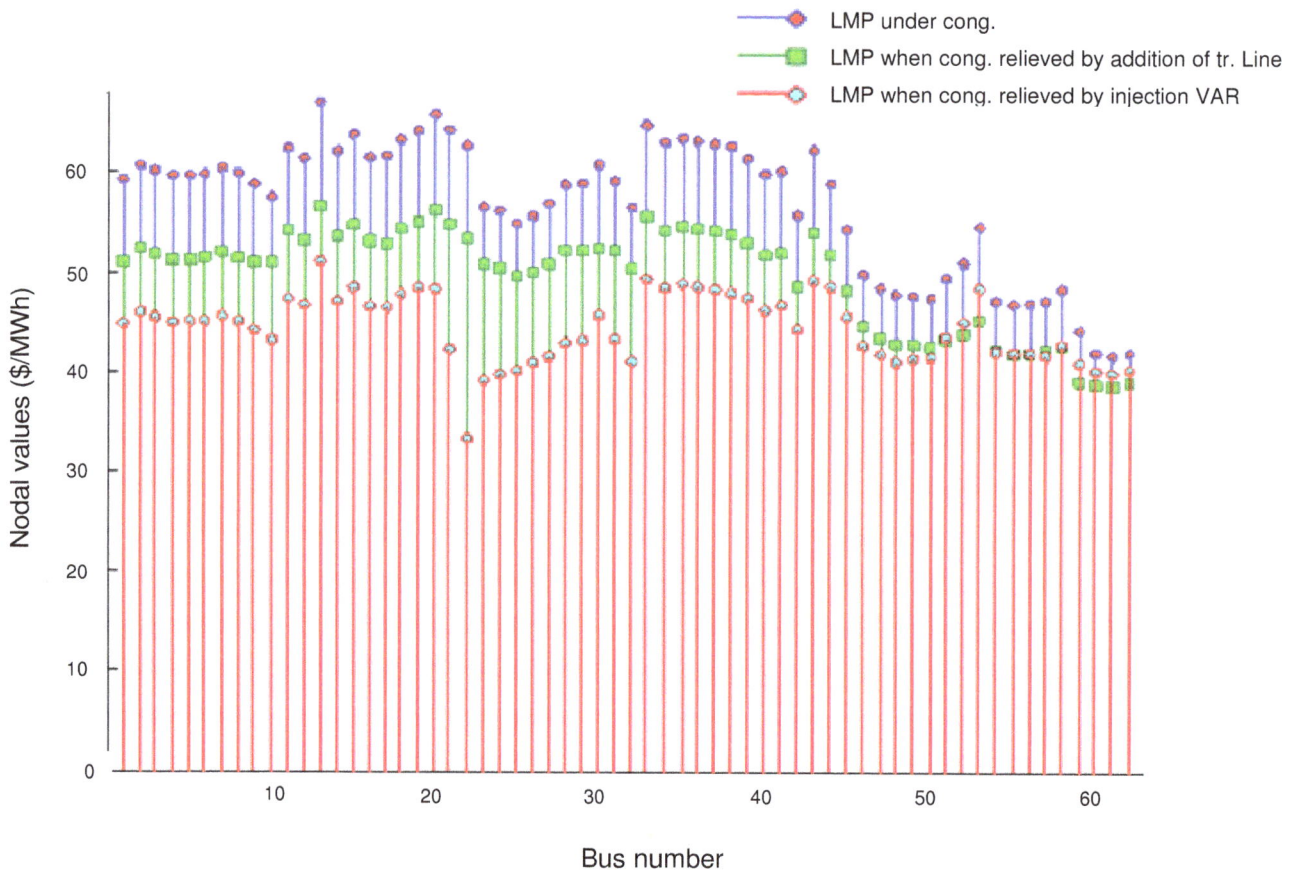

**Figure 3.** Comparison of LMPs – Increased in load condition.

addition of a transmission line and the installation of VAR devices. The contour representation of normal LMP values of the 62 bus system, covering the state of Tamil Nadu is given in Figure 4. Since the increase in load condition is the major cause of congestion, the variations in LMPs are given in the Figure 4. The areas of the state where drastic increase in LMPs are pronounced during congestion can be easily determined from the contour. The number of generating units in and around the Tamilnadu State capital (Chennai) is large and the congestion spots are also found to be crowded in the Chennai city and hence a zoomed view of Chennai contour

is produced.

### Transmission line contingency condition

In the Indian utility system, four transmission lines connected between the buses 11 to 12, 2 to 12, 3 to 12 and 23 to 32 were made out of service to carry out the contingency study and the corresponding optimal power flow solution was obtained. During the lines outage, it was observed that there was a drastic increase in LMP at the bus 13, 20, 21 and 33 and a notable increase at some

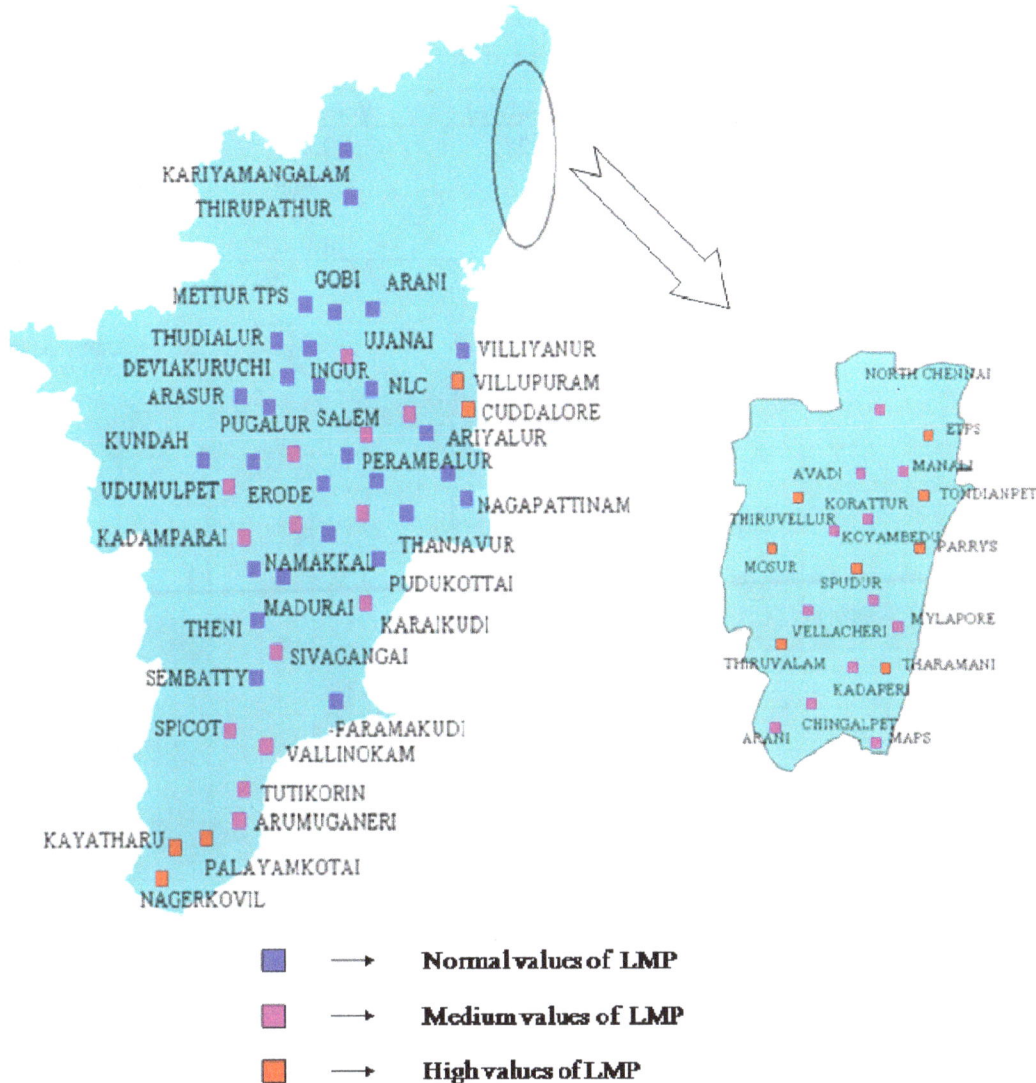

**Figure 4.** Contour representation of Tamil Nadu Bus system under congestion due to increased load.

buses. The relief method suggested for this case was the addition of reactive power source in the corresponding location to control the voltage violations. According to this proposal, the reactive power was injected in the voltage affected buses which served well in neutralizing the congestion. In this case, the reactive power was injected in buses 9, 10, 13, 34, 37, 48 and 59 and again OPF was carried out and results were given in the Table 2. Figure 5 shows the comparison of LMP values for abnormal condition of line outages and the reduced values of LMPs resulted from the installation of VAR devices.

**Generation failure condition**

To analyze the worst vulnerable condition, it is assumed that the generators at bus 12, 19 and 61 were made out

of order and optimal power flow solution was obtained. During this condition, it was observed that there was a drastic increase in LMP at the bus 13, 19, 20, 21, 33, 35, 43. According to the first method of congestion relief, four transmission lines were included to minimize the congestion. The transmission lines were added in the nodes of high LMP, at 13 to 15, 22 to 23, 26 to 30 and 53 to 54. After adding the transmission line OPF program was run again and the results were given in the Table 3. In case, if immediate relief was needed and there was violation of voltage limits, then injection of reactive power in voltage affected buses served good in neutralizing the congestion. In this case, the reactive power was injected in buses 9, 13, 21, 34, 37, 38, and 53 and again OPF was run and results were given in Table 3. Figure 6 shows the comparison of LMP values for abnormal condition of generation failure and the reduced values of

**Table 2.** Transmission line contingency condition.

| Bus no. | Line outage | | Injection of reactive power | |
|---|---|---|---|---|
| | Voltage (p.u) | LMP ($/MWh) | Voltage (p.u) | LMP ($/MWh) |
| 13 | 0.982 | 48.322 | 1.059 | 47.603 |
| 20 | 0.996 | 47.071 | 1.008 | 46.384 |
| 21 | 0.991 | 46.274 | 1.001 | 45.742 |
| 33 | 1.029 | 46.128 | 1.036 | 45.837 |

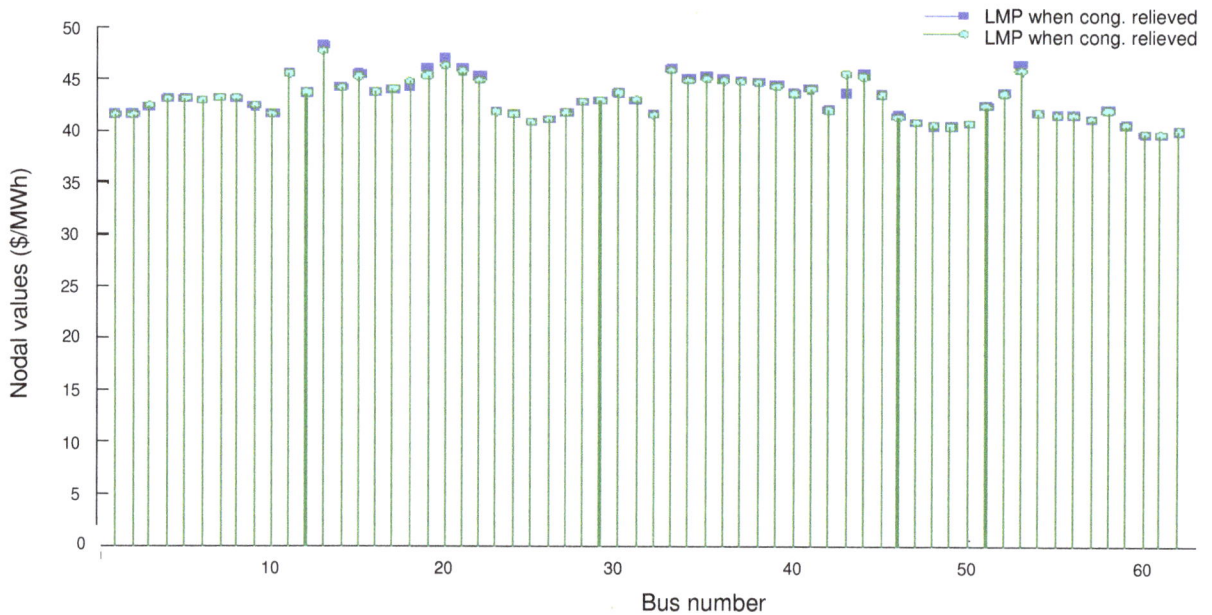

**Figure 5.** Comparison of LMPs – Transmission line contingency condition.

**Table 3.** Generation failure condition.

| Bus no. | Generation failure | | Line added | | Injection of reactive power | |
|---|---|---|---|---|---|---|
| | Voltage (p.u) | LMP ($/MWh) | Voltage (p.u) | LMP ($/MWh) | Voltage (p.u) | LMP ($/MWh) |
| 13 | 0.979 | 52.065 | 0.997 | 49.766 | 1.013 | 51.277 |
| 19 | 1.002 | 50.332 | 1.018 | 48.597 | 1.033 | 49.731 |
| 20 | 0.975 | 51.057 | 0.999 | 49.058 | 1.030 | 50.235 |
| 21 | 0.973 | 49.789 | 1.003 | 47.826 | 1.043 | 49.210 |
| 33 | 1.012 | 50.200 | 1.024 | 48.527 | 1.033 | 49.671 |
| 35 | 1.036 | 49.960 | 1.046 | 47.435 | 1.046 | 48.640 |
| 43 | 1.017 | 49.275 | 1.025 | 47.742 | 1.024 | 48.873 |

LMPs resulted from the addition of a transmission line and the installation of VAR devices.

## Conclusion

The transition from monopolistic to a competitive deregulated market though found to be more advantageous, encountered certain drawbacks, such as Congestion and difficulty in pricing. In this work, the Locational Marginal Pricing (LMP) was proved to be an effective solution in overcoming the above said barriers of deregulation. The LMPs are computed for the Indian utility system under normal and contingency conditions. Increase in LMP holds to be a good signal for identifying the Congested locations. Later, the congestion component of LMP is suggested to be used in congestion relief methods, such as addition of transmission line and injection

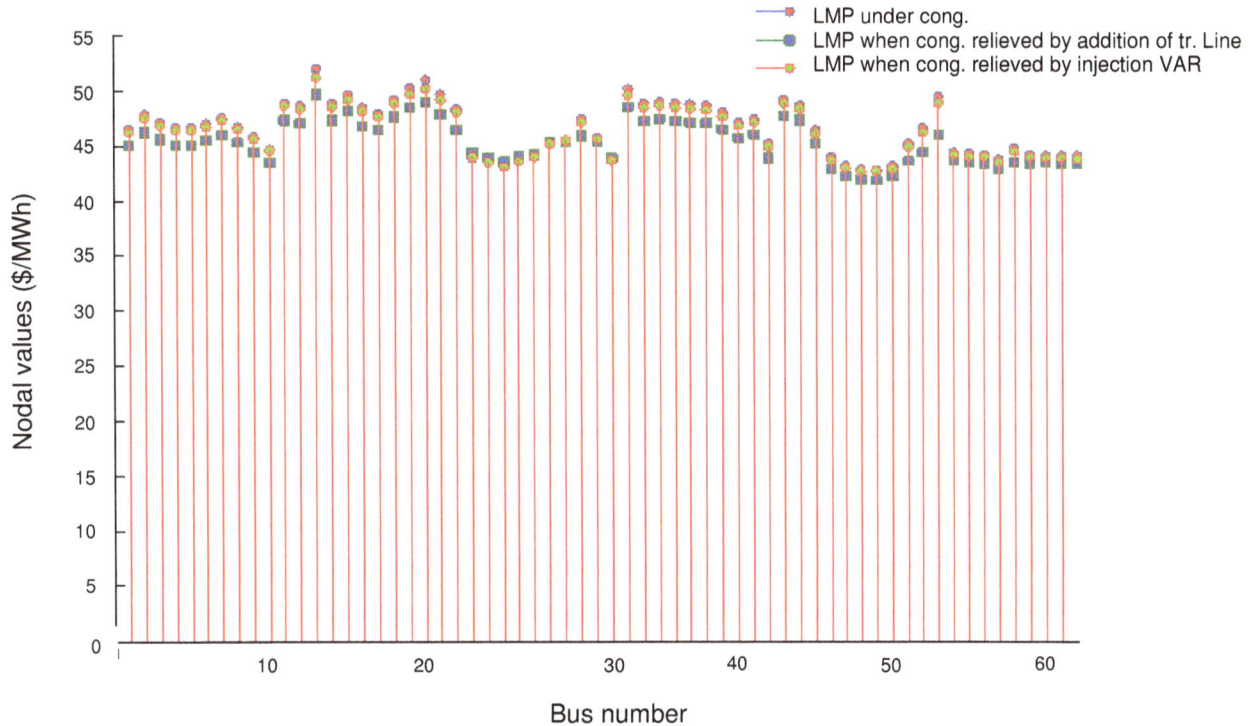

**Figure 6.** Comparison of LMPs - Generation failure condition.

of VAR sources. These methods proved well in relieving the system from congestion and have brought LMPs within limits.

**Nomenclature:** $P_{gk}$ - Power generated (MW), $P_{dk}$ - Power delivered (MW), $S_k$ - \$/MWh offers being made on an amount of generation of $P_{gk}$ over 1 h, $P_{loss}$ - Total power loss in the system (MW), $\Delta F_j$ - The change in flow on circuit $j$ (MW), $\Delta P_k$ - The change in real power injection at bus $k$ (MW), $t_{jk}$ - Sensitivity factor, $\lambda$ - Loss coefficient, $\mu$ - Congestion coefficient, $N$ - No. of generators, $M$ - No. of circuits of power flow.

## REFERENCES

Amarasinghe LYC (2008). 'Determination of Network Rental Components in a Competitive Electricity Market', IEEE Trans. Power Syst.,, 23(3): 1152-1161.

Briefing Note on the Proposed Market Arrangements in Electricity (MAE)' Prepared by the Commission for Energy Regulation (CER).

Chen L, Suzuki H, Wachi T, Shimuram Y (2002). 'Components of Nodal Prices for Electric Power Systems. IEEE Trans. Power Syst,, 17(1): 41-49.

Daniel C, Christopher LD (2007). 'Characterization of Feasible LMPs: Inclusion of Losses and Reactive Power'. IEEE Trans. Power Syst., 19: 3.

Enzo ES, Shmuel SO (2007). 'Aligning Generators' Interests with Social Efficiency Criteria for Transmission Upgrades in an LMP Based Market'. IEEE Trans. Power Syst.

Eugene L, Tongxin Z, Gary R, Payman S (2004). 'Marginal Loss Modeling in LMP Calculation'. IEEE Trans. Power Syst., 19: 2.

Fangxing L, Rui B (2007). 'DCOPF-Based LMP Simulation: Algorithm, Comparison with ACOPF, and Sensitivity'. IEEE Trans. Power Syst., 22: 4.

Fangxing L, Jiuping P, Henry C (2004). 'Marginal Loss Calculation in Competitive Electrical Energy Markets'. IEEE Internationa Conference on Electric Utility Deregulation, Restructuring and Power Technologies (DRPT2004) Hong Kong.

Fu Y, Zuyi L (2006). 'Different models and properties on LMP calculations,' in Proc. IEEE Power Engineering Society General Meeting.

Goncalves MJD, Zita AV (2003). 'Evaluation of transmission congestion impact in market power'. IEEE Bologna Power Tech Conf., 4: 6.

Hamoud G, Bradley I (2001). 'Assessment of Transmission Congestion Cost and Locational Marginal Pricing in a Competitive Electricity Market '. IEEE Trans. Power Syst., pp. 1468 -1470.

Jeffrey B, Jinxiang Z, Venkat B, Rana M (1999). 'Forecasting Energy Prices in a Competitive Market', IEEE Proc. Comput. Appl. Power, p. 42.

Keshi R, Saidi R, Narayana PP, Patel RN (2006). 'Congestion Management in Deregulated Power System using FACTS Devices'. IEEE Trans. Power Syst

Kim SG (2006). 'LMP as Market Signal for Reserve Supply in Energy and Reserve Integrated Market', IEEE Trans. Power Syst., 19: 2.

Kwok WC (2004). 'Standard Market Design for IS0 New England Wholesale Electricity Market: An Overview', IEEE International Conference on Electric Utility Deregulation, Restructuring and Power Technologies (DRPT2004) Hong Kong.

Locational Marginal Prices, [Online].Available: http://www.iso-ne.com

PJM Training Materials (LMP 101), PJM. [Available] http://www.pjm.com.

Scott MH, William WH (2000). Nodal and Zonal Congestion

Management and the Exercise of Market Power'. IEEE Trans. Power Syst.

Shariati H, . Askarian AH, Javidi MH, Razavi F (2008). ' Transmission Expansion Planning Considering Security Cost under Market Environment', DRPT-2008, pp. 1430-1434.

Silpa P (2007).' Power market Analysis tool for Congestion Management', thesis submitted to Department of Computer science and Electrical Engineering, Morgantown, West Virginia.

Srivastava SC, Verma RK (2000) 'Impact of FACTS devices on transmission pricing in a de-regulated electricity market', Electric Utility Deregulation and Restructuring and Power Technologies, Proceedings. DRPT Int. Conf,, pp. 642 -648.

Tina O, George G (2007). 'A General Formulation for LMP Evaluation'. IEEE Trans. Power Syst., 22: 3.

Xie K, Jiang CM, Zhang Z, Xie X, Xu XF, Ma X, Sun DI (2006). 'Application of the Locational marginal pricing model in North China grid: A preliminary study', International Conference on Power System Technology.

# 4

# Permanent magnet flux estimation method of vector control SPMSM using adaptive identification

## Abdoulaye M'bemba Camara*, Yosuke Sakai and Hidehiko Sugimoto

Department of Electrical and Electronics Engineering, University of Fukui, Fukui 910-8507, Japan.

The variation of permanent magnet flux deteriorates the performance of a torque controlled (TC) system. Without the torque sensor, the magnet flux information is indispensable for controlling the torque of the surface permanent magnet synchronous motor (SPMSM) with vector control. The magnet flux depends on variations of temperature inside of the motor. With the increase of stator winding temperature, the magnet temperature increases and the magnet flux decreases. Therefore, the magnet flux is not treated constantly and thus becomes a big issue to TC. So, the instantaneous value of the magnet flux is needed in any way. This paper proposes the estimation method of the magnet flux based on the adaptive identification used in the SPMSM with the vector control. Even at low speed, the influence of the stator resistance variation is not received easily because the proposed method has the stator resistance estimation function. The effectiveness of the proposed method is demonstrated by experiments.

**Key words:** SPMSM, magnet flux, vector control, stator resistance estimation, adaptive identification, mathematical model.

## INTRODUCTION

The surface permanent magnet synchronous motor (SPMSM) drive system has been widely used for high efficiency application. In recent years, torque controlled (TC) systems have become one of the favored control schemes for induction machine. In this case, controlling amplitude of the armature currents is required. The same principle has been applied in interior permanent magnet synchronous motors (Rahman et al., 2003; Tang et al., 2003).

Some AC servo motors used for injection molding machines (Yoshiharu et al., 1999) require torque control with less than 1% error. Such motors are SPMSMs in field system that has Neodym magnet, an high energy product, but torque feedback system is needed to detect torque with torque sensor. With the increase of the stator winding temperature, the magnet temperature increases and then the magnet flux decreases; and as a result, the torque decreases proportionally.

In Takeshit et al. (1997) and Beccera et al. (1991), an observer of the stator resistance variation is added to decrease the negative effects of the stator resistance variation. However, the stator resistance of the SPMSM is affected by many factors of the environment. Thus, it is difficult to estimate the resistance variation accurately on line, and the magnet flux of the motor cannot be estimated accurately with temperature increase. Some researchers have developed observers using the motor model (Han et al., 2000; Fukumoto et al., 2007; Hayashi et al., 2005). A suitable design of the observers produces a high level of insensitivity to parameter variations, but the observers are sensitive to noise measurement; the operation load seems to be large and their stability analysis is difficult because of many expressions.

During the low speed operation, the contact time between machine and plastic, injection molding machines, as an example, is important. That is why it is necessary to estimate the magnet flux during this period because much current would flow and cause a very important variation of temperature in the armature winding of SPMSM. Since the torque is necessary in the injection molding machine during the operation, it is constant even

*Corresponding author. E-mail: aziz71@u-fukui.ac.jp.

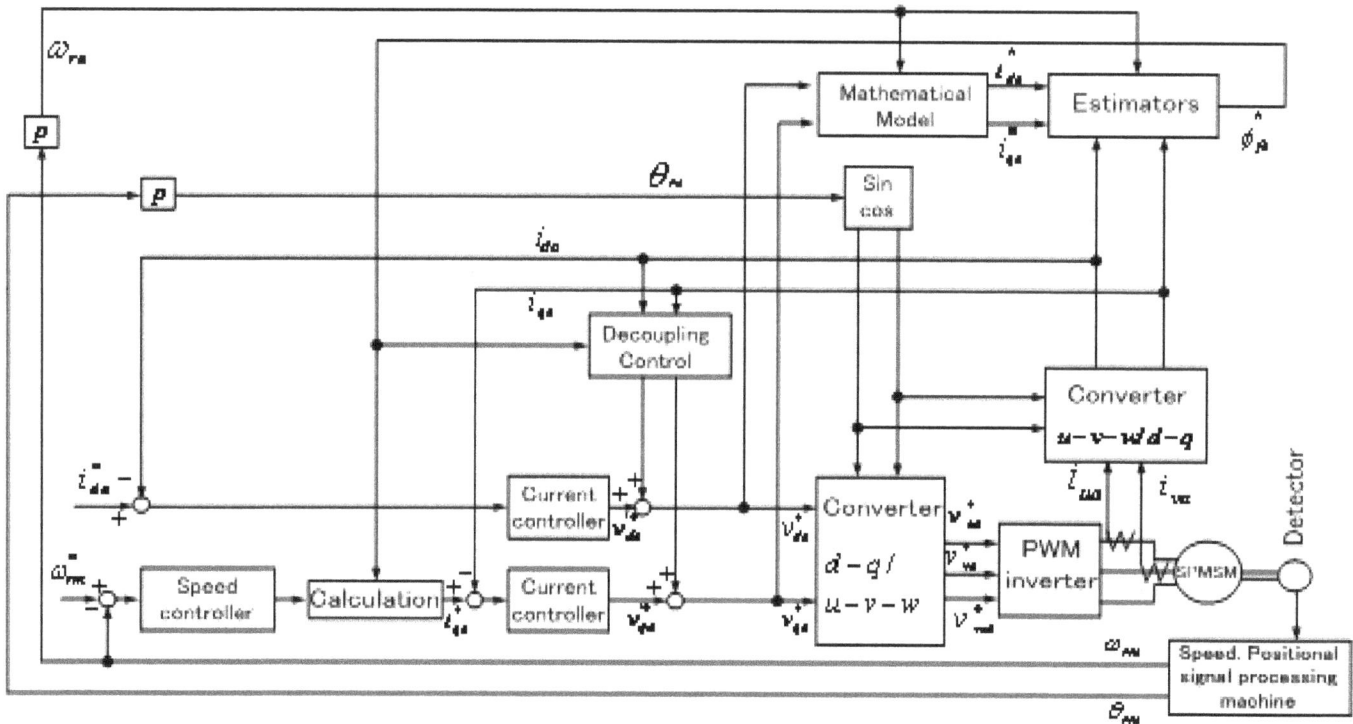

**Figure 1.** Block diagram of the control scheme.

at low speed; therefore, the magnet flux estimation becomes necessary for torque control systems.

The research about the temperature sensor is done, but the research about high accuracy of torque control without torque sensor is not active.

We propose a new torque control system without torque sensor controlling armature current with error less than 1% by using adaptive identification of magnet flux with temperature variation. The salient features of the new method are as follows:

(1) The magnet flux and the armature winding resistance are estimated by integrating the magnet flux estimation error with the armature winding resistance estimation error, leading to the derivation of the linearization error state equation of a real machine and using of a mathematical model.
(2) The estimation of the magnet flux is not influenced easily from estimation of the armature winding resistance.
(3) The design of the magnet flux estimator is simple; actually the design of bandwidth of closed loop transfer function between value and estimator of magnet flux is simple equation first–degree.
(4) For adjustment of moving average at low speed to estimate magnet flux.

In this thesis, the linearization error margin of state equation from state equation of a real machine and the mathematical model and the estimator is composed by using it. The stability is examined by bode diagram, and

the utility is confirmed by the experiment.

## MATERIALS AND METHODS

Mathematical model and linearization error equation of state

### Composition of the proposed system

Figure 1 shows the composition of the proposed system. The instruction value is shown as * in Figure 1. The voltages, $v_{da}$, $v_{qa}$ are given as real SPMSM; and mathematical model as an input, from which the mathematical model estimates, $\hat{\phi}_{fa}$ $\hat{R}_a$ the quantities, $\Delta\phi_{fa}, \Delta R_a$ are causes of the current estimation error. And these may become 0 by using the estimation of the output current.

### Mathematical model

The state equation of SPMSM in (d, q) coordinates is shown in Eq. (1). Moreover, Eq. (2) shows the mathematical model in Figure 1 by which the estimated current is described; $\hat{i}_{da}$ and $\hat{i}_{qa}$ are estimated by using $i_{da}, i_{qa}, v_{da}, v_{qa}, \hat{\phi}_{fa}, R_a$

$$\begin{bmatrix} L_{ti} & 0 \\ 0 & L_{ti} \end{bmatrix}\frac{d}{dt}\begin{bmatrix} i_{da} \\ i_{qa} \end{bmatrix}=\begin{bmatrix} -R_a & \omega L_{ti} \\ -\omega L_{ti} & -R_a \end{bmatrix}\begin{bmatrix} i_{da} \\ i_{qa} \end{bmatrix}+\begin{bmatrix} v_{da} \\ v_{qa} \end{bmatrix}-\omega_{re}\phi_{fa}\begin{bmatrix} 0 \\ 1 \end{bmatrix} \dots\dots (1)$$

$$\begin{bmatrix} L_d & 0 \\ 0 & L_d \end{bmatrix} \frac{d}{dt} \begin{bmatrix} \hat{i}_{da} \\ \hat{i}_{qa} \end{bmatrix} = \begin{bmatrix} -\hat{R}_a & \hat{\omega}_{\phi_a} \\ -\hat{\omega}_{\phi_a} & -\hat{R}_a \end{bmatrix} \begin{bmatrix} \hat{i}_{da} \\ \hat{i}_{qa} \end{bmatrix} + \begin{bmatrix} v_{da} \\ v_{qa} \end{bmatrix} - \hat{\omega}_{re}\hat{\phi}_{fa} \begin{bmatrix} 0 \\ 1 \end{bmatrix} + \begin{bmatrix} g_{11} & g_{12} \\ g_{21} & g_{22} \end{bmatrix} \begin{bmatrix} i_{da} - \hat{i}_{da} \\ i_{qa} - \hat{i}_{qa} \end{bmatrix}$$

$$\dots\dots\dots\dots\dots\dots\dots\dots\dots\dots\dots\dots\dots\dots (2)$$

$R_a$ : stator winding resistance $L_a$ :d,q axes inductances, $\phi_{fa}$ permanent magnet flux.

$v_{da}, v_{qa}$ :d,q axes voltages, $i_{da}, i_{qa}$ :d,q axes currents ,

$\omega_{re}$ :synchronous angular frequency.

In this work, the current of the armature and the parameter with the sign, ^ show "estimation" and the estimated parameter are $\phi_{fa}$ , $R_a$ . Moreover, $g_{11}, g_{12}, g_{21}, g_{22}$ are mathematical model gains.

### Linearization of estimation error of state equation

The following state equation for estimation error is obtained by multiplying matrix inverse of inductance matrix to each side of the Eqs (1) and (2):

$$\frac{d}{dt} \begin{bmatrix} i_{da} - \hat{i}_{da} \\ i_{qa} - \hat{i}_{qa} \end{bmatrix} = \begin{bmatrix} A_{11} & A_{12} \\ A_{21} & A_{22} \end{bmatrix} \begin{bmatrix} i_{da} - \hat{i}_{da} \\ i_{qa} - \hat{i}_{qa} \end{bmatrix} + \begin{bmatrix} B_{11} & B_{12} \\ B_{21} & B_{22} \end{bmatrix} \begin{bmatrix} \Delta\phi_{fa} \\ \Delta R_a \end{bmatrix}$$

$$= Ae_{ia} + Bu \dots\dots\dots\dots\dots\dots\dots\dots\dots\dots (3)$$

Where:

$$A_{11} = -\left(\hat{R}_a + g_{11}\right)\Big/L_a, A_{12} = \left(\omega_{re}L_a - g_{12}\right)/L_a, A_{21} = \left(-\omega_{re}L_a + g_{21}\right)/L_a$$

$$A_{22} = -\left(\hat{R}_a + g_{22}\right)\Big/L_a, B_{11} = 0, B_{12} = -i_{da}/L_a, B_{21} = -\omega_{re}/L_a, B_{22} = -i_{qa}/L_a$$

$$e_{ia} = \begin{bmatrix} i_{da} - \hat{i}_{da} \\ i_{qa} - \hat{i}_{qa} \end{bmatrix}, u = \begin{bmatrix} \Delta\phi_{fa} \\ \Delta R_a \end{bmatrix}$$

The above linear equation consists of the armature current, the armature voltage, the turning angle speed, and the armature winding resistance. The linearization makes the first separate equilibrium point (hereafter, it is shown that subscript 0 with lower right is an equilibrium point) and changes mathematical model gain when approximating. Each equilibrium point of the Eqs (1) and (2) is assumed to take the same value:

$$s\begin{bmatrix} i_{da} - \hat{i}_{da} \\ i_{qa} - \hat{i}_{qa} \end{bmatrix} = \begin{bmatrix} A_{110} & A_{120} \\ A_{210} & A_{220} \end{bmatrix} \begin{bmatrix} i_{da} - \hat{i}_{da} \\ i_{qa} - \hat{i}_{qa} \end{bmatrix} + \begin{bmatrix} B_{110} & B_{120} \\ B_{210} & B_{220} \end{bmatrix} \begin{bmatrix} \Delta\phi_{fa} \\ \Delta R_a \end{bmatrix}$$

$$= A_0 e_{ia} + B_0 u \dots\dots\dots\dots\dots\dots\dots\dots\dots\dots\dots\dots (4)$$

The mathematical model gain of Eq. (5) is chosen.

$$g_{11} = g_{22} = g_a, g_{12} = \omega_{re}L_a, g_{21} = -\omega_{re}L_a \dots\dots\dots\dots\dots (5)$$

The characteristic equation of the mathematical model is shown in Eq. (6):

$$(sI - A_0) = \begin{bmatrix} s + \dfrac{\hat{R}_{a0} + g_{a0}}{L_a} & 0 \\ 0 & s + \dfrac{\hat{R}_{a0} + g_{a0}}{L_a} \end{bmatrix}$$

$$\dots\dots\dots\dots\dots\dots\dots\dots\dots (6)$$

The mathematical model gain, $g_a$ , which Eq (6) stabilizes, is chosen. Moreover, the linearization estimation error becomes as follows:

$$\begin{bmatrix} i_{da} - \hat{i}_{da} \\ i_{qa} - \hat{i}_{qa} \end{bmatrix} = (sI - A_0)^{-1} B_0 \begin{bmatrix} \Delta\phi_{fa} \\ \Delta R_a \end{bmatrix}$$

$$= \begin{bmatrix} 0 & \dfrac{1}{s + \dfrac{\hat{R}_{a0} + g_{a0}}{L_a}}\left(-\dfrac{i_{da0}}{L_a}\right) \\ \dfrac{1}{s + \dfrac{\hat{R}_{a0} + g_{a0}}{L_a}}\left(-\dfrac{\omega_{re0}}{L_a}\right) & \dfrac{1}{s + \dfrac{\hat{R}_{a0} + g_{a0}}{L_a}}\left(-\dfrac{i_{qa0}}{L_a}\right) \end{bmatrix} \begin{bmatrix} \Delta\phi_{fa} \\ \Delta R_a \end{bmatrix}$$

$$= P_0(s) \begin{bmatrix} \Delta\phi_{fa} \\ \Delta R_a \end{bmatrix} \dots\dots\dots\dots\dots\dots\dots\dots (7)$$

From where the transmission function matrix $P_0(s)$ is:

$$P_0(s) = \begin{bmatrix} 0 & \dfrac{1}{s + \dfrac{\hat{R}_{a0} + g_{a0}}{L_a}}\left(-\dfrac{i_{da0}}{L_a}\right) \\ \dfrac{1}{s + \dfrac{\hat{R}_{a0} + g_{a0}}{L_a}}\left(-\dfrac{\omega_{re0}}{L_a}\right) & \dfrac{1}{s + \dfrac{\hat{R}_{a0} + g_{a0}}{L_a}}\left(-\dfrac{i_{qa0}}{L_a}\right) \end{bmatrix} \dots\dots\dots\dots\dots (8)$$

The inversion of $P_0(s)$ is:

$$P_0^{-1}(s) = \frac{\tilde{P}_0(s)}{\det(P_0)}$$

$$P_0^{-1}(s) = \begin{bmatrix} \left(s + \dfrac{\hat{R}_{a0} + g_{a0}}{L_a}\right)\left(\dfrac{L_a i_{qa0}}{\omega_{re0}i_{da0}}\right) & \left(s + \dfrac{\hat{R}_{a0} + g_{a0}}{L_a}\right)\left(-\dfrac{L_a}{\omega_{re0}}\right) \\ \left(s + \dfrac{\hat{R}_{a0} + g_{a0}}{L_a}\right)\left(-\dfrac{L_a}{i_{da0}}\right) & 0 \end{bmatrix} \dots\dots\dots (9)$$

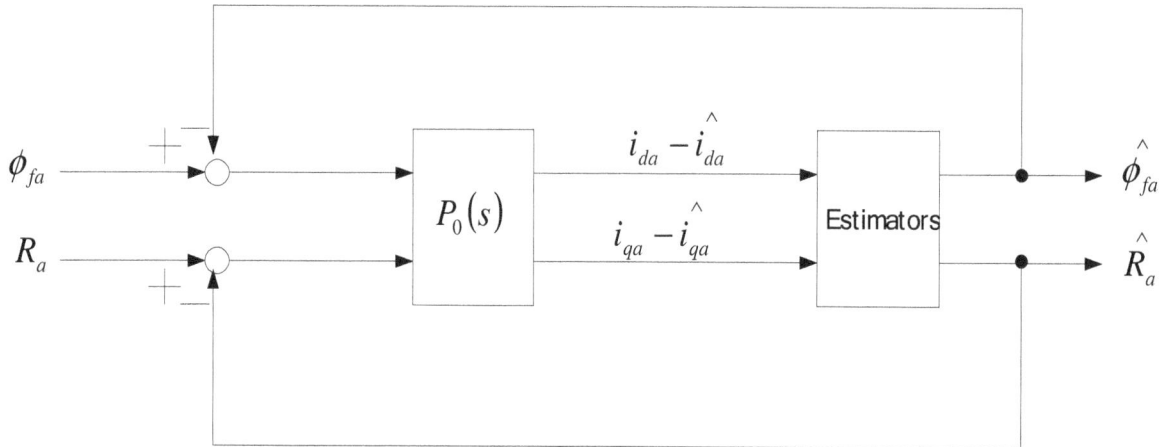

**Figure 2.** Construction of estimation system.

### *Influence of parameter errors*

Controlling torque of SPMSM is usually done with torque sensor. However, we can reduce the cost by making it sensorless. The estimation of the magnet flux, which depends on temperature, is therefore necessary and indispensable. The magnet type of the tested motor is Nd-Fe-B and the temperature coefficient is about $0.11 \% / {}^{0} C$. When the temperature of the motor changes to $\pm 75^{\circ}C$, the variation of the permanent magnet flux is within $\mp 9\%$ and the armature winding resistance changes to about $\pm 24\%$. In order to be related, occasionally the magnet flux is required in order to make a direct relation between the temperature and the magnet flux. With this proposition, the armature winding resistance and the magnet flux, which depend on temperature, are estimated; it is something which assures improvement of torque efficiency by the fact that the magnet flux estimator is used for control.

### COMPOSITION OF ESTIMATOR

#### *Construction of estimation system*

The construction of an estimation system is shown in Figure 2 and it showed the I

$$
\begin{bmatrix} \hat{\phi}_f \\ \hat{R}_a \end{bmatrix} = P_0^{-1}(s) \begin{bmatrix} i_{da} - \hat{i}_{da} \\ i_{qa} - \hat{i}_{qa} \end{bmatrix}
$$
................................(10)

And the estimator is:

$$
\begin{bmatrix} \hat{\phi}_f \\ \hat{R}_a \end{bmatrix} = \begin{bmatrix} \left(1+\frac{1}{s}\frac{\hat{R}_a+g_a}{L_a}\right)\left(\frac{K_{p1}L_a i_{qa}}{\omega_{re} i_{da}}\right) & \left(1+\frac{1}{s}\frac{\hat{R}_a+g_a}{L_a}\right)\left(-\frac{K_{p2}L_a}{\omega_{re}}\right) \\ \left(1+\frac{1}{s}\frac{\hat{R}_a+g_a}{L_a}\right)\left(\frac{K_{r1}L_a}{i_{da}}\right) & 0 \end{bmatrix} \begin{bmatrix} i_{da}-\hat{i}_{da} \\ i_{qa}-\hat{i}_{qa} \end{bmatrix}
$$
...... (11)

Here, $\omega_{re}$ and $i_{da}$ exist in the denominator when paying attention to the element of the first column of the first row. This makes $i_{da}$ an important parameter, because if it is zero we will not able to estimate the magnet flux and the stator winding resistance. The same consequences will happen when the speed becomes zero, that is $\omega_{re} = 0$. However, from the viewpoint of speed resolution, estimation is not possible even at speed near to zero, that is, below the nominal speed.

In both Eqs (7) and (9), we get:

$$
\begin{bmatrix} \hat{\phi}_{fa} \\ \hat{R}_a \end{bmatrix} = \begin{bmatrix} \dfrac{K_{p20}}{s} & \dfrac{\left(- K_{p10} + K_{p20}\right)i_{qa0}}{\omega_{re0}s} \\ 0 & \dfrac{K_{r10}}{s} \end{bmatrix} \begin{bmatrix} \Delta\phi_{fa} \\ \Delta R_a \end{bmatrix}
$$
...............(12)

Where $K_{p10}, K_{p20}$ and $K_{r10}$ represent each integrator gain in Eq. (12). The second element of the first row becomes zero with $K_{p10} = K_{p20}$. We thus have the following formulation (Figure 2):

$$
\begin{bmatrix} \hat{\phi}_{fa} \\ \hat{R}_a \end{bmatrix} = \begin{bmatrix} \dfrac{K_{p20}}{s} & 0 \\ 0 & \dfrac{K_{r10}}{s} \end{bmatrix} \begin{bmatrix} \Delta\phi_{fa} \\ \Delta R_a \end{bmatrix}
$$
........................... (13)

$$
\begin{bmatrix} \hat{\phi}_{fa} \\ \hat{R}_a \end{bmatrix} = \begin{bmatrix} \dfrac{K_{p20}}{s + K_{p20}} & 0 \\ 0 & \dfrac{K_{r10}}{s + K_{r10}} \end{bmatrix} \begin{bmatrix} \phi_{fa} \\ R_a \end{bmatrix}
$$
...................... (14)

These two equations are preferable when they are stable and are used to adjust the integrator gain to the satisfaction derived from value response. This is because the Eq. (14) is derived from the Eq. (13). The advantage of this method is robustness of estimation, because the influence of the armature winding resistance

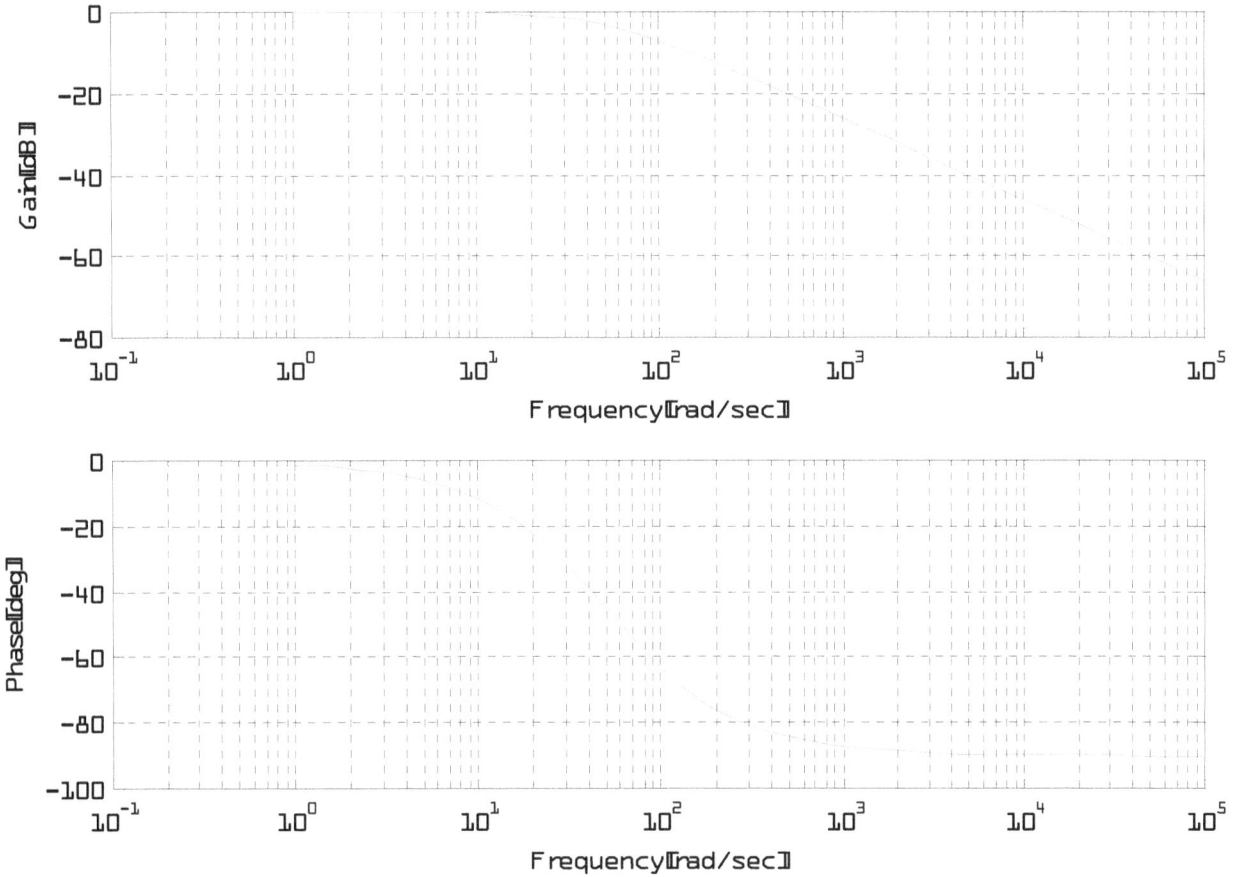

**Figure 3.** Bode diagram of $\hat{\phi}_{fa} \big/ \phi_{fa}$ .

estimation error $\Delta R_a$ is not easily received in the magnet flux estimation. The above two equations for the open-loop and the closed-loop transfer functions in the Eq. (13), (14) respectively work only at high-speed from where we can estimate the magnet flux. However, Eq. (1) can be used at speed near to zero because we cannot estimate the magnet flux and use it. First of all, we need to find when the speeds rotate faster than resolution. At that point, the open-loop transfer function expressed by Eq. (13) can estimate the magnet flux

Figure 3 shows the characteristic of the stability and it is realized at $K_{p20}$ = 50 . When a current condition is applied, it is arranged like Eqs. (13) and (14). The gain and phase differences between the input and output of the estimation system are a simple pole. The following observations can be made from Figure 3.
-That for a simple real pole, the piecewise linear asymptotic Bode plot for magnitude is at 0dB until the break frequency; and then drops at 20dB per decade (the slope is -20dB/decade).
-The phase plot is at 0° until one tenth the break frequency, and then drops linearly to -90º at ten times the break frequency.

### Improvement of the speed resolution

The problem of low-speed rotating is recorded as follows. The output $\hat{\phi}_{fa}$ of the estimator might become unstable when it is more low-speed than Eq. (15). The Eq. (15) becomes low-speed when the torque is output as for the injection molding machine. Therefore, it is necessary to estimate the magnet flux until the speed is near to zero. Under this condition, there is a method of improving speed resolution with increase of the number of moving average samples. The calculated example of the improvement of the speed resolution $\Delta \omega_{rm}$ (mechanical speed) is recorded as follows.

$$\Delta \omega_{rm} = \frac{2\pi}{n_p \bullet t_c \bullet N} [\text{rad/sec}]$$ ........................ (15)

Here $n_p$ [pulse/rev ] : the encoder pulse number $t_c$ [µsec] : operational period, $N$ number of moving average samples. $n_p = 4000$ [pulse/rev ], $t_c = 204.8$ [µsec] , $N = 64$ .

Eq. (15) shows the improvement of the speed resolution with an increase in the number of samples of moving averages. That is necessary because even at speed near to zero, we can detect the speed, and the flux estimation becomes possible. The reason to use Eq. (15) is to obtain the operation accuracy even when there is a sudden change in speed. At this point, we need to know where the estimation should stop. This is because the flux estimation

**Table 1.** Rating of tested motor.

| Rated power | | 1.5kW |
| --- | --- | --- |
| Rated current | | 8.6 A |
| Rated speed | | 2000rpm |
| Number of pole-pairs | $p$ | 3 |
| Torque constant | $\phi_{fa}$ | 0.1946Wb |
| Armature winding resistance | $R_a$ | 0.5157 $\Omega$ |
| Armature winding self-inductance | $L_a$ | 2.452mH |
| Moment of inertia | $J$ | $0.00525\ kg \cdot m^2$ |

cannot be done at speed slower than the resolution speed.

**Simulation and experimental conditions**

A test system was composed of a Digital Signal Processor DSP (TMS320C31-5kHz) control system (Texas instruments), a 3-phase PWM inverter and a 1.5 kW SPMSM. The operation cycle is 200 $\mu s$; the career frequency of the PWM inverter that drove the evaluation machine was assumed to be 5 KHz. The load machine used is the induction motor of 1.5kW and moment of inertia is 0.009 $kg \cdot m^2$. The detected current and voltage are fed to the input of DSP and then the voltage order was calculated. The voltage instruction from DSP is converted into the PWM signal, and then the short-circuit prevention time is added by FPGA which generates output to the circuit of the drive at the gate. Signal carrier (triangular wave) cycle was assumed to be 204.8 $\mu s$ using the triangular wave comparison method for the generation of the PWM signal. Hall CT (HAS-50S: LEM) was used for the current detector. The voltage proportional to the current from hall CT is output, and the voltage signal is converted into the digital signal with 16 bit A/D converter (AD976:AnalogDevices). DC power voltage $E_{DC}$ of the inverter is detected with 12 bit A/D converter (AD7864: Analog Devices) connected through the partial pressure machine. Voltage type PWM inverter is composed of the power-module and the circuit of the drive at the gate. IGBT-IPM (6MBP30RH060: Fuji Electric Co., Ltd.) was used for the power-module. The direct current voltage power supply of the inverter has vector control of faction 2.2kW of the three-phase circuit 200V type inverter (FRN2.2VG7S-2: Fuji Electric Co., Ltd.) that controls the torque; and DC linked the load machines. The pulse number output from the encoder used this time is 1000 pulses per rotation.

As for the voltage detection error margin $\pm 1/(2^{13}) = \pm 1/8192$, the delay of the voltage feed back loop becomes $300\mu s$, which is 1.5 times at sampling period $204.8\mu s$.

Rating of tested motor used in this experiment and simulation are in Table 1.

As for injection molding machine, the torque is particularly important especially at low speed and the estimation of the magnet flux at low speed becomes necessary.

With the increase of the stator winding temperature, the stator winding resistance increases and the magnet flux decreases as shown in Figure 4. The magnet flux estimation with robustness was confirmed by the simulation based on the condition described in

Table 1. This is because stator winding resistance is not estimated. The simulation has been made through the language of technical computing (MATLAB) that omitted the PWM inverter.

The estimation of magnet flux in the simutation started at 5 s. Also, during 10 s, the temperature in the real machine would not vary as shown in the simulation.

The simulation results in Figure 4 show that when the temperature increases, the armature winding resistance increases and the magnet flux decreases.

**RESULTS**

Figures 5, 6 showed the simulation results of estimated magnet flux with 25% load at 20 and 200 rpm when the temperature as well as the stator winding resistance does not change. The initial value is fixed at 0.295 Wb and the estimation started at 0.8 s.

The result of the flux estimation at 20 rpm with no-load by the experiment is recorded in Figure 7. The result is similar to that of the simulation- when the temperature and stator resistance do not change.

The estimation of flux at 100% load is easy to realize at low speed during the experiment but it is more difficult to realize the result at 25% load. Therefore, it showed the result of experiment at 25% load during 20 and 200 rpm (Figures 8 and 9).

This experiment shows that, the estimation of magnet flux was effective because the voltage sensor is used in the experimental operation and hence the estimation error is 0.3%, less than 0.5%. Also, these results showed the performance of the experiment compared to the results of simulation is similar (having the same condition).

**DISCUSSION**

The magnet flux information is important for controlling the torque of the SPMSM with vector control. We propose in this work a new method for estimating the magnet flux of the SPMSM with vector control. The proposed method showed good estimated performance by designing and

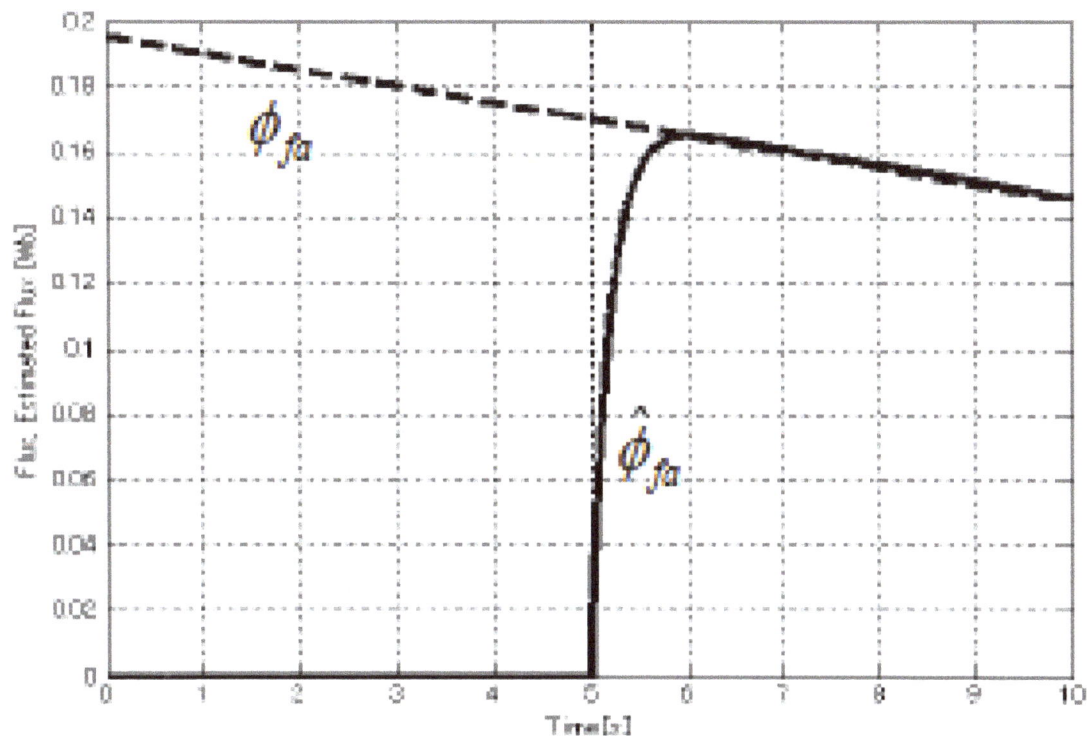

**Figure 4.** Simulation results for the flux and estimated flux.

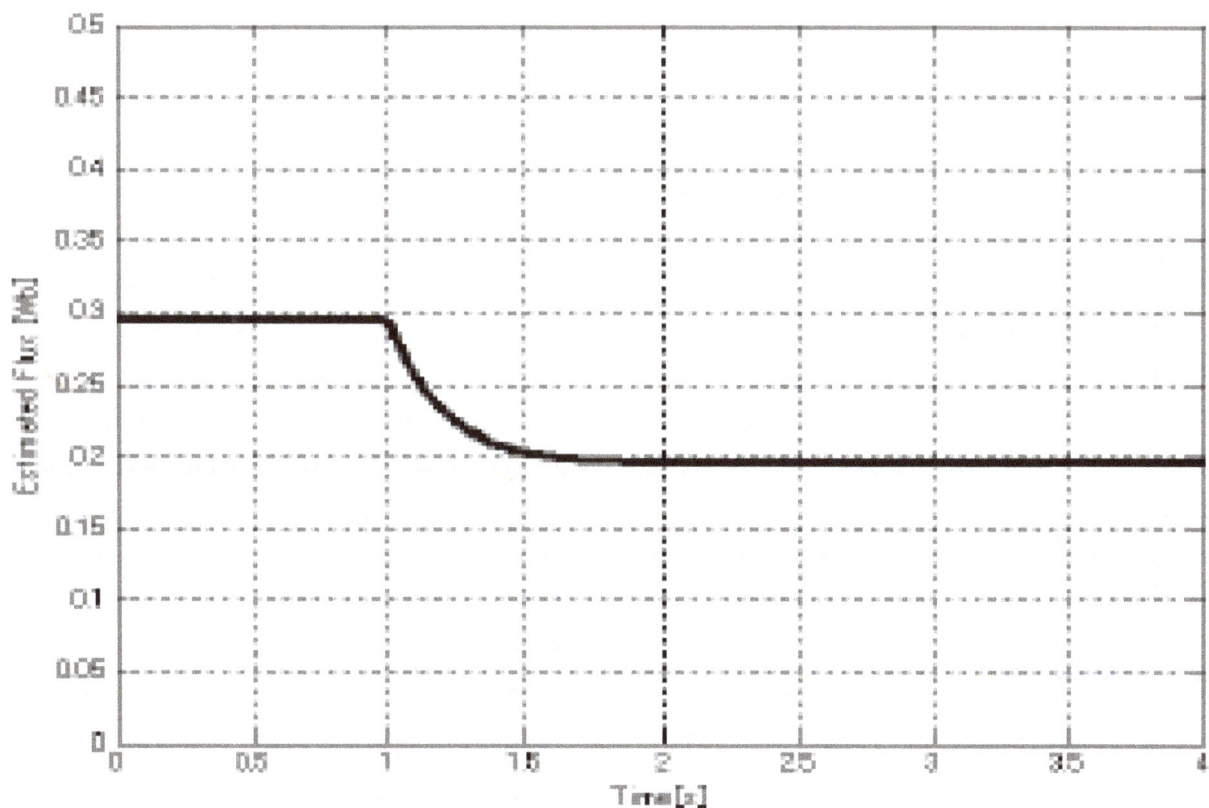

**Figure 5.** Simulation result for estimated magnet flux (25% load, 20 rpm).

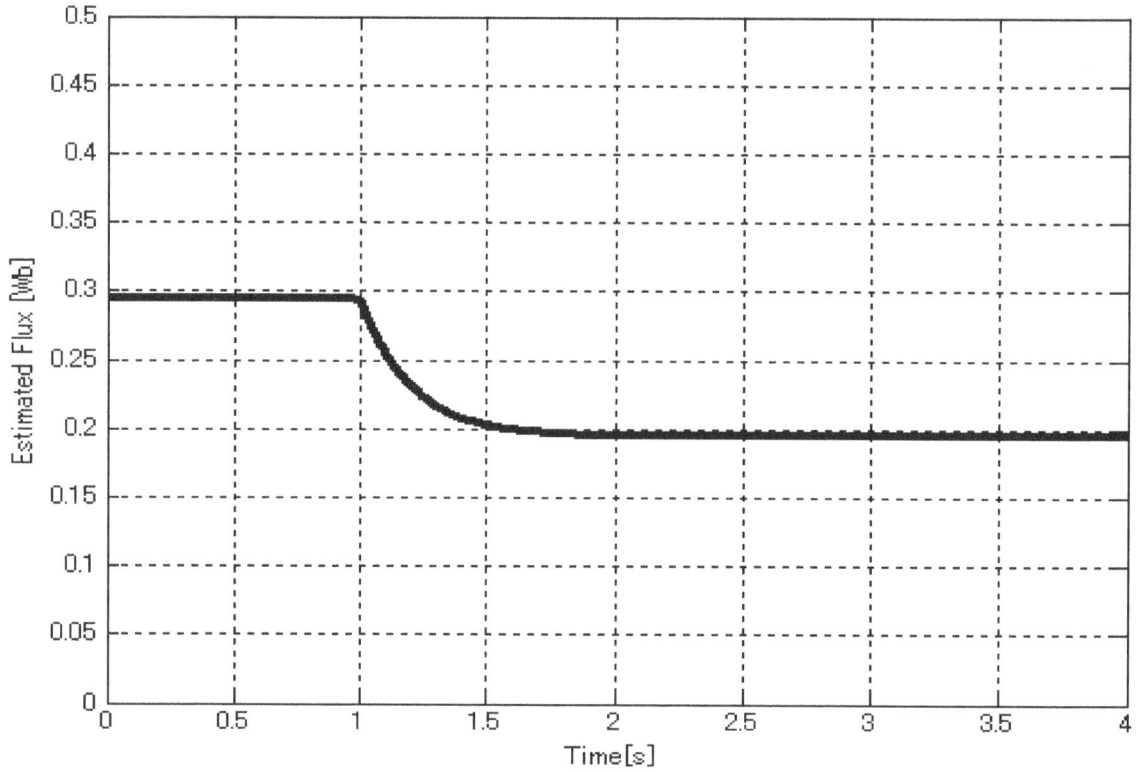

**Figure 6.** Simulation result for estimated for magnet flux (25% load, 200 rpm).

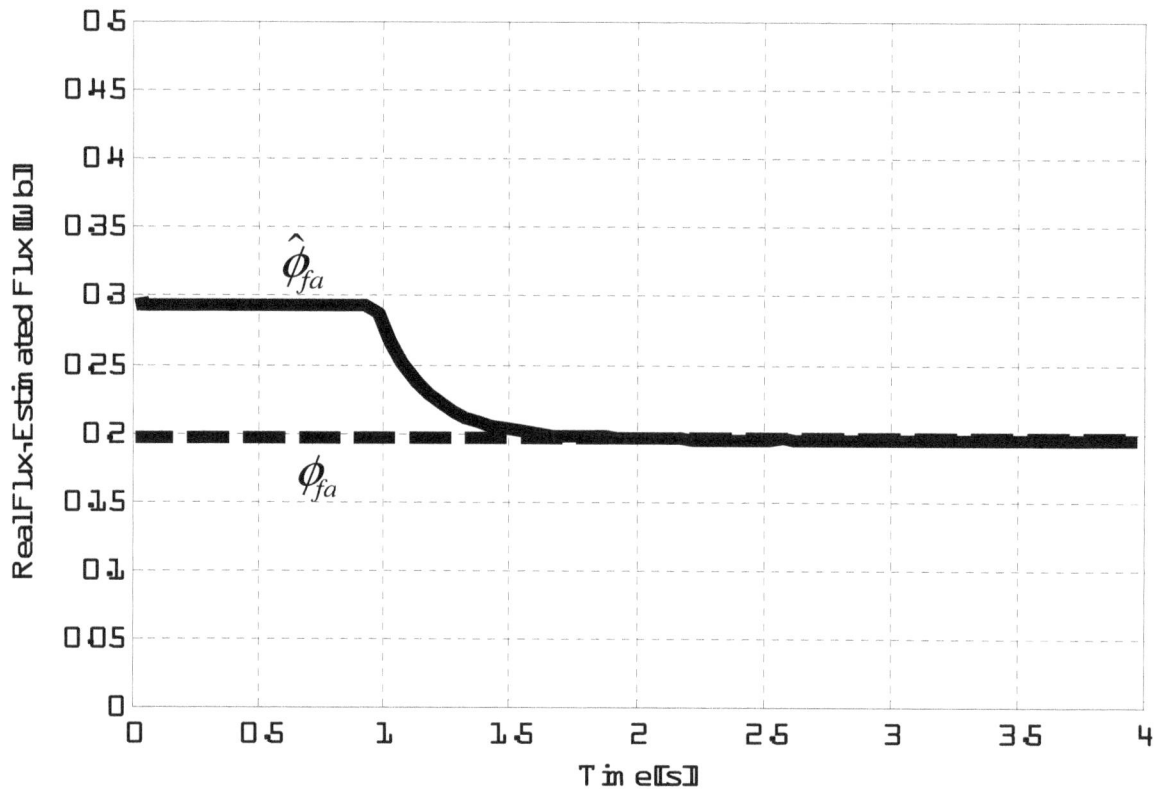

**Figure 7.** Experimental results for estimated of magnet flux (no-load, 20 rpm ).

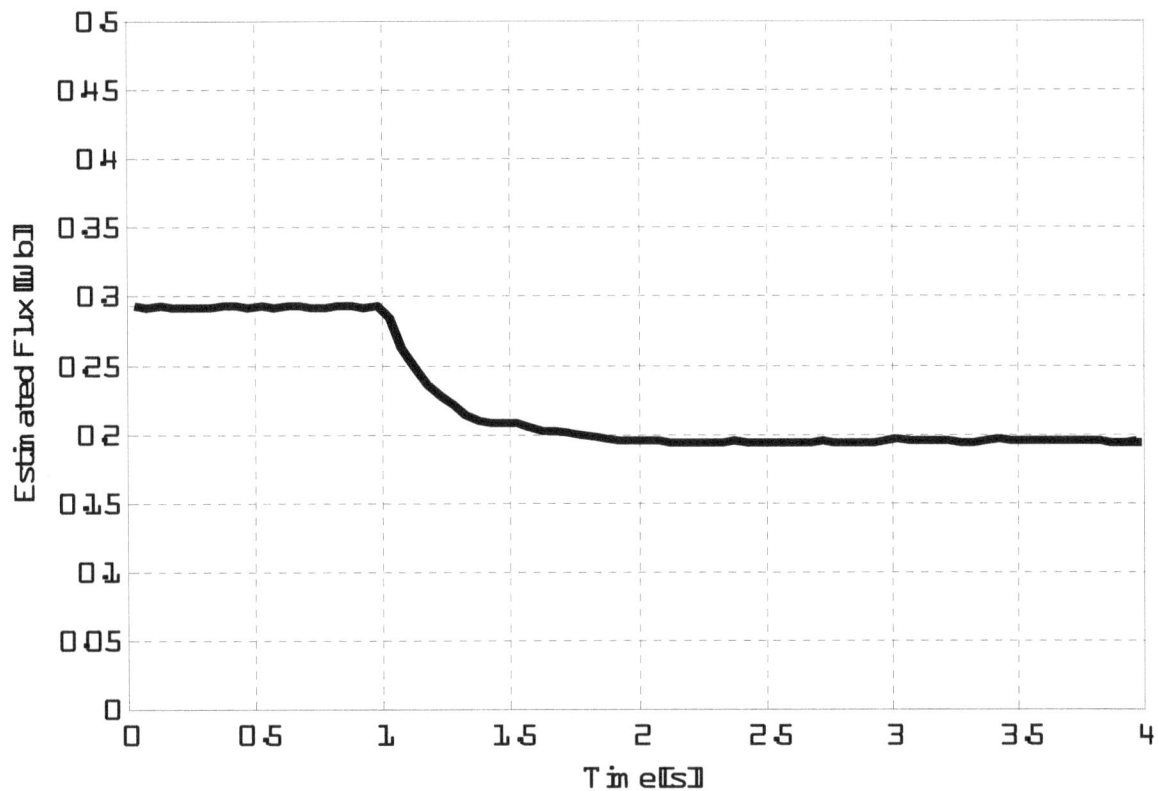

**Figure 8.** Experimental results for estimated of magnet flux (25% load, 20 rpm).

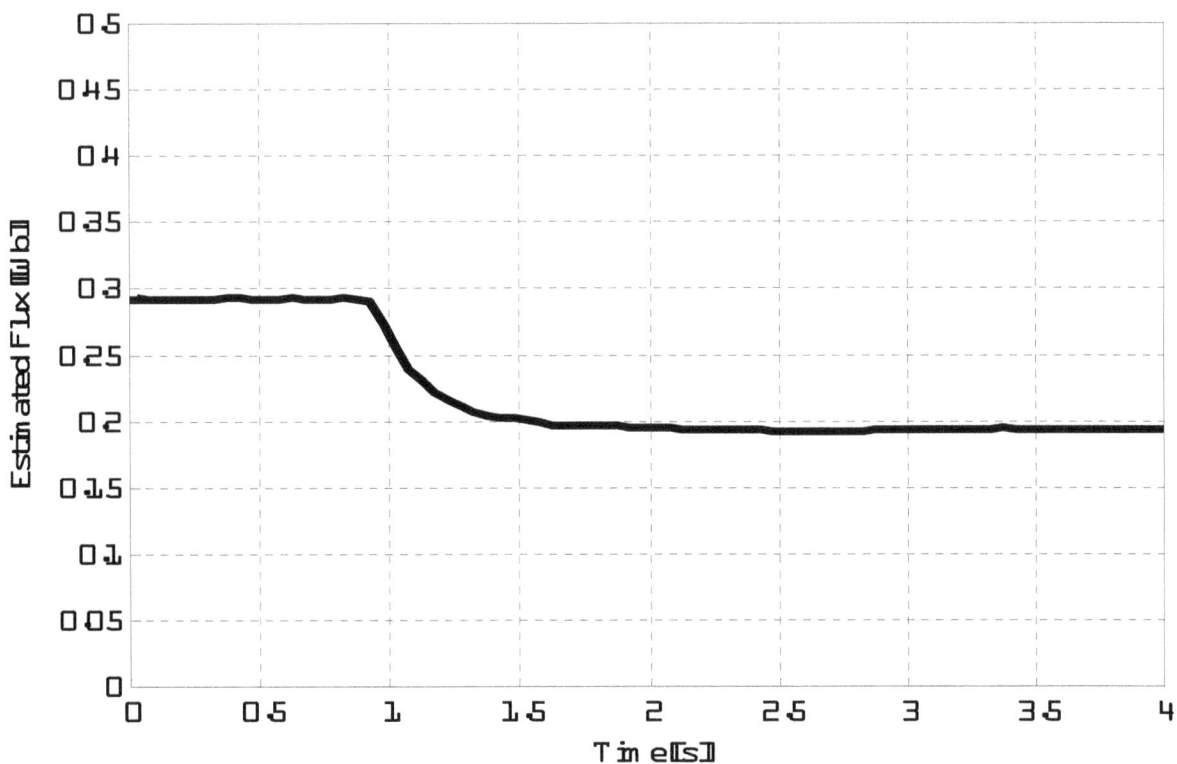

**Figure 9.** Experimental results for estimated of magnet flux (25% load, 200 rpm).

simulating the estimator of the magnet flux from linearization error state equation and also the robustness of the estimated magnet flux is demonstrated by noninterference of the stator winding resistance variation caused by the increase of the temperature in SPMSM. Currently, the condition of the experiment and starting flux estimation were not the same value. However, in future, we like to use the same value for both cases.

## Conclusion

Since the torque is necessary in the injection molding machine during operation, it is constant even at low speed; and so the magnet flux estimation becomes necessary for torque control systems without torque sensor during the variation of stator winding resistance. Therefore, this paper presents good estimated performance by designing and simulating the estimator of the magnet flux with vector control using adaptive identification from linearization error state equation and also presents the robustness demonstrated by noninterference of the stator winding resistance variation caused by the increase of the temperature in SPMSM.

## ACKNOWLEDGEMENT

The authors are thankful to Hidehiko Sugimoto of the Department of Electrical and Electronics Engineering for his sincere help.

## REFERENCES

Beccera RC, Jahns TM, Ehsani M (1991). Four quadrant sensorless brushless ECM drive. IEEE Power Electronics Conference and Exposition, Conference, pp. 202-209.
Fukumoto HT, Hayashi HY (2007). Performance Improvement of the IPMSM Position Sensor-less Vector Control System by the On-line Motor Parameter Error Compensation and the Practical Dead-time compensation. Power Conversation Conference-Nagoya PCC, (07): 314-321.
Han YS, Choi JS, Kim YS (2000). Sensorless PMSM drive with a sliding mode controlbased adaptive speed and stator resistance estimator. IEEE Trans Magnet., 36(5): 3588-91:3588-3591.
Hayashi Y, Kajino D, Kurita T, Hamane H, Fukumoto T (2005). Analysis and Design of an IPMSM Speed Sensor-less Vector Control System Using an Adaptive Observer. The International Power Electronics Conference. pp. 2205-2212.
Rahman MF, Zhong L, Haque E Md, Rahman MA (2003). A Direct Torque Controlled Interior Permanent-Magnet Synchronous Motor Drive Without a speed Sensor. IEEE Trans on energy conversion, 18(1): 17-22.
Takeshit T, Ichikawa M, Lee JS, Matui M (1997). "Back EMF Estimation-BasedSensorless Salient – Pole Brashless DC Motor Drives", Trans. IEE of Japan, 117(1): 98-104.
Tang L, Zhong L, Rahman MF, Hdu Y (2003). A Novel Direct Torque Control for Interior Permanent-Magnet Synchronous Machine Drive With Low Ripple in Torque and Flux- A Speed-Sensorless Approach. IEEE Trans. on Industry Applications, 39 (6): 1748-1756.
Yoshiharu I, Shunsuke MA, Masao K (1999). Force Control with Servomechanism in fully Electric Injection Molding Machine. The Japan Society for Precision Engineering, 65(4): 542-548.

# 5

# Artificial neural networks applied to DGA for fault diagnosis in oil-filled power transformers

**Mohammad Golkhah[1]\*, Sahar Saffar Shamshirgar[2] and Mohammad Ali Vahidi[3]**

[1]Department of Electrical Engineering University of Manitoba, Winnipeg, Canada.
[2]Islamic Azad University of Sciences and Researches, Tehran, Iran.
[3]K. N. Toosi University of Technology, Tehran, Iran.

Dissolved Gas Analysis (DGA) is a popular method to detect and diagnose different types of faults occurring in power transformers. This objective is obtained by employing different interpretations of dissolved gases in the mineral oil insulation of such transformers. This paper engages these interpretations and applies appropriate Artificial Neural Networks (ANN) to classify the different faults. Each interpretation method needs special neural network to determine the occurred fault. Three ANNs are applied to this aim. The classification results and some typical examples are presented to validate the networks.

**Key words:** DGA, duval triangle, ANN, power transformer faults.

## INTRODUCTION

Faults in power transformers can significantly decline the longevity of mineral oil insulation of those transformers. It is essential to detect and eliminate the occurred fault very soon preventing any jeopardous results. Insulating mineral oils under faults release gases which dissolve in the oils. The distribution of these gases relates to the type of fault. Analysis of the dissolved gases can result in very useful information in the maintenance programs. The advantages of dissolved gas analysis can be briefly stated as (DiGiorgio, 1996):

(i) Advance warning of developing faults.
(ii) Determining the improper use of units.
(iii) Status checks on new and repaired units.
(iv) Convenient scheduling of repairs.
(v) Monitoring of units under overload.

There are different detection and interpretation methods (DiGiorgio, 1996; Duval, 2006). IEC and ANSI/IEEE

standards are among the most prestigious sources for the dissolved and free gas interpretations (ANSI/IEEE C57.104; IEC 60599). Each interpretation method has its own pros and cons. These methods will be shortly discussed and evaluated.

The term of 'fault gases' is used to hint the gases which are originated through the faults. These fault gases are Methane ($CH_4$), Ethane ($C_2H_6$), Ethylene ($C_2H_4$), Acetylene ($C_2H_2$), Hydrogen ($H_2$), Carbon monoxide (CO), Carbon dioxide ($CO_2$), and the non-fault gases are Nitrogen ($N_2$), and Oxygen ($O_2$).

In addition to the oil, insulating papers also provide some gases under faults. The percentage of released gases under different faults is stated in Table 1. Corona, pyrolysis (over heating), and arcing in the oil and pyrolysis in the cellulose are considered as different types of faults in Table 1.

As a result, each of these gases can individually represent type fault. Table 2 presents such a conclusion (Jakob, 2003; Lewand, 2003).

There are different methods to measure the value of fault gases of the oil. The total combustible gases (TCG) and gas blanket analysis are such methods which take a sample of the space above the insulating oil in the power

\*Corresponding author. E-mail: m.golkhah@gmail.com.

**Table 1.** Percentage of each released gas under different faults.

| Fault type | $H_2$ (%) | $CO_2$ (%) | CO (%) | $CH_4$ (%) | $C_2H_6$ (%) | $C_2H_4$ (%) | $C_2H_2$ (%) |
|---|---|---|---|---|---|---|---|
| Corona in oil | 88 | 1 | 1 | 6 | 1 | 0.1 | 0.2 |
| Pyrolysis in oil | 16 | TRACE | TRACE | 16 | 6 | 41 | TRACE |
| Arcing in oil | 39 | 2 | 4 | 10 | TRACE | 6 | 35 |
| Pyrolysis in cellulose | 9 | 25 | 50 | 8 | TRACE | 4 | 0.3 |

**Table 2.** Interpretation based on a single released gas amount.

| Gases | Indication |
|---|---|
| Hydrogen | Partial discharge, heating, arcing |
| Methane ,Ethane, Ethylene | "Hot metal" gases |
| Acetylene | Arcing |
| Carbon oxides | Cellulose insulation degradation |

transformers (DiGiorgio, 1996). TCG has the advantage of high speed analysis and continues monitoring but it is not able to collect noncombustible gases such as Carbon dioxide, Nitrogen, and oxygen. The gas blanket analysis is capable of sampling both combustible and noncombustible fault gases.

In general, both of the mentioned methods suffer from some disadvantages. Indeed, these methods can not be engaged to detect fault gases in transformers which are full of oil and do not contain any gas blanket above their insulating oil. Furthermore, since the faults are often originated from the bottom of the oil, it takes time to the released gases to saturate the oil at first and then penetrate in the gas blanket. Therefore, the total time of the analysis will be significantly augmented.

Dissolved Gas Analysis (DGA) is the most popular informative method to this aim. In this method, a sample of oil containing dissolved fault gases is taken from the oil of the unit; then the fault gases are detached from the sample. Eventually, each gas is separated from the others and the value of each gas is derived in part per million level (ppm). The main advantage of DGA is the quick detection of the gases right after occurrence of a fault. All these methods provide the value of fault gases in the oil. Now it is required to interpret the attained values to determine the type of the occurred fault.

There are some interpretation methods which classify the faults according to the obtained gases values. Artificial neural networks are employed to solve these pattern classifications for three popular interpretation methods in this paper.

## DORNENBURG PLOT

This earlier IEEE method plots two different ratios in two axes. Three different faults, Thermal, arcing, and corona, can be detected by using this method.

A multilayer perceptron neural network is designed to simulate Dornenburg interpretation. Construction of this network is presented in Figure 1. This configuration contains 10 neurons in the first layer and three neurons in the last one. Inputs nodes are ratios of $C_2H_2/C_2H_4$ and $CH_4/H_2$ and three outputs represent three types of faults. Each output node is assigned to a special type of fault hence the neuron which is high in its output indicates that which fault is occurred.

Transfer function of all the neurons of the two layers is the step function. When the input of a step function is negative, the output becomes zero and correspondingly the output is unity when the input is at least zero. The first layer is designed to make all the decision boundaries and the second one plays an OR rule to create three different classes of the three faults. Each input is applied to all the neurons of the first layer by a weight. All of the neurons include biases. Abbreviated notation of this network is also presented in Figure 2 (Hagan, 1996)

Weight and bias matrixes are evaluated as (1) to (4):

$$W^1 = \begin{bmatrix} 1 & -1 & 0 & 1 & 0 & -1 & 0 & 1 & 0 & 0 \\ 0 & 0 & 1 & 0 & -1 & 0 & 1 & 0 & -1 & 1 \end{bmatrix} \quad (1)$$

$$b^1 = \begin{bmatrix} 0 & 2 & -1 & -2 & 0.07 & 5.83 & 0 & -5.84 & 1 & -0.07 \end{bmatrix}^T \quad (2)$$

$$W^2 = \begin{bmatrix} 1 & 1 & 1 & 0 & 0 & 0 & 0 & 0 & 0 & 0 \\ 0 & 0 & 0 & 1 & 1 & 1 & 1 & 0 & 0 & 0 \\ 0 & 0 & 0 & 0 & 0 & 0 & 0 & 1 & 1 & 1 \end{bmatrix}^T \quad (3)$$

$$b^2 = \begin{bmatrix} -2.5 & -3.5 & -2.5 \end{bmatrix}^T \quad (4)$$

This network has been simulated in the Matlab software and the classification problem has been solved. A large amount of random inputs have been applied as inputs and Figure 3 has been obtained. Red areas are corresponding to thermal faults, green areas represent

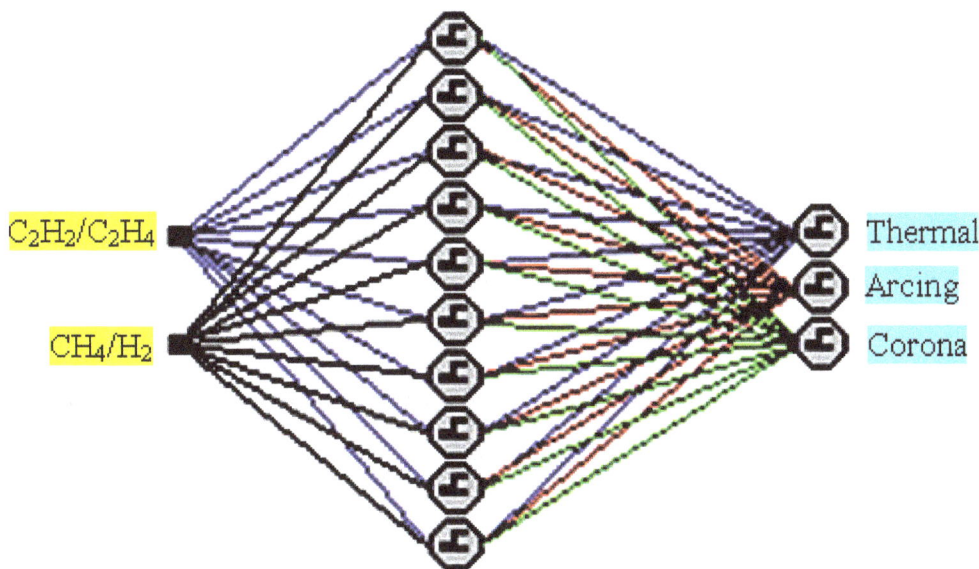

**Figure 1.** Multilayer perceptron for Dornenburg method.

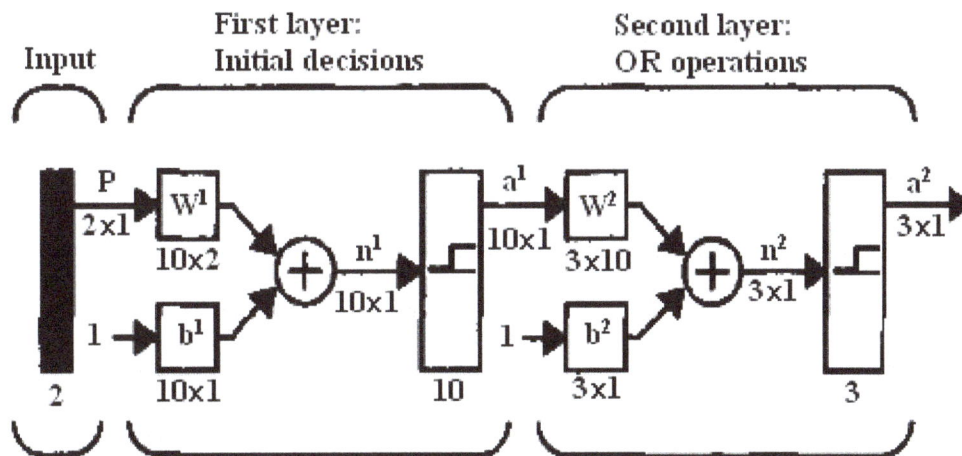

**Figure 2.** Configuration of the three layer perceptron.

arcing faults, and blue areas indicate corona faults.

## VALIDATION OF THE NETWORK

Some experimental data and the type of fault have been presented in (Jakob, 2003). To validate the results of the proposed neural network, these data have been engaged. All of the values are in ppm.

The data presented in Table 3 have been obtained by the method of DGA under normal operation of the power transformer (Jakob, 2003). The proposed neural network is employed to judge about the condition. Figure 4

indicates that the network correctly selects the normal condition.

One year later, this unit was tested once again. The obtained data indicated that the unit was under thermal runaway condition. Table 4 represents the data. The neural network notices that the unit is under heating fault (Figure 4).

Engineers removed the unit from the power system to repair. The unit was tested again after installation. The data of Table 5 and Figure 4 prove that the unit was under normal condition.

Red areas are corresponding to thermal faults, green areas represent arcing faults, and blue areas indicate

Domenburg plot

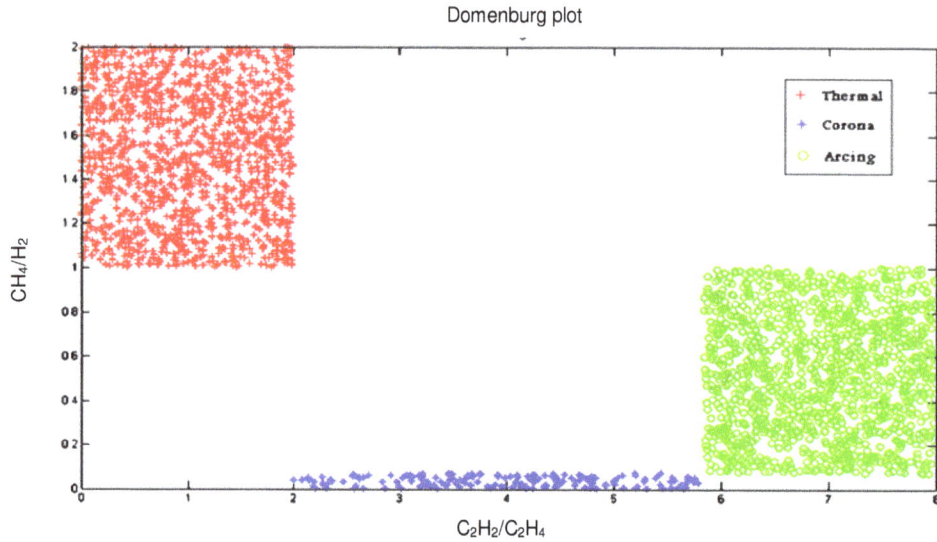

**Figure 3.** Dornenburg plot attained by two layer perceptron neural network.

**Table 3.** Experimental data on February 25, 1993.

| Date | $C_2H_2$ | $CH_4$ | $C_2H_6$ | $C_2H_4$ | $H_2$ | CO | $CO_2$ |
|------|----------|--------|----------|----------|-------|-----|--------|
| 02/25/93 | 0 | 5 | 1 | 4 | 34 | 71 | 350 |

Domenburg plot

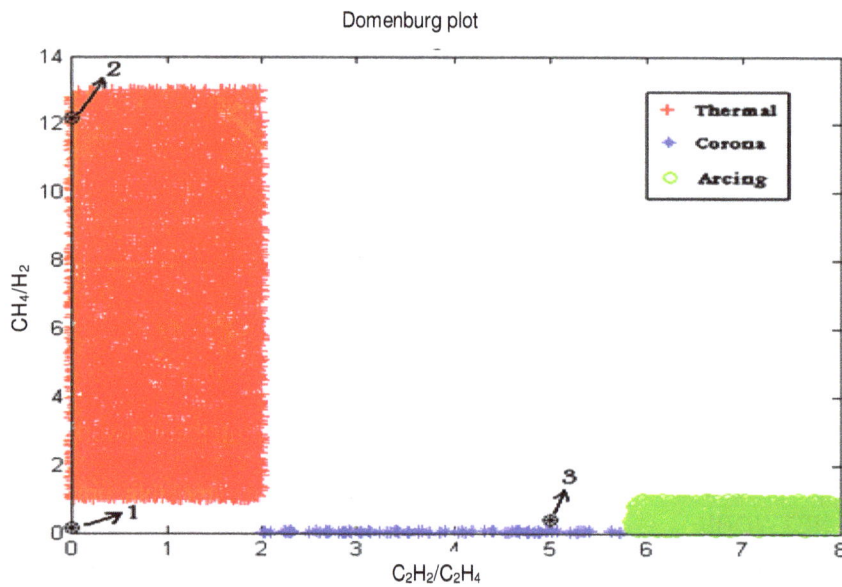

**Figure 4.** Three condition plot of the unit; 1, 3: Normal, 2: Thermal fault.

corona faults. All of the other areas show normal condition. Therefore point 1 indicates normal condition. Point 2 is situated in red areas hence the transformer is operating under overheat condition and it is required to remove the transformer and eliminate the occurred fault or repair the unit. After repairing the transformer, it should

**Table 4.** Experimental data on February 25, 1994.

| Date | $C_2H_2$ | $CH_4$ | $C_2H_6$ | $C_2H_4$ | $H_2$ | CO | $CO_2$ |
|------|------|------|------|------|------|------|------|
| 02/25/94 | 44 | 1812 | 576 | 3143 | 149 | 33 | 645 |

**Table 5.** Experimental data after repairs on February 27, 1994.

| Date | $C_2H_2$ | $CH_4$ | $C_2H_6$ | $C_2H_4$ | $H_2$ | CO | $CO_2$ |
|------|------|------|------|------|------|------|------|
| 02/27/94 | 44 | 1812 | 576 | 3143 | 149 | 33 | 645 |

**Table 6.** C. E.G. B. fault gas ratios developed by Rogers.

| Ratio | Range | Code |
|-------|-------|------|
| | ≤ 0.1 | 5 |
| | > 0.1 < 1 | 0 |
| $CH_4/H_2$ | ≥ 1 < 3 | 1 |
| | ≥ 3 | 2 |
| $C_2H_6/CH_4$ | < 1 | 0 |
| | ≥ 1 | 1 |
| $C_2H_4/C_2H_6$ | < 1 | 0 |
| | ≥ 1 < 3 | 1 |
| | ≥ 3 | 2 |
| $C_2H_2/C_2H_4$ | < 0.5 | 0 |
| | ≥ 0.5 < 3 | 1 |
| | ≥ 3 | 2 |

**Table 7.** C. E. G. B. diagnostics developed by Rogers.

| | Code | | | Diagnosis |
|---|---|---|---|-----------|
| 0 | 0 | 0 | 0 | Normal |
| 5 | 0 | 0 | 0 | Partial discharge |
| 1,2 | 0 | 0 | 0 | Slight overheating < 150°C |
| 1,2 | 1 | 0 | 0 | Slight overheating 150 - 200°C |
| 0 | 1 | 0 | 0 | Slight overheating 200 - 300°C |
| 0 | 0 | 1 | 0 | General conductor overheating |
| 1 | 0 | 1 | 0 | Winding circulating currents |
| 1 | 0 | 2 | 0 | Core and tank circulating currents, overheated joints |
| 0 | 0 | 0 | 1 | Flashover, no power follow through |
| 0 | 0 | 1,2 | 1,2 | Arc, with power follow through |
| 0 | 0 | 2 | 2 | Continuous sparking to floating potential |
| 5 | 0 | 0 | 1,2 | Partial discharge with tracking (note CO) |
| $CO_2 / CO > 11$ | | | | Higher than normal temperature in insulation |

be installed and tested by DGA and related equipments. It is done and point 3 proves that the new condition is normal and the unit can satisfy the network requirements.

## ROGERS METHOD

Central Electric Generating Board (CEGB) of Great Britain has employed a method developed by Rogers, IEEE method (Duval, 2006), in which four ratios of fault gases are calculated to generate a four digit code presenting in Table 6 and 7. Table 6 illustrates circumstance of developing the digits and Table 7 describes the fault diagnosis assigning to each of the digits.

A competitive neural network has been developed and proposed in the Matlab software to simulate the Rogers method. This network is presented at Figure 5.

Indeed this type is a Hamming network by two layers. The weights of the first layer are desired prototypes. All the inputs, the four ratios plus $CO_2/CO$, are compared to the first layer weights and the hamming distances are calculated. The less is the hamming distance, the more is the output of the neuron which has a linear transfer function. The outputs of the first layer then become the inputs of second layer, competitive layer. The second layer contains recurrent neurons in which the outputs represent one time less than the inputs. Each output of the second layer is back propagated to its input by a weight equal to unity however all the other outputs feed the input of that neuron by a "-ε" weighted loop. "ε" is much less than unity and should be less than $\frac{1}{S-1}$ where S is the number of neurons in the first layer. It is important to note that the second layer has the same number of neurons in the first one.

After following the outputs of the first layer into the second one and passing a few iterations, the neuron which has had the biggest initial value wins the competition, global winning neuron. The winning neuron has unity on its output while all of the other neurons are zero on their outputs. As a result an input which is more near to one of the weights of the first layer, will take all the other inputs, so called winner-takes all (WTA). 'D' block represents a time delay. To prevent drawing a complex diagram, which will nor be readily readable, abbreviate notation of this network is only presented. For simplification, the second layer can be replaced by a competitive layer and since all of the biases are zero and the output of a linear transfer function is equal to its input, Figure 5 can be redrawn as Figure 6.

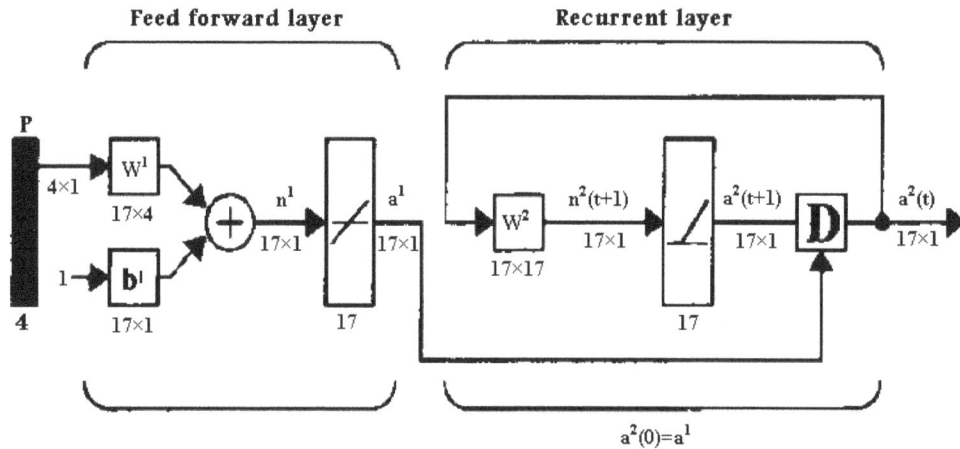

**Figure 5.** Hamming network for Rogers method.

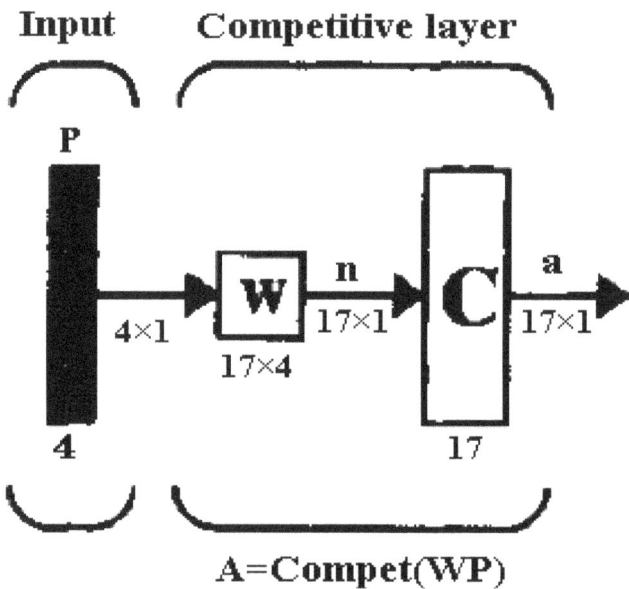

**Figure 6.** Hamming network of Figure 5 with a competitive layer.

This network has been simulated in the Matlab software. The weights of the first layer are the codes stated in table X hence it is not required to rewrite here. The simulator prompts this network while $CO_2/CO<11$. As this ratio outmatches 11, the network is interrupted and the output is set to a value indicating that the temperature of the insulation is higher than the normal value.

**VALIDATION OF THE NETWORK**

A bushing soaked in oil has been tested by DGA (Jakob, 2003). Table 8 represents the obtained data. John Stead

**Table 8.** Bushing overwhelmed on oil under partial discharge.

| Gas | Value in ppm |
|---|---|
| Hydrogen | 19132 |
| Oxygen | 4041 |
| Nitrogen | 50767 |
| Carbon monoxide | 537 |
| Methane | 1256 |
| Carbon dioxide | 1459 |
| Ethylene | 11 |
| Ethane | 409 |
| Acetylene | 0.2 |

has stated on his presented paper at the 1996 Doble Conference that this unit has been under partial discharge condition.

These data were applied to the neural network. The network produced codes: [5 0 0 0] which demonstrates the correctness of the decision, partial discharge fault. As another instance, suppose Table 9 presented in (Lewand, 2003). The unit is subjected in high temperature overheating of the oil. Applying these data to the proposed neural network eventuated codes [0 1 2 0] which means that the temperature of the insulation is higher than normal.

**DUVAL TRIANGLE**

The dual triangle was first developed in 1974 (Duval, 2006). It uses only three hydrocarbon gases (CH4, $C_2H_2$, and $C_2H_4$). The three sides of the triangle are expressed in triangular coordinates (X, Y, Z) representing the relative proportions of $CH_4$, $C_2H_4$ and $C_2H_2$, from 0 to

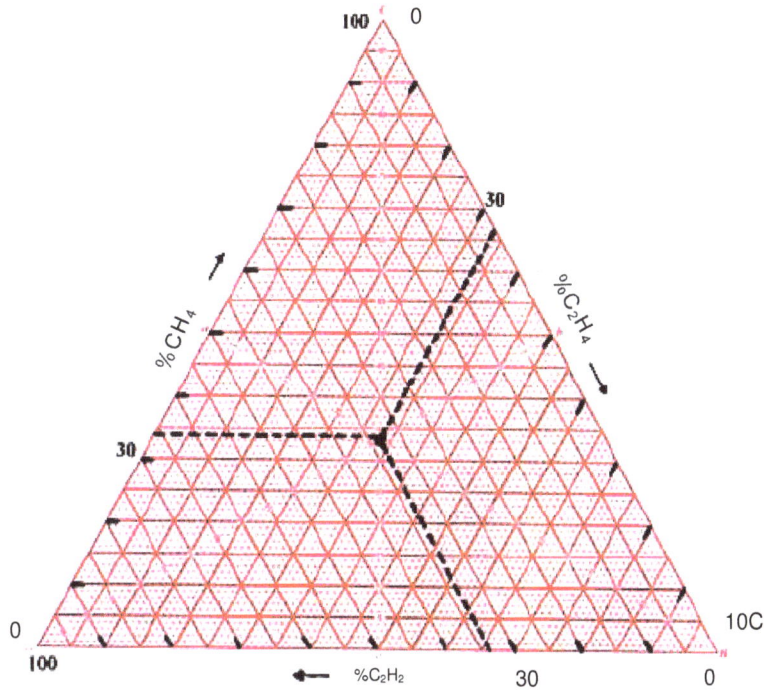

**Figure 7.** Example of triangular graphical plot.

**Table 9.** Value of fault gases under high temperature, McGraw Edison Transformer, 400 MVA, 345 KV, 1969.

| Gas | Value in ppm |
| --- | --- |
| Hydrogen | 7040 |
| Methane | 17700 |
| Ethane | 4200 |
| Ethylene | 21700 |
| Acetylene | 165 |
| Carbon Monoxide | 67 |
| Carbon Dioxide | 1040 |

**Table 10.** Faults detectable by Duval triangle.

| Symbol | Fault |
| --- | --- |
| PD | Partial discharge |
| D1 | Discharges of low energy |
| D2 | Discharges of high energy |
| T1 | Thermal fault, T <300°C |
| T2 | Thermal fault, 300<T<700°C |
| T3 | Thermal fault, T>700°C |
| DT | Mixtures of electrical and thermal faults |

100% for each gas.

In order to display a DGA result in the triangle, one must start with the concentrations of the three gases, $(CH_4) = A$, $(C_2H_4) = B$, and $(C_2H_2) = C$, in ppm.

First calculate the sum of these three values: $(CH_4 + C_2H_4 + C_2H_2) = S$, in ppm, then calculate the relative proportion of the three gases, in %:

$X = \%CH_4 = 100(A/S)$,  $Y = \%C_2H_4 = 100(B/S)$,
$Z = \%C_2H_2 = 100(C/S)$.

X, Y and Z are necessarily between 0 and 100%, and $(X+Y+Z)$ should always 100%. Plotting X, Y and Z in the triangle provide only one point in the triangle.

For example, if the DGA results are A=B=C=100 ppm, X=Y=Z=33.3%, which corresponds to only one point in the centre of the triangle, as indicated in Figure 7. Duval triangle can diagnose the fault types of Table 10. These faults are shown in Figure 8.

Michel Duval found his proposed method the most suitable. He has presented Table 11 to demonstrate his claim.

A three layer perceptron has been proposed here to simulate Duval triangle. This network is presented in Figure 9.

This neural network has been simulated in the Matlab software. Many random inputs have been applied to the network to indicate its performance. Figure 10 presents the results.

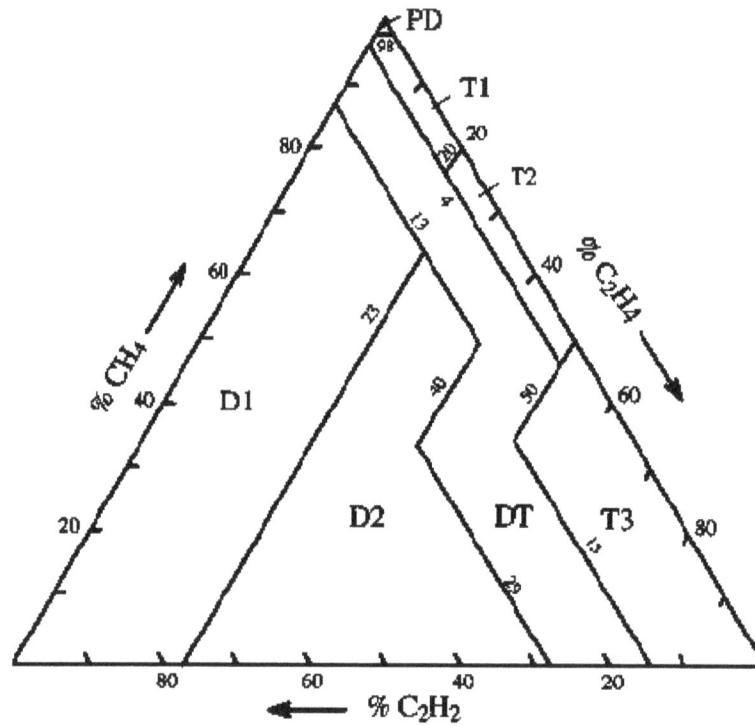

**Figure 8.** Fault dispersal on Duval triangle.

**Table 11.** Comparing faults of diagnostic methods by Duval.

| Diagnostic method | % Unresolved diagnoses | % Wrong diagnoses | %Total |
|---|---|---|---|
| Key gases | 0 | 58 | 58 |
| Rogers | 33 | 5 | 38 |
| Dornenburg | 26 | 3 | 29 |
| IEC | 15 | 8 | 23 |
| Triangle | 0 | 4 | 4 |

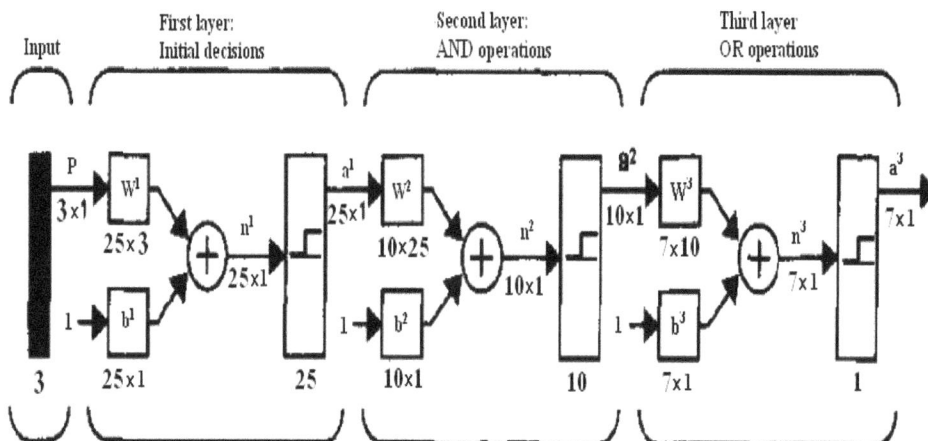

**Figure 9.** Three layer neural network for Duval triangle.

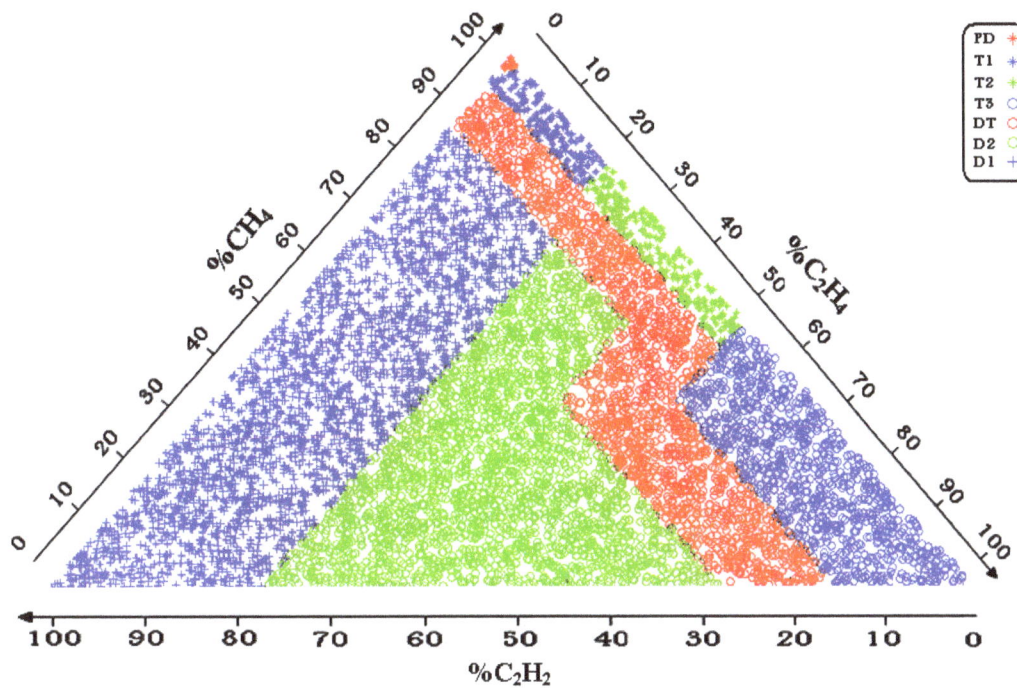

**Figure 10.** Duval triangle plot obtained by the three layer perceptron neural network.

**Table 12.** Examples of DGA cases (concentrations in percent).

| Fault | CH$_4$ | C$_2$H$_4$ | C$_2$H$_2$ |
|-------|--------|------------|------------|
| PD    | 99     | 1          | 0          |
| D1    | 38     | 12         | 50         |
| D2    | 15     | 50         | 35         |
| T2    | 69     | 30         | 1          |
| T3    | 20     | 75         | 5          |

It is comprehended from Figure 10 that the proposed neural network can successfully classify the seven faults of Duval method. It is important to note that the input is three dimensional and a conversion has been applied to a two dimensional plot. For instance, When C$_2$H$_2$=CH$_4$= C$_2$H$_4$=%33, X (horizontal axis) will be equal to:

$$X = 100 - \left( \%C_2H_2 + \frac{Y}{\tan(\frac{\pi}{3})} \right)$$

Correspondingly, Y (vertical axis) will be:

$$Y = \% CH_4 \times \cos(\frac{\pi}{6}) \quad (6)$$

## VALIDATION OF THE NETWORK

Michel Duval has engaged some experimental data of DGA to indicate the correctness of his triangle (Duval, 2006). These data are presented in Table 12.

All of the cases in Table 12 have been presented to the proposed neural network and Figure 11 indicates the results. All the five points corresponding to the faults of Table 12 have been plotted in Figure 11. Circumstance of drawing the points on such a plot is also shown by thin lines connected to the points. By a glance on the figure, it can be understood that the faults are correctly classified.

## CONCLUSION

Appropriate design of artificial neural networks can help simulate the interpretation methods of fault diagnoses in power transformers. Three well-known methods were engaged and a neural network was designed for each of them in this paper. Validation results for the proposed networks prove that they can predict the occurred faults correctly.

As a matter of fact, interpretation methods of fault gases are theoretic and it is required to employ artificial intelligences such as neural networks to realize them. Therefore, once DGA detects the value of all the fault gases in the insulating mineral oil of a transformer, a

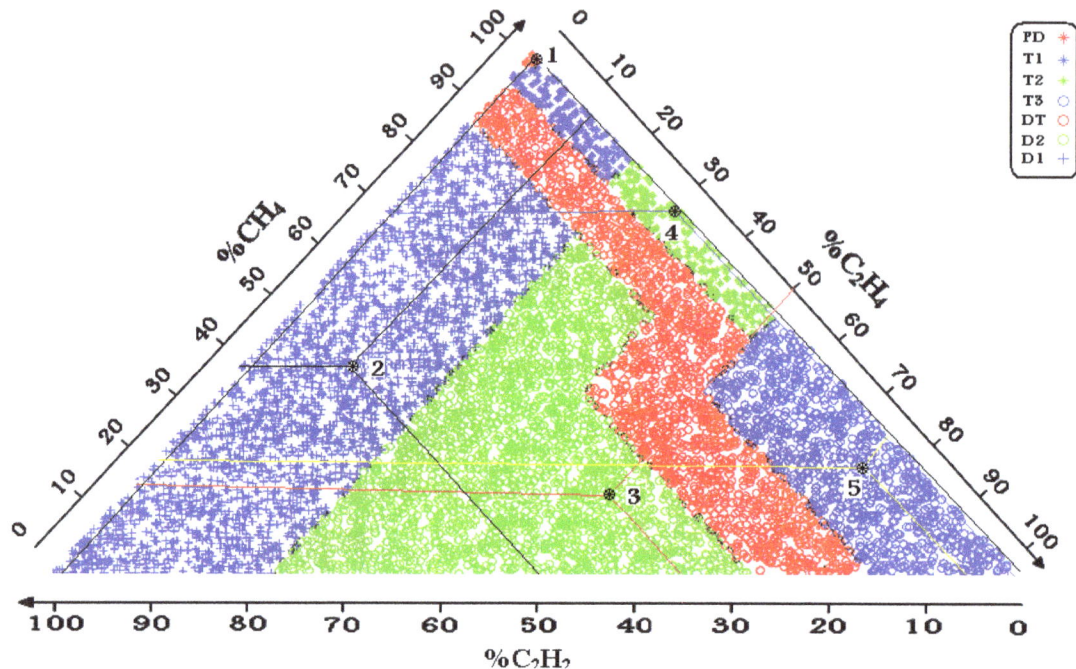

**Figure 11.** Decision making about the faults of table XV by the neural network.

neural network allocating to a desired interpretation method is selected. Eventually, the designed neural network can be employed to real-time decision making of any fault resulting in continues monitoring of that unit. Each neural network has its own characteristics and it is not possible to make comparisons in most cases; hence, for each type of the fault interpretation method, an appropriate network can be previously defined.

## REFERENCES

DiGiorgio JB (1996-2005). Dissolved Gas Analysis of Mineral Oil Insulating Fluids. NTT copyrighted material.

Duval M (2006). Dissolved Gas Analysis and the Duval Triangle. Fifth AVO New Zealand International Technical Conference.

Jakob (2003). Dissolved Gas Analysis – Past, Present and Future. Weidmann Electrical Technology, Technical Library.

Lewand LR (2003). Using Dissolved Gas Analysis to Detect Active Faults in Oil-Insulated Electrical Equipment. Doble Engineering Company, Practicing Oil analysis Magazine, Issue Number: 200303.

Hagan MT, Demuth HB, Beale M (1996). Neural Network Design. PWS Publishing Company.

ANSI/IEEE C57.104. IEEE Guide for the Interpretation of Gases Generated in Oil-Immersed Transformers –Description. IEEE Standard.

IEC 60599. Mineral oil-impregnated electrical equipment in service - Guide to the interpretation of dissolved and free gases analysis. IEC Standard.

# 6

# Fundamentals and literature review of wavelet transform in power quality issues

**Lütfü SARIBULUT[1], Ahmet TEKE[2], Mohammad BARGHI LATRAN[2] and Mehmet TÜMAY[2]**

[1]Department of Electrical-Electronics Engineering, Adana Science and Technology University, Adana/TURKEY.
[2]Department of Electrical-Electronics Engineering, Çukurova University, Adana/TURKEY.

**The effects of power quality (PQ) disturbances on power utilities and power customers are increasing day by day with widespread use of power electronics devices. These disturbances should be detected rapid and accurate. Wavelet transform (WT) has become a popular method to detect, classify and analyze various types of PQ disturbances in power systems. In this study, the fundamentals of WT, its distinctions from Fourier transform and the results of an extensive literature review of published studies are presented. The results of the literature review reveal that the detection of PQ disturbances using WT is a very powerful method and has been used in PQ mitigation devices effectively (that is, Custom Power, Flexible AC Transmission System, Flexible, Reliable and Intelligent Energy Delivery System).**

**Key words:** Wavelet transform, power quality disturbances, literature review, feature extraction.

## INTRODUCTION

The term wavelet means a small wave. In addition, a wave refers the condition that this function is an oscillatory. The other term of Wavelet transform (WT) is the mother wavelet. It implies that the functions used in the decomposition and reconstruction processes are derived from one main function or the wavelet. In other words, the mother wavelet is a prototype for generating the other wavelet functions. The term translation ($\tau$) is used in the same sense as it was used in short time Fourier transform (STFT). It is related to the location of the window, as the window is shifted through the signal. It is visualized in Figure 1 (Amara, 1995; Robi, 2012).

The application of WT has become a popular for some time under the various names of multirate-sampling (Quadrature Mirror Filters) QMF in electrical engineering (Sarkar et al., 1998). WT possesses many desirable properties that are useful in engineering, economics and

finance. The ability of wavelet analysis deals with both the stationary and non-stationary data, their localization in time and the decomposing and analyzing the punctuation in variable signal. While a review of the possible future contributions of WT is provided to the engineering discipline, there are explored two ways in which wavelets might be used to enhance the empirical toolkit of our profession in engineering (Crowley, 2005). These are summarized in the following:

**Time scale versus frequency:** An examination of data sets to assess the presence and ebb and the flow of frequency components is potentially valuable (Crowley, 2005).

**Time scale decomposition:** Recognition components of the signal can be possibly found at disaggregate (scale)

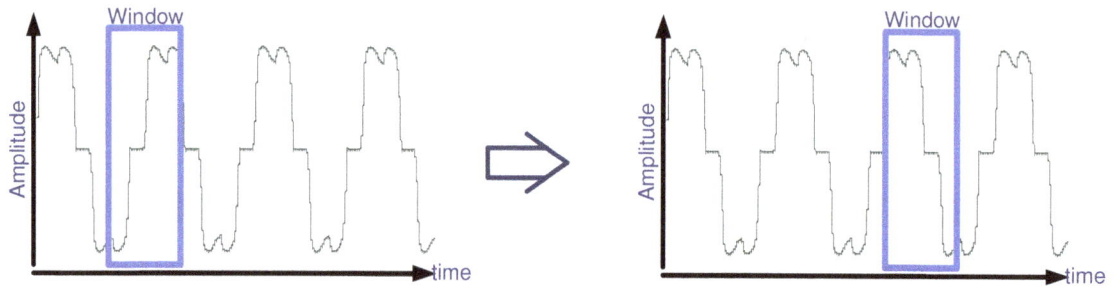

**Figure 1.** Window is shifted through the signal.

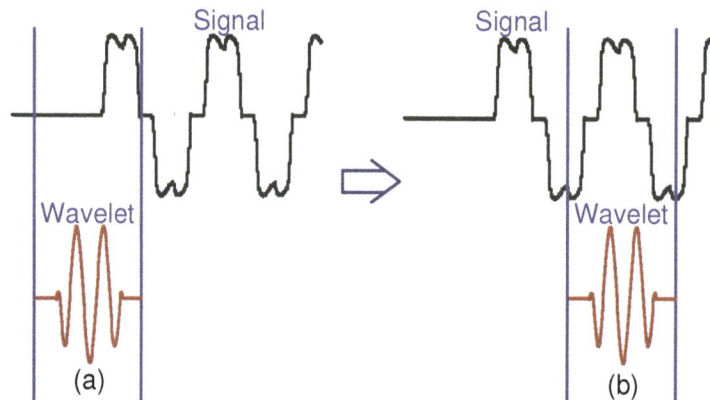

**Figure 2.** Wavelet function with it's shifted and dilated versions.

level rather than an aggregate level forecasting: disaggregate forecasting, establishing global versus local aspects of series (Crowley, 2005).

In this study, the basic principles of WT especially Continuous Wavelet Transform (CWT) and Discrete Wavelet Transform (DWT), the properties of most used Haar and Daubechies wavelets are presented. The similarity and contrast points of WT with the Fourier Transform (FT) are also given in this study. The recent studies with the subject of WT applications in power quality (PQ) issues are examined in details in the last part of the work.

## CONTINUOUS WAVELET TRANSFORM

Unlike FT, CWT possesses the ability to construct a time-frequency representation of a signal that offers very good time and frequency localization. In other words, CWT is used to divide a continuous-time function into the equal-time intervals and equal-frequency intervals. In CWT, the signal is multiplied with the wavelets, same as STFT and the transform is computed separately for different parts of the time-domain signal. The mathematical definition of

CWT is given in the following (Wikipedia, 2012).

$$\psi(\tau, s) = \frac{1}{\sqrt{|s|}} \int x(t) \, \psi^* \left( \frac{t - \tau}{s} \right) dt \qquad (1)$$

The scale ($s$) and $\tau$ parameters are the parameters of CWT functions. The function $\psi(\tau, s)$ named as mother wavelet is the transforming function. Let, $x(t)$ be the signal to be analyzed. The mother wavelet is chosen to serve as a prototype for all windows in the process. All the windows used for the dilated, compressed and shifted are the versions of the mother wavelet. There are many functions used for this purpose. Once the mother wavelet is chosen in CWT, the computation can be started any value of $s$ and computed for all values of $s$ (smaller and/or larger than 1). Such as, the procedure can be started from scale $s=1$ and it is continued with increasing the values of $s$. In this situation, the analyzing of the signal is started from the high frequencies and proceeded towards the low frequencies. This procedure is repeated for every value of $s$. Every computation for a given value of $s$ fills the corresponding single row of the time-scale plane (Robi, 2012). This process is summarized in Figure 2.

An accurate signal analysis is the proper analysis of

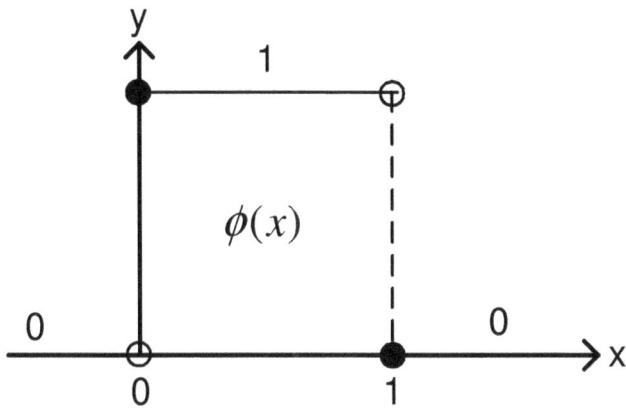

**Figure 3.** Scaling function of Haar wavelet.

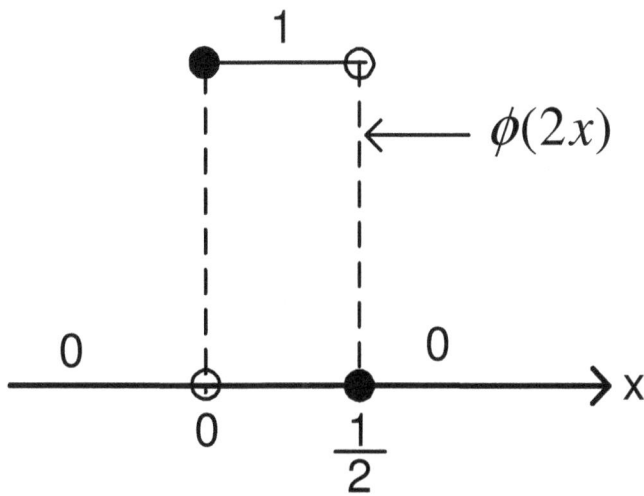

**Figure 4.** Graph of $\phi(2x)$.

each trace. FT is not a good tool for this purpose. It can only provide frequency information of the oscillations that comprise the signal. It gives no direct information about when an oscillation is occurred. Another tool for signal analysis is STFT. The full-time interval is divided into a number of small, equal-time intervals. These are individually analyzed by using FT. The result contains time and frequency information. However, there is a problem with this approach. The equal-time intervals are not adjustable. When the time intervals are very-short-duration, high-frequency bursts are hard to detect (Sarkar, 1998; Morlet, 1982a, b).

CWT is a good tool for analysis the signal included the harmonics both the time-domain and the frequency-domain. The signal, that its frequency is changed with time, is detected easily when the time intervals are very-short-duration. The most common wavelets are summarized in the following.

## Haar Wavelet

There are two functions that play a primary role in the wavelet analysis. These are the scaling function ($\phi$) and wavelet ($\psi$). These functions generate a family of functions that can be used to break up or reconstruct the signal. To emphasize the marriage involved in building this family, $\phi$ is sometimes called as father wavelet and $\psi$ as mother wavelet (Boggess, 2001). The simplest wavelet analysis is based on the Haar scaling function illustrated in Figure 3.

The Haar scaling function is defined as

$$\phi(x) = \begin{cases} 1, & if \ 0 \leq x \leq 1 \\ 0, & elsewhere. \end{cases} \tag{2}$$

The function $\phi(x$-$k)$ is obtained by shifted versions of $\phi(x)$ to the right by $k$ units (assuming $k$ is positive). Let, $V_0$ be the space of functions in the form of $\phi(x$-$k)$. It is given in the following.

$$\sum_{k \in z} a_k \phi(x-k) \tag{3}$$

where $k$ can range over any finite set of positive or negative integers.

The thinner blocks are required to analyze the high-frequency signals. The function $\phi(2x)$, whose width is half of $\phi(x)$, is illustrated in Figure 4.

The simplest $\psi(x)$ of Haar wavelet is the function that consists of two blocks and it can be written as (4).

$$\begin{aligned} \psi(x) &= \phi(2x) - \phi(2(x-1/2)) \\ &= \phi(2x) - \phi(2x-1) \end{aligned} \tag{4}$$

Haar wavelets have one particularly desirable property that they are zero everywhere except on a small interval. If the function is closed everywhere outside of a bounded interval, it has compact support. If a function has compact support, the smallest closed interval on which the function has non-zero values is called the support of the function (Aboufadel et al., 1999). If $k \neq 0$ and $\psi(x)$ is orthogonal to $\phi(x)$, then the compact support of $\psi(x)$ and the compact support of $\phi(x$-$k)$ do not overlap in the Cartesian coordinate system.

So, $\int_{-\infty}^{\infty} \psi(x)\phi(x-k)dx = 0.$ Consequently, $\psi(x)$

belongs to $V_1$ and it is orthogonal to $V_0$. Then, $\psi(x)$ is called Haar wavelet. It is given in Figure 5.

In general, $n^{th}$ generation of daughters will have $2^n$ wavelets and it is defined as (5).

$$\psi_{n,k}(t) = \psi(2^n t - k), \qquad 0 \leq k \leq 2^n - 1 \tag{5}$$

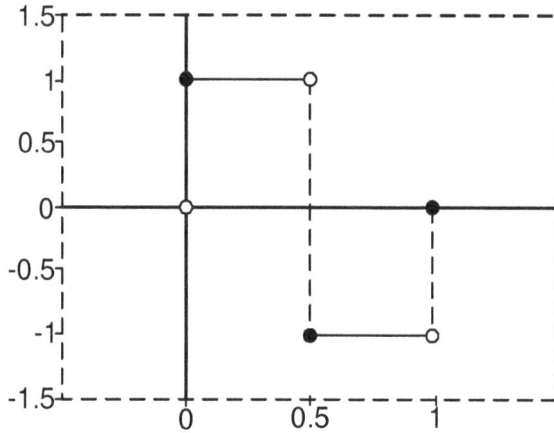

**Figure 5.** Graph of Haar wavelet $\psi(x)$.

## Daubechies Wavelet

Ingrid Daubechies in 1987 sought a wavelet family that had compact support and some sort of smoothness. The scaling function of her discovery was quite a feat and was greeted with enthusiasm. There are three requirements for the following example of Daubechies wavelets.

(i) The scaling function has compact support that $\phi(x)$ is zero outside of the interval $0<t<3$.
(ii) The orthogonality condition is the heart of any multi-resolution analysis.
(iii) The regularity condition is related with the smoothness of the scaling function.

The dilation equation of Daubechies wavelet is given in the following.

$$\phi(t) = \frac{1+\sqrt{3}}{4}\phi(2t) + \frac{3+\sqrt{3}}{4}\phi(2t-1)$$
$$+ \frac{3+\sqrt{3}}{4}\phi(2t-1) + \frac{1-\sqrt{3}}{4}\phi(2t-1) \quad (6)$$

Daubechies wavelets have the compact support, orthogonality and regularity conditions. These properties are to be required in constructing the scaling function, if the accuracy is desired. Hence, the mother wavelet and all children can be approximated as well. This indicates that it is not specifically necessary to know a formula for the scaling function in order to work with wavelets. Other wavelet families are developed in this way (Aboufadel et al., 1999; Addison, 2002).

## DISCRETE WAVELET TRANSFORM

DWT is the form of CWT that uses the discrete values of

the signal in the time-domain. The mathematical expression of DWT is given in the following.

$$\psi_{m,n}(t) = \frac{1}{\sqrt{a_0^m}}\left(\frac{t - nb_0 a_0^m}{a_0^m}\right) \quad (7)$$

where the integers $m$ and $n$ control the wavelet dilation and translation respectively, $a_0$ is a specified fixed dilation step parameter which should be greater than one and $b_0$ is the location parameter which should be greater than zero (Addison, 2002).

The control parameters $m$ and $n$ contain the set of positive and negative integers. WT of a continuous signal $x(t)$ gives the decomposition wavelets by using the discrete form of (1). In other words, the scale-location of the signal in the time-frequency-domain is determined by using (8).

$$T_{m,n} = \int_{-\infty}^{\infty} x(t) a_0^{-m/2} \psi\left(a_0^{-m}t - nb_0\right) dt \quad (8)$$

where the values of $T$ are the scale-location grid of the signal in the time-frequency-domain according to indexes $m$ and $n$.

For DWT, the values of $T_{m,n}$ are known as wavelet coefficients or detail coefficients. Common choices for DWT parameters, $a_0$ and $b_0$ are selected as two and one, respectively. The term "power of two" is emerged for the practical application. This power of two is known as the dyadic grid arrangement for dilation and translation steps. The dyadic grid is perhaps the simplest and most efficient discretization for practical purposes and lends itself to the construction of an orthonormal wavelet basis (Aboufadel et al., 1999; Addison, 2002).

$$\psi_{m,n}(t) = 2^{-m/2}\psi\left(2^{-m}t - n\right) \quad (9)$$

By choosing an orthonormal wavelet basis $\psi_{m,n}(t)$, the original signal can be reconstructed with utilizing the wavelet coefficients $T_{m,n}$ by using inverse discrete wavelet transform (IDWT) given in the following.

$$x(t) = \sum_{m=-\infty}^{\infty} \sum_{n=-\infty}^{\infty} T_{m,n}\, \psi_{m,n}(t) \quad (10)$$

## ADVANTAGES AND DISADVANTAGES OF WT WITH FFT

Many transforms have been devised by engineers and mathematicians such as the Laplace transform, the Hilbert transform, $Z$-transform and FT. However, FT is the most popular among these transforms. It presents lots of limitations in many areas of study. Only one of them is present at a time. The frequency-domain gives no information about the time of signal generation or the time at which the frequencies are occurred (Bousaleh et al., 2009). The points of similarity and contrast between WT and FT are given in the following (Sarıbulut, 2012).

(i) The matrix of inverse transform is the transpose of the original for both FFT and DWT. Both transforms can be viewed as a rotation in a function space to a different domain. For FFT, this new domain contains basis functions included in the sinus and cosine functions. For WT, this new domain contains more complicated basis functions called wavelets, mother wavelets or analyzing wavelets (Amara, 1995; Sarkar et al., 1998).

(ii) The basis functions are localized in frequency, making mathematical tools such as power spectra (how much power is contained in a frequency interval) and scale grams useful at picking out frequencies and calculating power distributions (Amara, 1995).

(iii) The most interesting dissimilarity between these two kinds of transforms is that the individual wavelet functions are space whereas localized. However, the sine and cosine functions of FT are not space whereas localized. This localization feature makes many functions and operators along with wavelets' localization of frequency when it is transformed into the wavelet domain (Amara, 1995).

(iv) An advantage of WTs is that the windows vary. In order to isolate signal discontinuities, some very short basis functions would be required. At the same time, some very long basis functions would be required in order to obtain the detailed frequency analysis (Sarkar et al., 1998).

## LITERATURE REVIEW OF WAVELET TRANSFORM IN POWER QUALITY ISSUES

The latest studies related with WT are summarized as literature survey in the following:

A conventional synchronous fundamental DQ-frame algorithm is modified in filtering process (Firouzjah et al., 2009). As an improvement, the Fourier-based low-pass filter is replaced with a windowing-wavelet method. The adopted windowing aspect is Hamming window to reduce conventional rectangular window effect in the frequency-domain. The disturbance classification schema is performed with a wavelet-neural network (Uyar et al., 2008). It performs feature extraction and classification algorithm composed of a wavelet feature extractor based on norm entropy and classifier based on a multi-layer perceptron.

A transmission-line fault location model based on an Elman recurrent network has been presented for balanced and unbalanced short circuit faults (Ekici et al.,

2009). WT is used for selecting distinctive features about the faulty signals. The fault signals are analyzed by using DWT which splits the signal into detail and approximation coefficients. The energies of detail coefficients fed neural networks to estimate fault location. The recognition of PQ disturbances is studied (Kaewars et al., 2008). The multi-wavelet-based neural classifier is used to automatically detect, localize and classify the transient disturbance type for high accuracy and low usage time. PQDs classification based on WT and self-organizing learning array system is proposed (He et al., 2006). WT is utilized to extract feature vectors for various PQ disturbances based on the multi-resolution analysis. Then, these feature vectors are applied to a solar system for training and testing.

An integrated approach for the PQ data compression by using the Spline WT and ANN is presented (Meher et al., 2004). Its performance is assessed in terms of compression ratio, mean square error and percentage of energy retained in the reconstructed signals. A compression technique for power disturbance data via DWT and the wavelet packet transform is introduced (Hamid et al., 2004). The compression technique is performed through the signal decomposition up to a certain level, threshold of wavelet coefficients and signal reconstruction. A fuzzy-wavelet packet transform based PQ indices are introduced (Morsi et al., 2009). Online real-time detection and classification of voltage events in power systems are presented (Pérez et al., 2008). Wavelet analysis is used for detection and estimation of the time-related parameters of an event and the extended Kalman's filtering is used for confirmation of the event and for computation of the voltage magnitude during the vent.

An algorithm based on the wavelet-packet transform for the analysis of harmonics in power systems is given (Barros et al., 2008). The proposed algorithm decomposes the voltage/current waveforms into the uniform frequency bands. It uses a method to reduce the spectral leakage due to the imperfect frequency response of the used wavelet filter bank. The use of WT and the computational intelligence techniques are presented to quantify voltage short-duration variation in electric power systems (Machado et al., 2009). WT is used to determine the event duration and to obtain the characteristic curve that is related with the signal norm as the function of the number of cycles for a waveform without disturbance.

A fault location procedure for distribution networks based on the wavelet analysis of the fault-generated traveling waves is presented (Borghetti et al., 2008). CWT is applied to the voltage waveforms recorded during the fault in correspondence of a network bus. An automatic classification of different PQD by using WT and fuzzy k-nearest neighbor based classifier is proposed (Panigrahi et al., 2009). The training data samples are generated using parametric models of the PQ disturbances. The features are extracted using some of the statistical measures on the WT coefficients of the disturbance signal when decomposed up to the fourth level. A method for detection and classification of PQ disturbances is proposed (Masoum et al., 2010). The distorted waveforms (PQ events) are generated based on the IEEE 1159 standard, captured with a sampling rate of 20 kHz and de-noised using DWT to obtain signals with the higher signal-to-noise ratio.

An implementation of DWT using LC passive filters for operating a DVR system is presented (Saleh et al., 2008). The proposed implementation is based on designing Butterworth LC passive filters with cutoff frequencies that are identical to cutoff frequencies of DWT associated digital filters. An integrated model for recognizing PQDs using a novel wavelet multiclass-support vector machine is presented (Lin et al., 2008). Various disturbance events were tested for proposed method and the wavelet-based multilayer-perceptron neural network was used for comparison. To enhance the performance of WT-based monitoring systems and to improve the classification accuracy of WT based classifiers, two standards statistical procedures have been proposed (Dwivedi et al., 2009). PQ waveform data has been performed by the correct implementation of wavelet threshold using the supreme functional of the Brownian bridge process.

A non-invasive methodology to evaluate and classify the electrical system failures is presented (Sartori et al., 2010). It is based on the electrical system magnetic signature recognition by using the wavelet signal decomposition and the variance spectrum evaluation, respectively. A simple and effective technique to de-noise PQ waveform data for enhanced detection and time localization of PQ disturbances is proposed by using inter and inter-scale dependencies of wavelet coefficients (Dwivedi et al., 2010). The proposed method operates on WT coefficients that are obtained from wavelet multi-resolution decomposition of PQ waveform data and exploits the local structure of wavelet coefficients within a sub-band as well as across the scales to produce noise-free coefficients.

A new algorithm based on WT to detect the starting point, ending point and the magnitude of the voltage sag is developed (Gencer et al., 2010). DWT is used to detect fast changes in the voltage signals which allow time localization of differences frequency components of a signal with different frequency wavelets. WT is used to reformulate the recommended PQ levels (Morsi et al., 2010). The non-stationary waveforms are analyzed for the smart grid. An effective technique is proposed by using inter and inter-scale dependencies of wavelet coefficients to de-noise the waveform of PQ data for enhanced detection and time localization of PQ disturbances (Dwivedi et al., 2010).

An algorithm for continuous real-time power waveforms capture, compression, and transmission is presented (Tse et al., 2012). The integer-lifting WT with integer

mapping and an adaptive threshold scheme are used for truncating wavelet coefficients and compressing the non-stationary component, respectively. Automatic classifications of PQ events and disturbances by using WT and support vector machine (SVM) are presented (Eristi et al., 2012). WT is used in order to obtain the distinctive features of event signals. PQ event types are determined by using SVM classifier. At the last stage of intelligent recognition system, types of PQ disturbances regarding each fault event are identified by doing a further analysis. An approach for the recognition and classification of PQ disturbances by using WT and wavelet based SVM is presented (Moravej et al., 2010). The proposed method employs WT techniques to extract the most important and significant feature from details and approximation waves. The obtained severable feature vectors are used for training SVMs to classify the PQ disturbances.

The use of near perfect reconstruction filter banks on the harmonic analysis of power system waveforms is investigated (Taskovski et al., 2012). A dual-purpose wavelet-based transient detection module to perform the fault detection and the voltage regulation through conditional threshold selection are described (Lakshmi et al., 2013). An effective boundary element numerical algorithm combining with WT and generalized minimum residual method for the solution of the electric field problem in substations are proposed (Deng et al., 2012). A wavelet-based methodology for the characterization of voltage sags is described (Costa et al., 2013). Frequency-Shifting wavelet decomposition via the Hilbert transform is introduced for PQ analysis introduced (Tse et al., 2012). The proposed algorithm overcomes the spectra leakage problem in DWT. A hybrid model is proposed for monthly runoff prediction by using wavelet transform and feed forward neural networks (Okkan, 2012). DWT and Levenberg-Marquardt optimization algorithm based feed forward neural networks are considered for the modeling study. The two hybrid AI-based models which are reliable in capturing the periodicity features of the process are introduced for watershed rainfall–runoff modeling (Nourani et al., 2011). In the first model, an ANN is used to find the non-linear relationship among the residuals of the fitted linear SARIMAX model. In the second model, wavelet transform is linked to the ANFIS concept and the main time series of two variables are decomposed into some multi-frequency time series by wavelet transform.

## CONCLUSIONS

Power quality disturbances might have large financial impacts on the customers of modern power industrial. New power electronics based devices can be used to mitigate these PQ disturbances. These disturbances in power systems should be detected, classified and analyzed with certain accuracy. The number of studies related with the applications of Wavelet Transform in these devices has been increasing. WT performs a superior performance in the applications of power systems. It extracts the distinct features of typical PQ disturbances to obtain dynamic parameters superior to traditional methods. A comprehensive review of WT in PQ issues has also been carried out to explore a broad perspective on their applications. The literature review reveals more precise in discriminating the type of transients, higher robustness and faster processing time for detecting and classifying voltage and current disturbing events.

## REFERENCES

Aboufadel E, Schlicker S (1999). Discovering wavelets. Wiley-Interscience, 1st edition.

Addison N (2002). The illustrated wavelet transform handbook. Taylor & Francis. 1st edition.

Barros J, Diego RI (2008). Analysis of harmonics in power systems using the wavelet-packet transform. IEEE Trans. Instrum. Meas. 57(1):63-69.

Boggess A, Narcowich FJ (2001). A first course in wavelets with Fourier analysis. Prentice Hall, 1st edition.

Borghetti A, Bosetti M, Di Silvestro M, Nucci CA, Paolone M (2008). Continuous-wavelet transform for fault location in distribution power networks: Definition of mother wavelets inferred from fault originated transients. IEEE Trans. Power Syst. 23(2):380-388.

Bousaleh G, Hassoun F, Ibrahim T (2009). Application of wavelet transform in the field of electromagnetic compatibility and power quality of industrial systems. International Conference on Advances in Computational Tools for Engineering Applications, pp. 284-289.

Costa FB, Driesen J (2013). Assessment of voltage sag indices based on scaling and wavelet coefficient energy analysis. IEEE Trans. Power Deliv. 28(12):336-346.

Crowley PM (2005). An intuitive guide to wavelets for economists. Bank of Finland. Res. Discussion Pap. 1:71.

Deng J, Hao Y, Chen H, Li L, Wang J, Zheng X, Wu H (2012). A wavelet transform boundary element method for electric field problem inside substations. IEEE Trans. Electromagn. C. 54(1):193-197.

Dwivedi UD, Singh SN (2009). Denoising techniques with change-point approach for wavelet-based power-quality monitoring. IEEE Trans. Power Deliv. 24(3):1719-1727.

Dwivedi UD, Singh SN (2010). Enhanced detection of power-quality events using intra and interscale dependencies of wavelet coefficients. IEEE Trans. Power Deliv. 25(1):358-366.

Ekici S, Yildirim S, Poyraz M (2009). A transmission line fault locator based on Elman recurrent networks. Appl. Soft Comput. 9(1):341-347.

Eristi H, Demir Y (2012). Automatic classification of power quality events and disturbances using wavelet transform and support vector machines. IET Gener. Transm. Distrib. 6(10):968-976.

Firouzjah KG, Sheikholeslamia A, Karami-Mollaeia MR, Heydaria F (2009). A predictive current control method for shunt active filter with windowing based wavelet transform in harmonic detection. Simul. Model. Pract. Theory 17(5):883-896.

Gencer Ö, Öztürk S, Erfidan T (2010). A new approach to voltage sag detection based on wavelet transform. Int. J. Elect. Power Energy Syst. 32(2):133-140.

Graps A (1995). An introduction to wavelets. IEEE Comput. Sci. Eng. 2(2):50-61.

Hamid EY, Kawasaki ZI (2002). Wavelet-based data compression of power system disturbances using the minimum description length criterion. IEEE Trans. Power Deliv. 17(2):460-466.

He H, Starzyk JA (2006). A self-organizing learning array system for power quality classification based on wavelet transform. IEEE Trans. Power Deliv. 21(1):286-295.

Kaewars S, Attakitmongcol K, Kulworawanichpong T (2008). Recognition of power quality events by using multiwavelet-based neural networks. Int. J. Elect. Power Energy Syst. 30(4):254-260.

Lakshmi VIK, Xiaomin L, Usama Y, Ramakrishnan V, Narayan CK (2013). A Twofold Daubechies-wavelet-based module for fault detection and voltage regulation in SEIGs for distributed wind power generation. IEEE Trans. Ind. Electron. 60(4):1638-1651.

Lin WM, Wu CH, Lin CH, Cheng FS (2008). Detection and classification of multiple power-quality disturbances with wavelet multiclass SVM. IEEE Trans. Power Deliv. 23(4):2575-2582.

Machado M, Bezerra RN, Pelaes UH, Oliveira EG, de Lima Tostes ME (2009). Use of wavelet transform and generalized regression neural network (GRNN) to the characterization of short-duration voltage variation in electric power system. IEEE Latin Am. Trans. 7(2):217-222.

Masoum MAS, Jamali S, Ghaffarzadeh N (2010). Detection and classification of power quality disturbances using discrete wavelet transform and wavelet networks. IET Sci. Meas. Technol. 4(4):193-205.

Meher SK, Pradhan AK, Panda G (2004). An integrated data compression scheme for power quality events using spline wavelet and neural network. Electric Power Syst. Res. 69(2-3):213-220.

Moravej Z, Abdoos AA, Pazoki M (2010). Detection and classification of power quality disturbances using wavelet transform and support vector machines. Electric Power Compon. Syst. 38:182-196.

Morlet J (1982). Wave propagation and sampling theory-Part II: Sampling theory and complex waves. Geophysics 47:222.

Morlet J, Arens G, Fourgeau E, Glard D (1982). Wave propagation and sampling theory-Part I: Complex signal and scattering in multilayered media. Geophysics 47:203.

Morsi WG, El-Hawary ME (2009). Fuzzy-Wavelet-based electric power quality assessment of distribution systems under stationary and non stationary disturbances. IEEE Trans. Power Deliv. 24(4):2099-2106.

Morsi WG, El-Hawary ME (2010). Novel power quality indices based on wavelet packet transform for non-stationary sinusoidal and non-sinusoidal disturbances. Electric Power Syst. Res. 80(7):753-759.

Nourani V, Kisi O, Komasi M (2011). Two hybrid artificial intelligence approaches for modeling rainfall-runoff process. J. Hydrol. 402(1):41-59.

Panigrahi BK, Pandi VR (2009). Optimal feature selection for classification of power quality disturbances using wavelet packet-based fuzzy k-nearest neighbour algorithm. IET Gener. Transm. Distrib. 3(3):296-306.

Pérez E, Barros J (2008). A proposal for on-line detection and classification of voltage events in power systems. IEEE Trans. Power Deliv. 23(4):2132-2138.

Polikar R (2012). Wavelet transform tutorials. (http://users.rowan.edu/~polikar/ wavelets/ wttutorial.html).

Saleh SA, Moloney CR, Rahman MA (2008). Implementation of a dynamic voltage restorer system based on discrete wavelet transforms. IEEE Trans. Power Deliv. 23(4):2366-2375.

Sarıbulut L (2012). Detection and mitigation methods of power quality disturbances. PhD Thesis, Çukurova University, Institute of Natural and Applied Science.

Sarkar TK, Su CA, Salazar-Palma R, Garcia-Castillo M, Boix LRR (1998). A tutorial on wavelets from an electrical engineering perspective. I. Discrete wavelet techniques. IEEE Antenn. Propag. Mag. 40(5):49-68.

Sartori CAF, Sevegnani FX (2010). Fault classification and detection by wavelet-based magnetic signature recognition. IEEE Trans. Magn., 46(8):2880-2883.

Taskovski D, Koleva L (2012). Measurement of harmonics in power systems using near perfect reconstruction filter banks. IEEE Trans. Power Deliv. 27(2):1025-1026.

Tse NCF, Chan JYC, Lau WH, Lai LL (2012). Hybrid wavelet and hilbert transform with frequency-shifting decomposition for power quality analysis. IEEE Trans. Instrum. Meas. 61(12):3225-3233.

Okkan U (2012). Using wavelet transform to improve generalization capability of feed forward neural networks in monthly runoff prediction. Sci. Res. Essays 7(17):1690-1703.

Tse NCF, Chan JYC, Wing-Hong L, Poon JTY, Lai LL (2012). Real-time power-quality monitoring with hybrid sinusoidal and lifting wavelet compression algorithm. IEEE Trans. Power Deliv. 27(4):1718-1726.

Uyar M, Yildirim S, Gencoglu MT (2008). An effective wavelet-based feature extraction method for classification of power quality disturbance signals. Electric Power Syst. Res. 78(10):1747-1755.

Wikipedia (2012). (http://en.wikipedia.org/wiki/Continuous_wavelet_transform).

# 7

# Non-polarizing beam splitter and antireflection coating design

Hayder A. Ahmed* and Arafat J. Jalil

Physics Department, College of Science, Basrah University, Basrah, Iraq.

The optical coating design of beam splitters and antireflection coatings that are non-polarizing, those that have the same reflection for both "s- and p-polarizations" at specified angles, is a challenge. This is because the effective indices of refraction for the media and coating layer materials have a different function of the angle of incidence for each polarization. This limits what can be achieved. Specific materials can be used to design reasonable non-polarizing coatings at only certain angles, whereas other materials can be made to serve for other angles. The choice of materials seems to be critical to the success of any particular requirement. Since the range of practical coating materials and substrates is limited for most applications, non-polarizing solutions seem to be quantized in reflectance and angle. This is illustrated with the aid of reflectance amplitude (circle) diagrams, and some design "rules-of-thumb" are provided.

**Key words:** Beam splitter, antireflection coating, reflectance amplitude diagrams.

## INTRODUCTION AND BASIC THEORY

Beam splitters are used to divide incident light into transmitted and reflected beams in a certain ratio over a broad range of wavelengths. To physically separate the three beams, these beam splitters are often required to operate at oblique angles of incidence. At oblique angles, unfortunately, the optical admittances for "s- and p-polarized" light are different according to LiLi (2003). When light travels from the first medium with a refractive index $n_0$ to the second medium with a refractive index $n_1$, the reflected and transmitted electric field amplitudes of the electromagnetic wave at the interface are governed by the Fresnel equations and given in Equations (1) and (2):

$$\begin{cases} r = \dfrac{\eta_o - \eta_1}{\eta_o + \eta} = |r|e^{i\varphi} \\ t = \dfrac{2\eta_o}{\eta_o + \eta} = |t|e^{i\varphi_t} \end{cases} \qquad (1)$$

$$\begin{cases} \eta_o = n_o \cos\theta_o \\ \eta_1 = n_1 \cos\theta_1 \end{cases}(s-pol.) \quad \begin{cases} \eta_o = n_o/\cos\theta_o \\ \eta_1 = n_1/\cos\theta_1 \end{cases}(p-pol.) \qquad (2)$$

$$n_o \sin\theta_o = n_1 \sin\theta_1 \qquad (3)$$

where r and t are also called the amplitude reflection and transmission coefficients, respectively. $\varphi$ and $\varphi_t$ are the phase changes on reflection and transmission. $\eta_o$ and $\eta_1$ are the optical admittances for the two media and are different for s- and p-polarized light. $\theta_o$ is the incident angle and $\theta_1$ is the refracted angle inside the second medium, and they are related to each other by Snell's law (Equation 3), and thus reflectance or transmittance of s- and p-polarized light tends to be different; it is rather difficult to design non-polarizing beam splitters that have the same reflectance and transmittance for both s- and p-polarized light. However, it has been mentioned above that a conventional thin film polarizing beam splitter (PBS) operating at angles smaller than the critical angle always reflects s-polarized light and transmits

*Corresponding author. E-mail: dr.hayder73@yahoo.com.

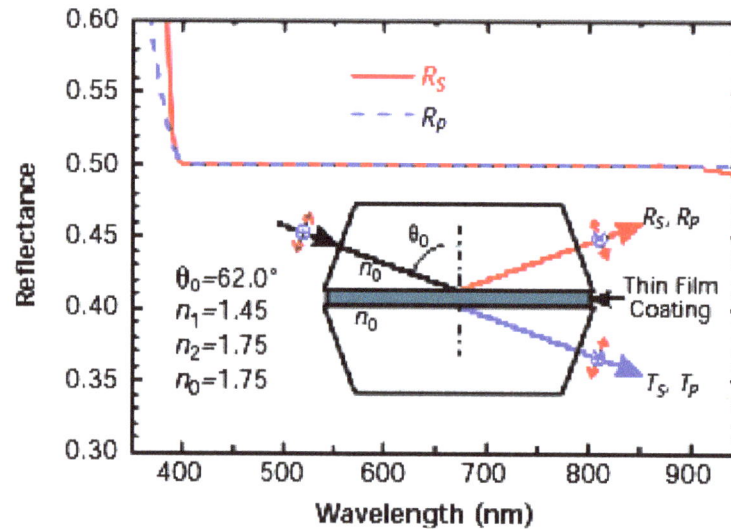

**Figure 1.** Calculated performance of a non-polarizing beam splitter operating beyond the critical angle (Macleod, 2001).

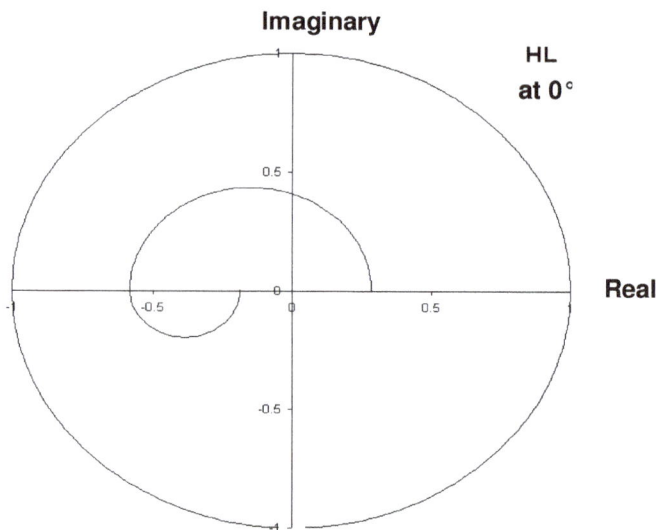

**Figure 2.** Reflectance amplitude diagram of a two QWOT layer coating at normal incidence of design: SHLA.

p–polarized light, while a PBS operating at angles greater than the critical angle always transmits s-polarized light and reflects p-polarized light. A transition angle must therefore exist at which the reflectance's for s- and p-polarized light are both equal to 50%. And indeed, this angle is above "but close to" the critical angle for the low refractive index layer. Macleod (2001) has described the design of such non-polarizing beam splitters based on admittance diagrams of frustrated total internal reflection (FTIR) layers. Figure 1 shows a design for a visible/near infrared beam splitter that operates from 400 to 900 nm.

The optical coating design of beam splitters and

antireflection coatings that are non-polarizing at a specific wavelength and angle of incidence are addressed here. This work is further confined to "non-immersed" designs as opposed to coatings surrounded by glass such as cube beam splitters which offer broader possibilities when they can be employed. Macleod (2001) comments that "The techniques which are currently available operate only over very restricted ranges of wavelength and angle of incidence (effectively over a very narrow range of angles)." The current results support this conclusion. The results of the techniques of Thelen (1988) show examples which are also of this nature and Baumeister (2004) shows another variety of examples along the same lines.

These limitations on angle and bandwidth occur because the effective indices of refraction for the media and coating layer materials have different functions of the angle of incidence for each polarization.

We use (filmstar free version) (http://www.ftgsoftware.com/design.htm) to draw the reflectance amplitude and phase diagram and for the comparison purpose with other experimental works. Figure 2 shows the reflectance amplitude and phase diagram (or "circle" diagram) (Willey, 2002) of a two quarter wave optical thickness QWOT layer coating at normal incidence of the following design: Substrate (S) HL Air (A), where L,M and H represent low, medium and high respectively. Such diagrams are closely related to admittance (Macleod, 2001; Willey, 2002) diagrams when rotated 180° about the origin.

Figure 2 shows how Figure 3 changes when the angle is 50° and the polarizations split. The s-polarization shifts to the left and the p to the right. The final reflectance of the coating at the end point to the right is greater for the p than for the s. Figure 4 shows the case at 89° as the

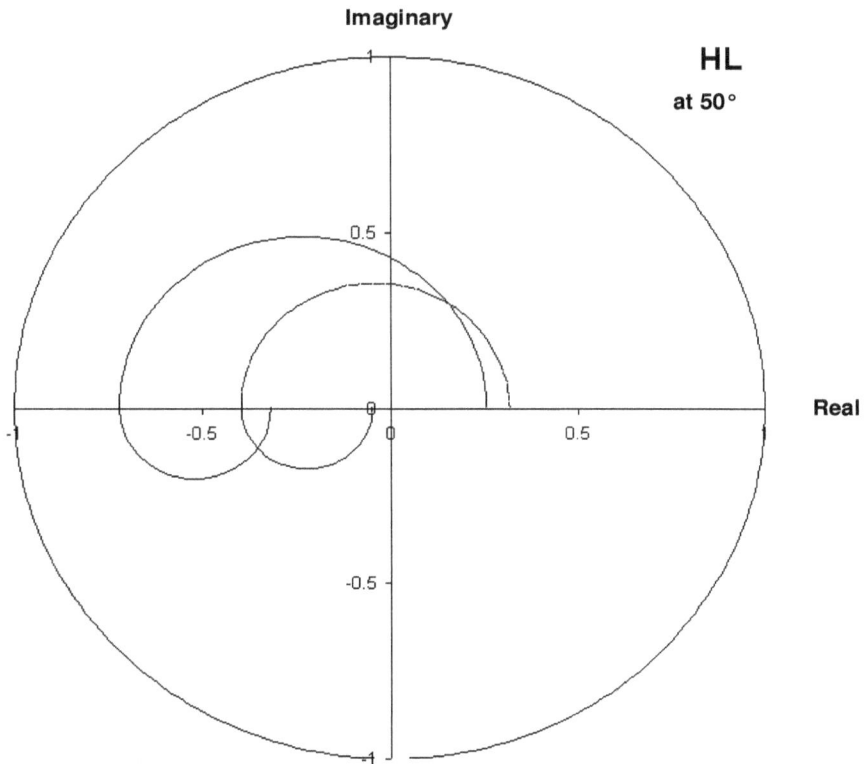

**Figure 3.** Change from Figure 2 when the angle is 50° and the polarizations split.

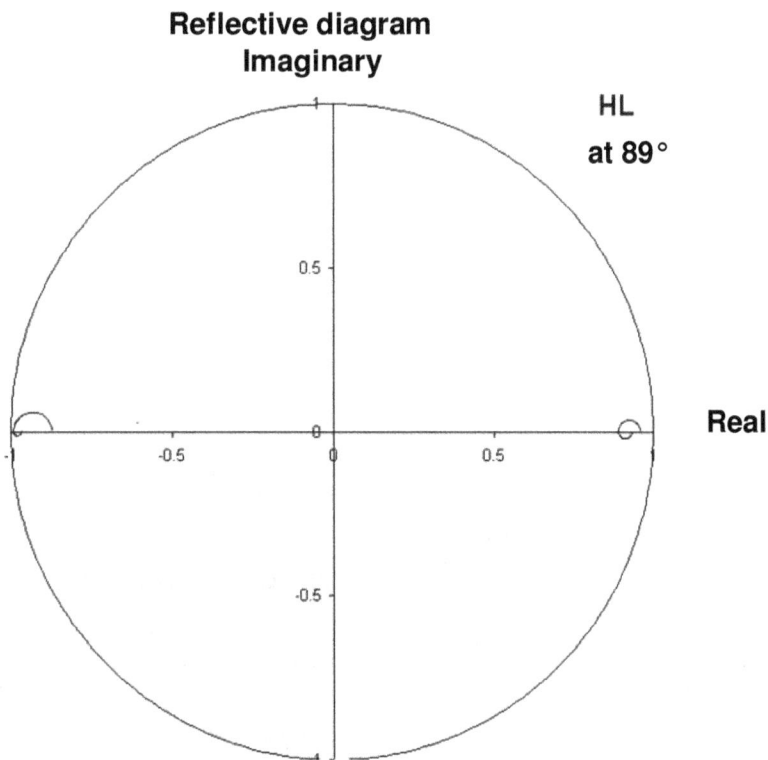

**Figure 4.** Same case as Figure 1 at 89° as 90° is approached. The s-polarization shifts to the left and the p to the right.

Reflective vs. angle for angle corrected QWOTS
Design: S HL A, where s=1.46, H=2.35 and L=1.45

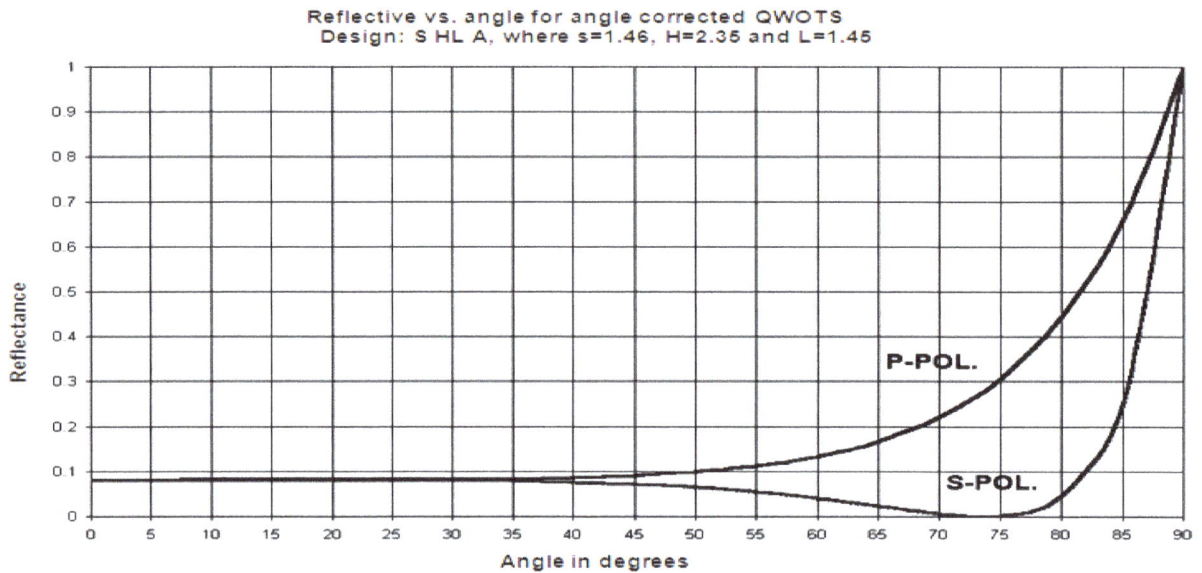

**Figure 5.** Reflectance of the SHLA design in s- and p-polarizations over 0 to 90°.

extreme angle is approached; both s and p approach 100% reflectance, but are 180° apart in phase. Figure 5 shows the reflectance of this coating configuration as a function of all angles of incidence from 0 to 90°.

The foregoing figures are cases where the thicknesses are adjusted to produce quarter wave optical thickness (QWOT) for each layer and index at the angle being evaluated. This are referred to as "angle matched" thicknesses and are discussed more in details in the referenced texts (Macleod, 2001; Thelen, 1988; Baumeister, 2004). It will be seen that two advantages accrue when only "matched" QWOT layers are used:

1. The change in reflectance with wavelength is minimum at the design wavelength because it goes through zero, and
2. The sensitivity to layer errors can be minimum.

### PROCEDURE

A systematic search was done of all possible QWOT coating designs having from one to five layers, where the substrate index was 1.46 and the coating material indices were L = 1.45, M = 1.65, and H = 2.35. Each design was evaluated from 0 to 90° with the necessary adjustments of thickness with angle. If the reflectances of the s- and p-polarizations intersect at other than 0 or 90°, there is a non-polarizing design with all QWOT's at that angle and reflectance. Figure 6 is an example of such a case where, in the design S HMHM A, the s and p reflectances intersect near 45°. Twelve intersecting designs were found for five layers or less. An exhaustive search of all possible QWOT designs with 6 and 7 layers was beyond the scope of this work, but all of the 7-layer designs which start with MH layers on the substrate were evaluated. Fourteen additional non-polarizing designs were found. These 26 designs are plotted in Figure 7 for their %R versus non-polarizing angle.

Table 1 shows these designs and resulting non-polarizing angles, %R, and dR/dA (rate of change of the reflectance difference between s and p versus angle in degrees) for the 26 cases found. Table 2 groups the designs together where they have similar or identical results. There are only 14 unique cases found for angle and reflectance. The choice between designs of similar results might be based on the number of layers or other physical property considerations, but the optical performances are the same.

The results in Figure 7 point to the possibility of beam splitter designs at 45° which have a reflectance between 20 and 30%. If a 50/50% beam splitter is needed related to this group, it would apparently be at angles in the range of 59 to 69°. It is conjectured that 50/50% all QWOT dielectric designs might be achievable with more than seven layers of these materials at 45°, over a small wavelength and angle range. The popular compromise for a nearly non-polarizing 50/50% beam splitter at 45° which is available from various component suppliers is a cube beam splitter (immersed coating) with about 45/45% over the visible or similar bandwidth. This is done with a silver layer that has 5 to10% absorbance and additional dielectric layers.

Figure 8 plots the change of percentage reflectance difference in the s- and p-polarizations with change of angle in degrees for the various intersecting cases examined. It can be seen that there is a change with increasing angle of incidence which appears to be unavoidable. This can be seen to be close to a function of $1/\cos^2(\text{angle})$. For example, at 45°, one can expect at least 0.2% split in reflectance of s and p for each degree of departure from the non-polarizing angle of the design.

### CHANGES WITH INDICES

The effects of varying indices were evaluated using DOE methodology (Schmidt and Launsby, 1994;DOE, 1997) over the ranges of 1.38 to 1.52 for L, 1.55 to 1.75 for M, and 2.2 to 2.5 for H for a representative design, S MHMHLHM A. Figures 9 and 10 show, in surface plot format, how the non-polarizing angles vary with L, M, and H, and Figures 11 and 12 show, in contour plot format, how the percentage reflectance (%R) at the non-polarizing angles vary with L, M and H.

**Reflective vs. angle for angle corrected QWOTS**
**Design: S HMHM A**

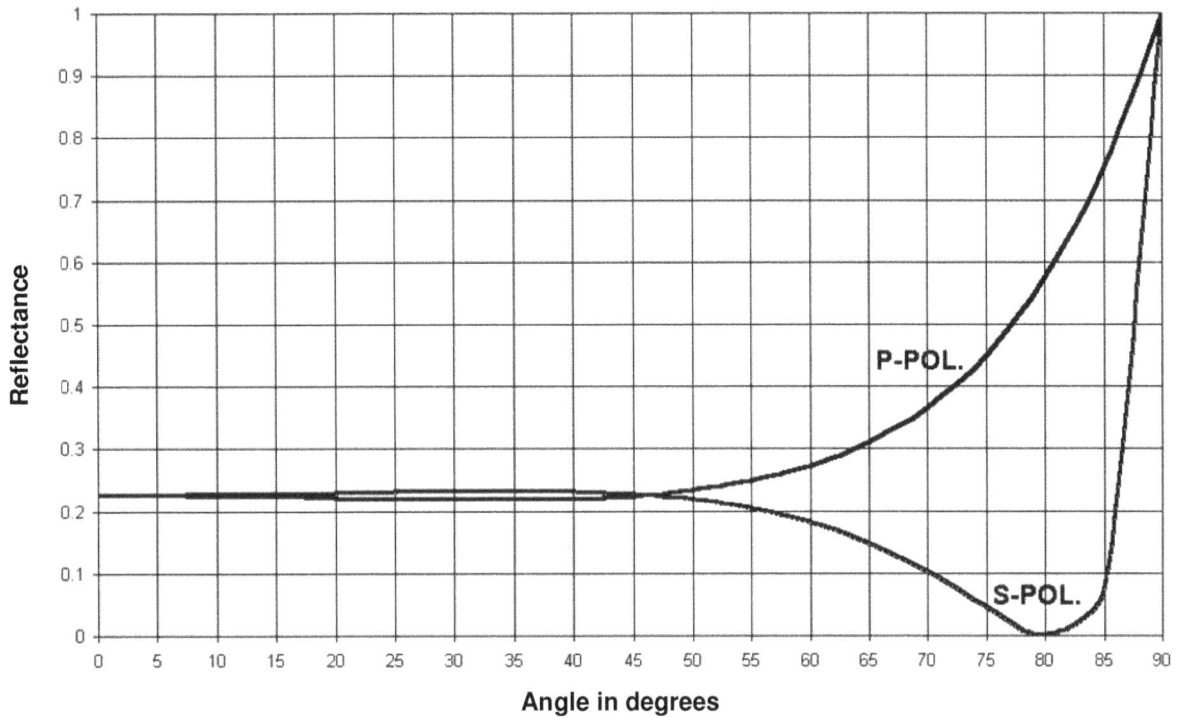

**Angle in degrees**

**Figure 6.** Reflectance versus angle where, in the design SHMHMA, the s and p reflectances intersect near 45°.

## Range of all QWOT solution within index bounds used

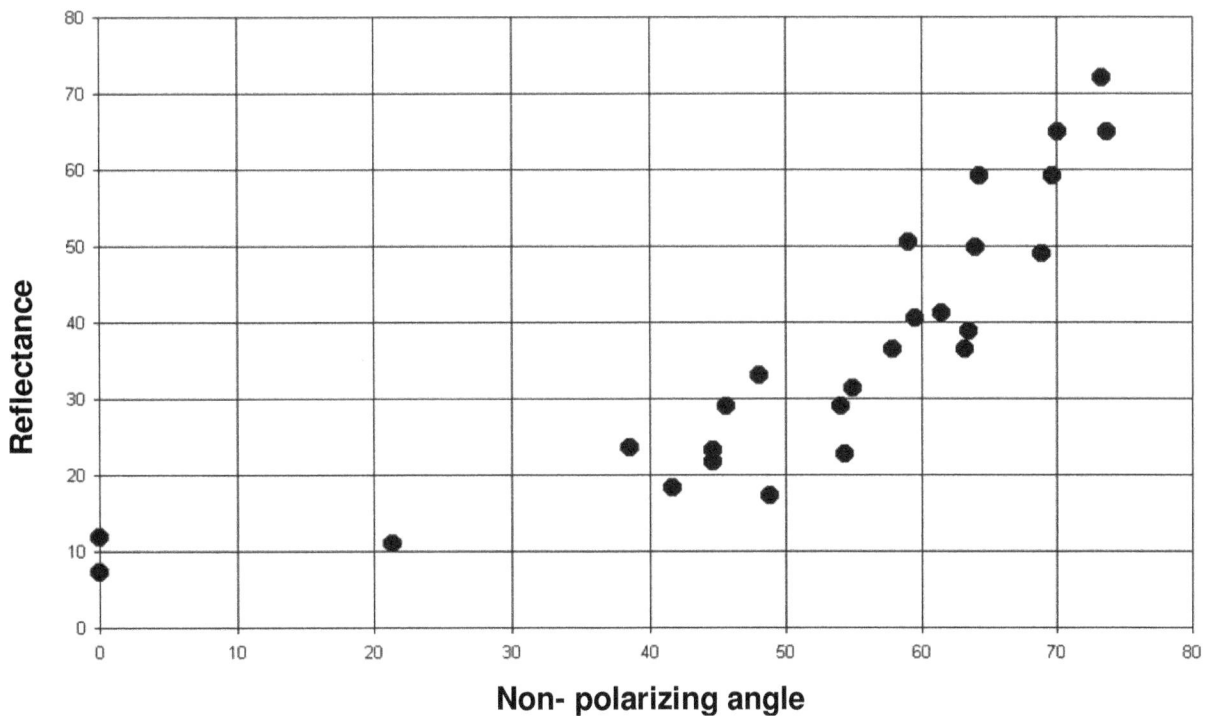

**Non- polarizing angle**

**Figure 7.** Reflectance versus non-polarizing angle of the 26 all-QWOT designs found.

**Table 1.** Designs, resulting non-polarizing angles, %R, and dR/ for the 26 all-QWOT designs found.

| Design | Angle | %R | dR/dA |
|---|---|---|---|
| HL | 27.57 | 8.15 | 0.026 |
| MLHL | 53.18 | 15.95 | 0.39 |
| MLHM | 27.57 | 8.15 | 0.026 |
| HLML | 53.18 | 15.95 | 0.39 |
| HLHL | 67.2 | 42.33 | 1.73 |
| HLHM | 59.48 | 32.47 | 0.712 |
| HMHL | 59.48 | 32.47 | 0.712 |
| HMHM | 45.67 | 22.68 | 0.342 |
| MHLML | 30.64 | 8.51 | 0.054 |
| MHLHL | 60.17 | 33 | 1.223 |
| MHLHM | 46.66 | 23.18 | 0.328 |
| MHMHL | 46.66 | 23.18 | 0.328 |
| MHLMLML | 54.5 | 16.5 | 0.53 |
| MHLMLHL | 68 | 42.9 | 1.13 |
| MHLMLHM | 60.3 | 32.9 | 0.96 |
| MHLHLML | 68 | 42.9 | 1.13 |
| MHLHLMH | 31 | 8.5 | 0.037 |
| MHLHLHL | 74.5 | 66.2 | 2.75 |
| MHLHLHM | 70.5 | 58.5 | 1.92 |
| MHLHMHL | 70.5 | 58.5 | 1.92 |
| MHLHMHM | 64 | 49.8 | 1.42 |
| MHMHLML | 60.3 | 32.9 | 0.96 |
| MHMHLHL | 70.5 | 58.5 | 1.92 |
| MHMHLHM | 64 | 49.8 | 1.42 |
| MHMHMHL | 64 | 49.8 | 1.42 |
| MHMHMHM | 55 | 40.4 | 0.68 |

This information can be used as guidance to adjust the indices of a design to achieve a specific reflectance at a given angle. For example, it was calculated that this S MHMHLHM A design with L = 1.52, M = 1.75 and H = 2.35 would have 33.06% R at 48.04°. It was found that adjusting this for 30%R at 45° caused the L and M to move toward each other in the region of 1.7. This then suggested that the one L-layer could be replaced by an M-layer to make a two-material only design. The S MHMHMHM A design could then be adjusted to 30%R at 45° when M = 1.6845 and H = 2.31. These might be achievable indices with the proper process parameters in a given wavelength region. Figures 13 and 14 show the reflectance versus wavelength and versus angle for this design. Figure 15 shows the reflectance amplitude diagram for this design at 45°.

**OTHER ADJUSTMENTS**

The choice of having only QWOT's in the cases shown above leads to the coating terminating with the s and p reflectance's on the 0° phase line of the circle diagram. This is the condition where the change in the difference between s and p reflectance with angle passes through zero and is minimal in that angular region. It is possible to sacrifice some of this minimal difference in s-p in exchange for less variation of the difference with wavelength. This then requires non-QWOT layers, and the final phase is not likely to be zero.

**ANTIREFLECTION COATINGS**

Some of the findings of the search procedure described above included cases where the reflectance was near zero over a broad angular range, such as 0 to 40°. When these were optimized from 0 to 45°, none proved to be better than the classic S MHHL A design (3-layer quarter-half-quarter wave (QHQ)) after it had been optimized for the broader angles. This, of course, results in a design with non-QWOT layers. Therefore, this present design approach has yet to add anything to the antireflection coating knowledge.

**CONCLUSION**

Specific materials can be used to design reasonable non-polarizing coatings at only certain angles, whereas other materials can be made to serve for somewhat different angles. The choice of materials seems to be critical to the success of any particular requirement. Since the range of practical coating materials and substrates is limited for most applications, non-polarizing solutions seem to be quantized in reflectance and angle to the same extent that the indices of the materials are

**Table 2.** Designs of Table 1 grouped together where they have similar or identical results.

| Design | Angle | %R | dR/dA |
|---|---|---|---|
| HL | 27.57 | 8.15 | 0.026 |
| MLHM | 27.57 | 8.15 | 0.026 |
| MHLML | 30.64 | 8.51 | 0.054 |
| MHLHLMH | 31 | 8.5 | 0.037 |
| HMHM | 45.67 | 22.68 | 0.342 |
| MHLHM | 46.66 | 23.18 | 0.328 |
| MHMHL | 46.66 | 23.18 | 0.328 |
| MLHL | 53.18 | 15.95 | 0.39 |
| HLML | 53.18 | 15.95 | 0.39 |
| HLHM | 59.48 | 32.47 | 0.712 |
| HMHL | 59.48 | 32.47 | 0.712 |
| MHLMLHM | 60.3 | 32.9 | 0.96 |
| MHMHLML | 60.3 | 32.9 | 0.96 |
| MHLHMHM | 64 | 49.8 | 1.42 |
| MHMHLHM | 64 | 49.8 | 1.42 |
| MHMHMHL | 64 | 49.8 | 1.42 |
| MHLMLHL | 68 | 42.9 | 1.13 |
| MHLHLML | 68 | 42.9 | 1.13 |
| MHLHLHM | 70.5 | 58.5 | 1.92 |
| MHLHMHL | 70.5 | 58.5 | 1.92 |
| MHMHLHL | 70.5 | 58.5 | 1.92 |

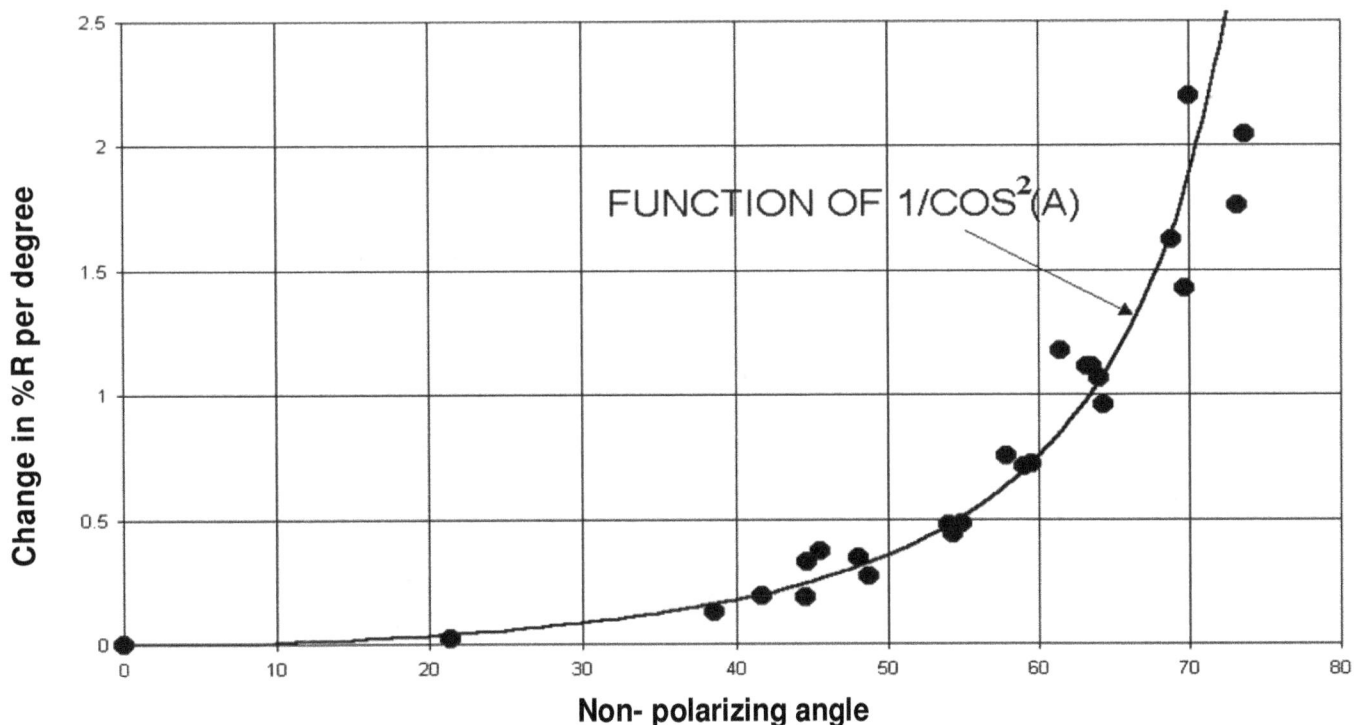

**Figure 8.** Percentage reflectance difference in the s and p-polarizations versus non-polarizing angle of the 26 designs.

**Surface plot of L vs. M**

Constants:   H = 2.35

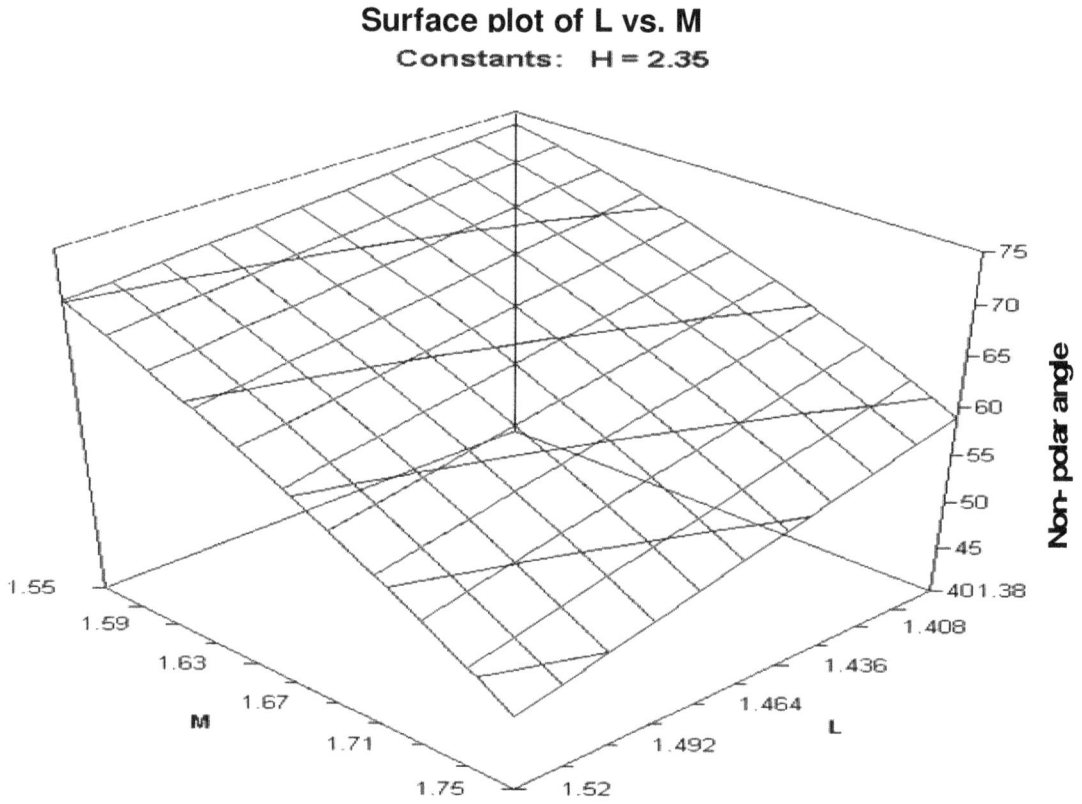

**Figure 9.** Surface plot of the variation of the non-polarizing angles with L and M.

**Surface plot of H vs. M**
**Constant: L=1.45**

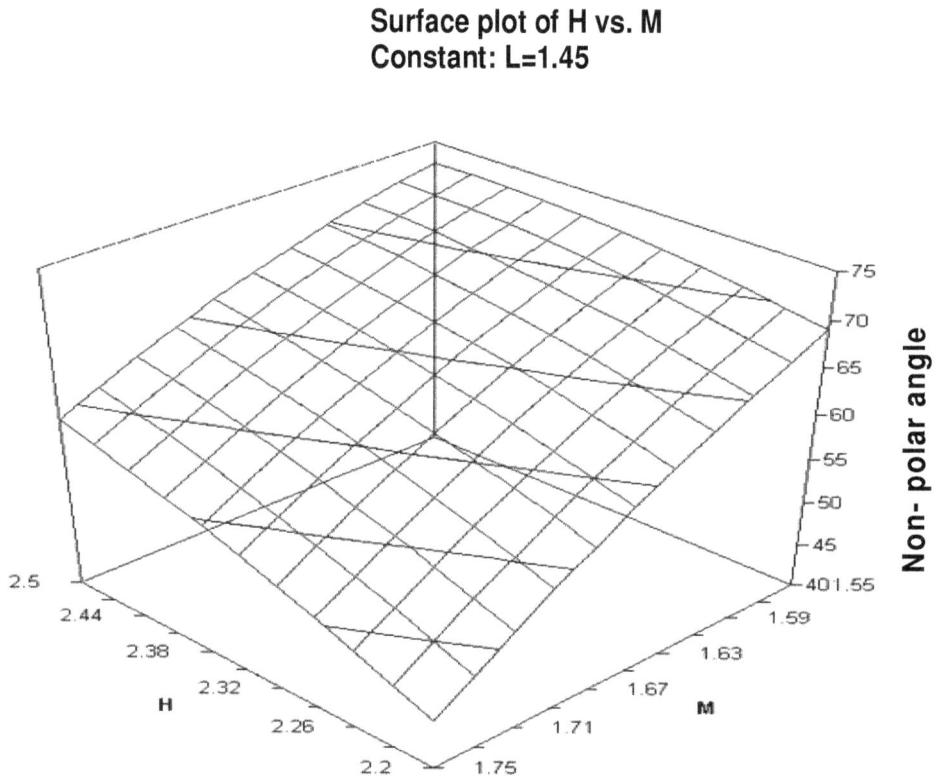

**Figure 10.** Surface plot of the variation of the non-polarizing angles with M and H.

**Contour plot of L vs. M**

**Constants:  H = 2.35**

**%R
MHMHLHM**

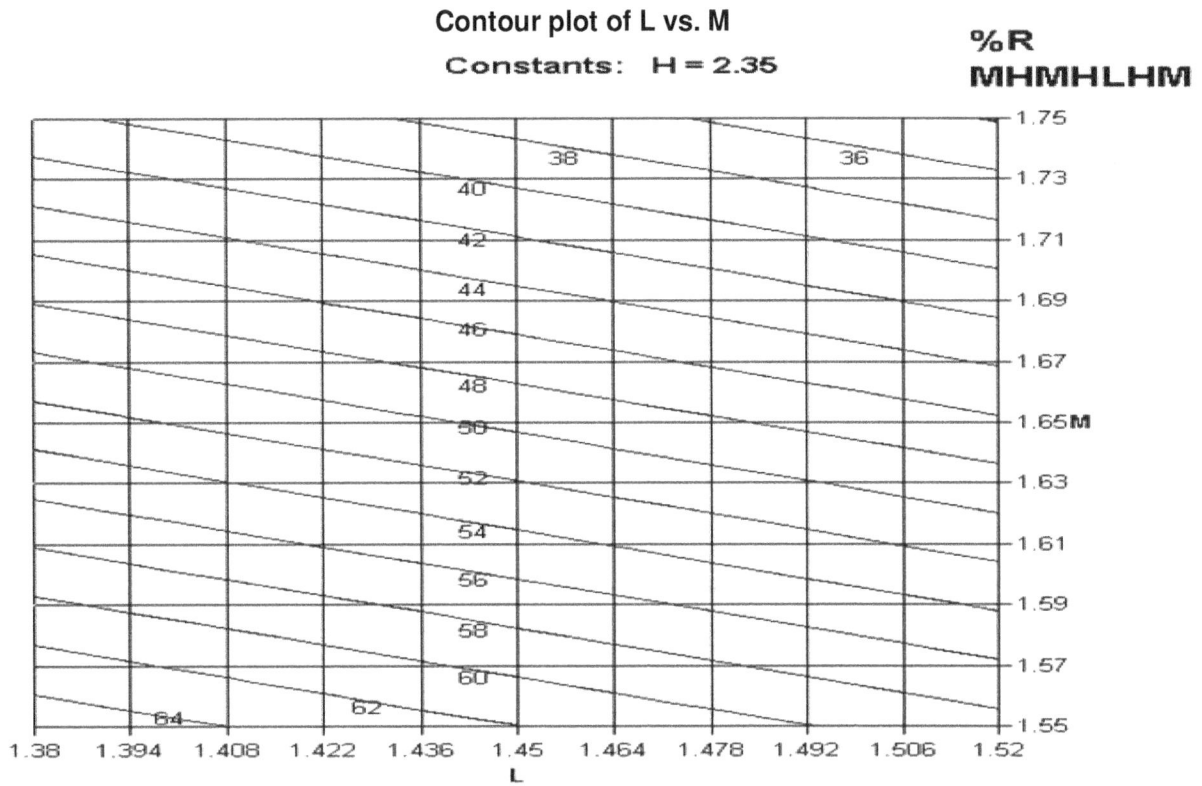

**Figure 11.** Contour plot of the variation of the percent reflectance (%R) at the non-polarizing angles with L and M.

**Contour plot of H vs. M**

**Constants:  L = 1.45**

**%R
MHMHLHM**

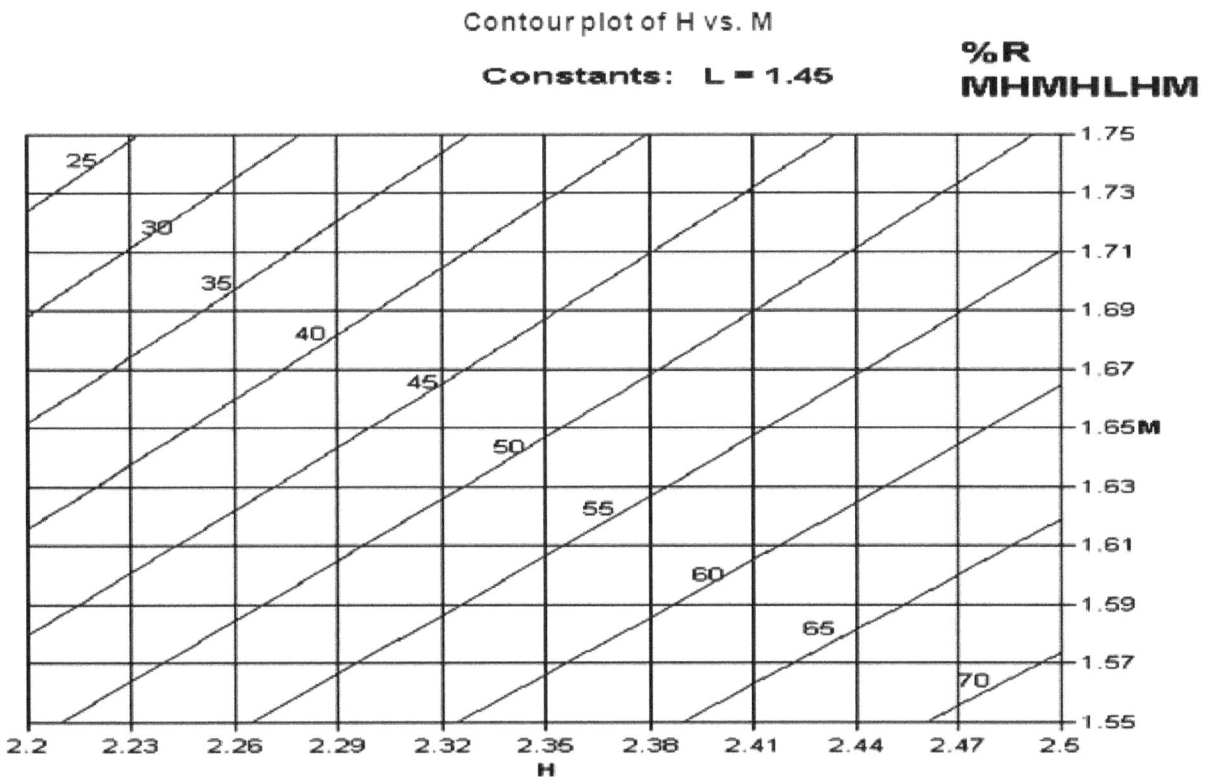

**Figure 12.** Contour plot of the variation of the percent reflectance (%R) at the non-polarizing angles with H and M.

**MHMHMHM 30% at 45°**

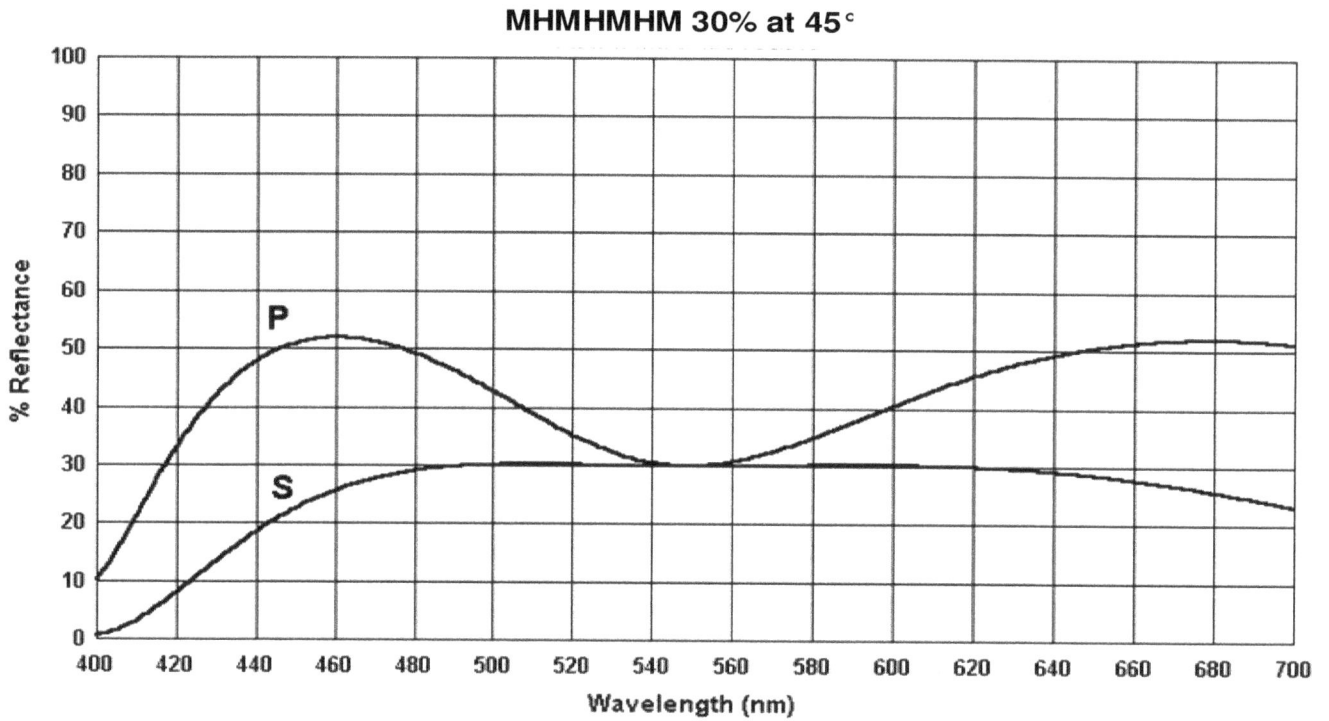

**Figure 13.** Reflectance versus wavelength for the design SMHMHMHMA.

**MHMHMHM 30% at 45°**

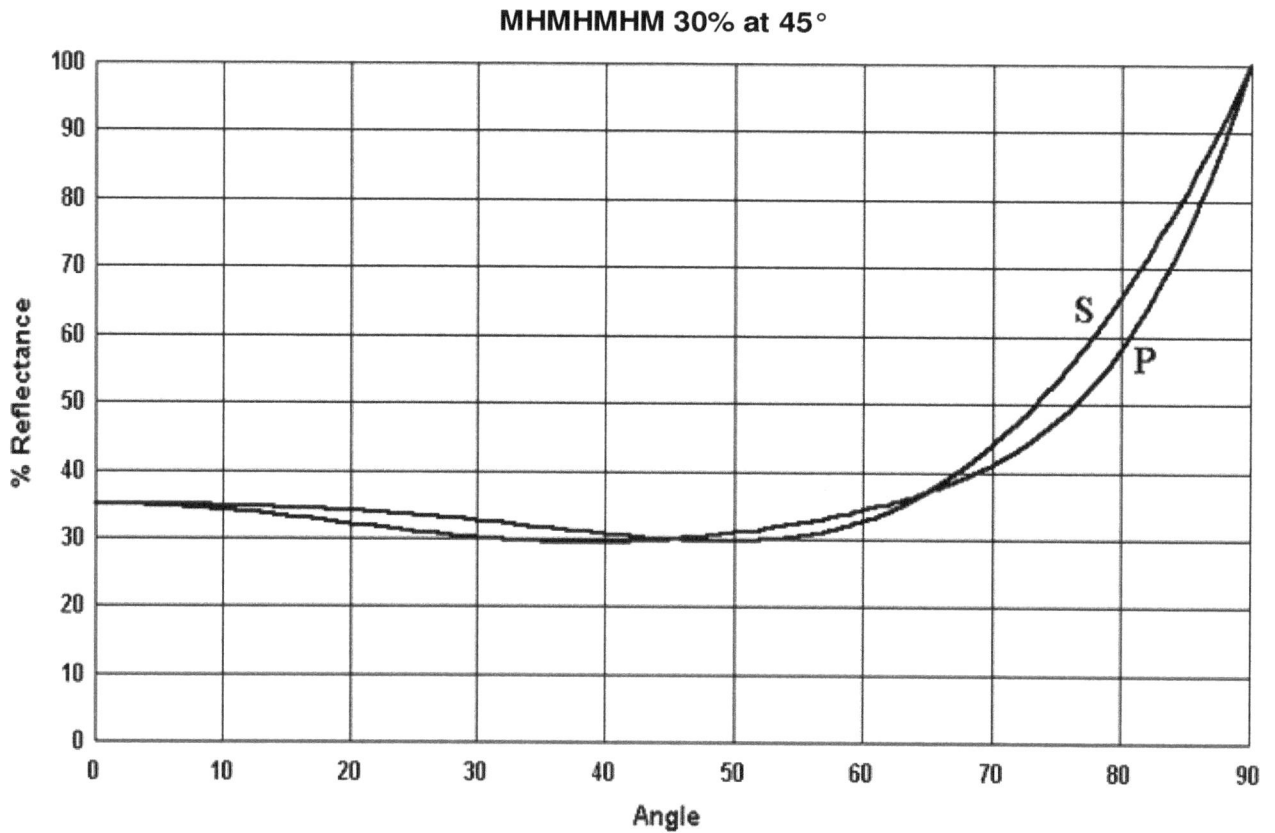

**Figure 14.** Reflectance versus angle for the design SMHMHMHMA.

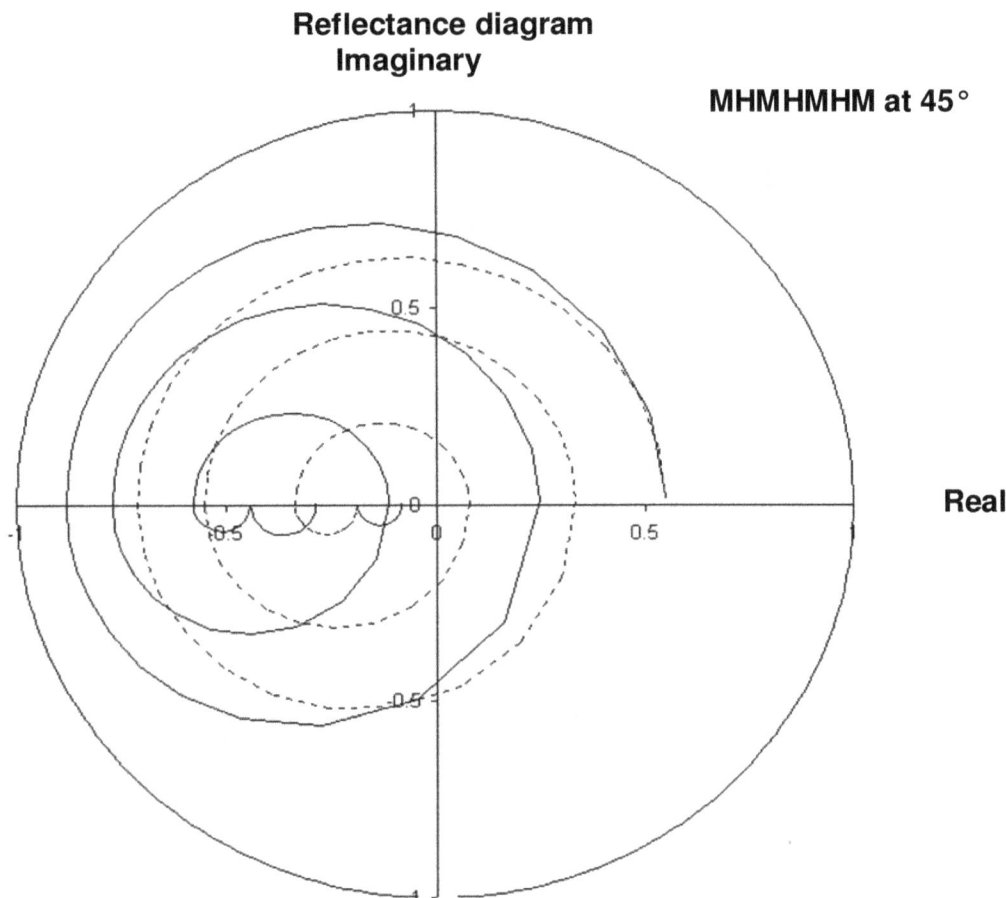

**Figure 15.** Reflectance amplitude diagram for this design at 45°. The sold line for s-polarization and the dotted line for the p- polarization.

quantized into narrow bands of index. This has been illustrated with the aid of reflectance amplitude (circle) diagrams and other plots. We use (filmstar free version) (http://www.ftgsoftware.com/design.htm) to draw the reflectance amplitude and phase diagram and for the comparison purpose with other experimental works.

Design rules-of-thumb for this family of coatings might be:

1. Determine the low, medium and high indices which your facility can reliably produce at the wavelength needed.
2. Search for the possible combinations of layers using these materials with QWOT solutions (intersections) near the desired angle and reflectance (be sure to include the proper "angle matched" thicknesses).
3. Adjustment the indices to gain the angle and reflectance as needed (within the ranges that can be produced.
4. Adjust the design with optimization as necessary, using non-QWOT layers to achieve the most acceptable compromise for the application.

**REFERENCES**

Angus Macleod H (2001). Thin-Film Optical Filters, 368, 3rd Ed., Inst. of Phys. Pub.
Baumeister PW (2004). Optical Coating Technology, 6-29 to 6-39, SPIE Press, Advanced Optical Thin Film Technology from: FTG Software Associates http://www.ftgsoftware.com/design.htm
DOE KISS, Ver. 97 for Windows, Air Academy Associates (and Digital Computations, Inc.), Colorado Springs, 1997.
http://www.optics.arizona.edu/opti380a/Opti380A/Lab2/Papers/OPNpaper.pdf.
Li Li (2003). The Design of Optical Thin Film Coatings With Total and Frustrated Total Internal Reflection, Optics and Photonics News, Optical Society of America, September 2003.
Schmidt SR, Launsby RG (1994). Understanding Industrial Designed Experiments, Air Academy Press, Colorado Springs.
Thelen A (1988). Design of Optical Interference Coatings, 114, McGraw-Hill.
Willey RR (2002). Practical Design and Production of Optical Thin Films, 2nd Ed., 8, Marcel Dekker.

# 8

# Reducing torsional oscillation and performance improvement of industrial drives using PI along with additional feedbacks

T. Ananthapadmanabha[1]*, A. D. Kulkarni[1], Benjamin A. Shimray[1], R. Radha[1]
and Manoj Kumar Pujar[2]

Department of Electrical and Electronics Engineering, The National Institute of Engineering, Mysore-08, Karnataka, India.
[2]Manoj Kumar Pujar Bangalore Transmission Zone, KPTCL Karnataka India.

Torsional oscillations in electrical drive systems with elastic shafts are a well known problem. In the industrial drive systems, a shaft torsional vibration or oscillation is often generated when a motor and a load are connected with a flexible shaft. Large inertias of the motor and load side and a long shaft create an elastic system. The motor speed is different from the load side and the shaft undergoes large torsional torque. In order to damp the torsional vibration effectively, the application of various feedbacks are necessary. The control structures of electrical drives working in the industry are usually based on linear PI controllers. This paper presents the design, analysis and comparison of the conventional PI-control to control structure using PI supported by various feedbacks for performance improvement of the Industrial drives system.

Key words: Torsional oscillation, 2- mass drive system, industrial drive system, transient response, mechanical elasticity, vibration suppression.

## INTRODUCTION

Without Drive control systems there could be no manufacturing, no vehicles, and no material handling. Modern industrial drive systems require relatively high dynamic properties. When the industrial drives are designed, the elasticity of the shaft is neglected. In the case of standard drive and low power servos such an assumption is reasonable; however, there is a large group of high power drives and other application such as rolling-mill drives, robot arms, servo systems, textile drives, throttle systems, conveyor belts, and deep-space antenna drives where characteristics of the mechanical part have to be included in the analysis (Szabat and Orlowska, 2007; Szabat and Orlowska, 2008; Zhang and Furusho, 2000), and the shaft elasticity must be taken into consideration. This type of assumption (neglecting elasticity) can lead to damaging oscillations (Valenzuela et al., 2005; Sugiura and Hori, 1996; Ji and Sul, 1995). This oscillation or torsional vibration decreases the

product quality and system reliability; the system can even lose stability.

The control problem of the two-mass system is originally derives from rolling-mill drives (Zhang and Furusho, 2000; Zhang, 1999; Valenzuela et al., 2005). Large inertias of the motor and rolls and a long shaft makes the drive system an elastic system. The non ideal characteristics of the shaft worsen the performance in practical industrial drive system. An analogous problem also appears in the paper and textile industry (Valenzuela et al., 2005) and rolling mills (Zhang et al., 2007).

To suppress the torsional oscillations, different control structures have been developed (Rached et al., 1994). The simplest method to avoid the system state variable oscillations till now relies on decreasing the dynamics of the control structures. However, this approach neglects the performance of the drive and is hardly ever utilized. If desirable control system performances are required, the application of the additional feedbacks from a selected state variable is necessary ((Szabat and Orlowska, 2007; Szabat and Orlowska, 2008).

A technique that can improve system performance

*Corresponding author. E-mail: drapn2008@yahoo.in.

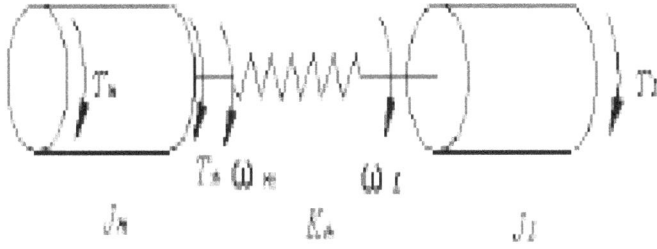

**Figure 1.** 2-mass system. Where: $\omega_m$: motor speed, $\omega_l$: load speed; $T_m$: motor torque; $T_l$: load torque; $T_{sh}$: shaft torque; $J_m$: motor inertia; $J_l$: load inertia; $K_{sh}$: spring coefficient (stiffness) of drive shaft.

exploit alternative tuning techniques for the classical cascade control structure (with a PI speed controller and the basic feedback from the motor speed), based on a suitable location of the close loop system poles. Three different pole locations with identical radius, damping coefficient, and real part were presented (Zhang and Furusho, 2000). Suggestions for the application of a proportional–integral–derivative controller are also made. But the derivative part D increased the inertia ratio of the system and virtually decreasing the moment of inertia of the motor. To improve the performances of industrial drive system, additional feedback loop from one selected state variable can be used. The additional feedbacks can be inserted to the electromagnetic-torque control loop or the speed control loop. Another modification of the control structure results from inserting the additional feedback from the shaft torque. This type of feedback was applied in (O'Sullivan et al., 2006). The damping of the torsional vibration is reported to be successful. This structure is less sensitive to measurement noises than the previous one since the derivative of the shaft torque does not exist. Use of additional feedback from the derivative of the load speed was proposed in (Zhang, 1999). This results in the same dynamical performance as for the previous control structure.

In recent development, nonlinear and soft computing control methods have attracted much attention. The application of the sliding or the fuzzy control will increases the robustness of the drive system to parameter variations. These techniques can allow obtaining better dynamical characteristics of the system, as compared to the classical ones, but they are not yet popular in industrial applications.

Control structures of electrical drives working in almost all the present industry are usually based on linear PI controllers. Thus, the main goal of this paper is to undergo a systematic analysis and present design guidelines for the speed control structures of the two-mass system using PI speed controller that is supported by different additional feedbacks. This work is similar to work carried out by Krzysztof Szabat and Teresa Orlowska-Kowalska (Szabat and Orlowska, 2007; Szabat and Orlowska, 2008) but the method of analysis and

results extracted are different.

## MODELING OF DRIVE SYSTEM

Even though the main drive system used in industrial systems are complex multi-mass system, in many cases, it can be roughly modeled by a two-mass system considering only the first resonant mode as shown in Figure 1.

The damping of the 2-mass system due to the friction is very small, so that it can be neglected without affecting the analysis accuracy. The following Laplace transfer functions are derived from the dynamic mechanical principles:

$$\omega_m = \frac{1}{J_m s}\left(T_m - T_{sh}\right) \tag{1}$$

$$\omega_l = \frac{1}{J_l s}\left(T_{sh} - T_l\right) \tag{2}$$

$$T_{sh} = \frac{K_{sh}}{J_l s}\left(\omega_m - \omega_l\right) \tag{3}$$

The stated equation of the system is given by

$$\overline{X} = AX + BU + ET_L \tag{4}$$

Where:

$$X = \begin{bmatrix} \omega_m & \omega_l & \omega_{sh} \end{bmatrix}^T$$

$$A = \begin{bmatrix} 0 & 0 & -\frac{1}{J_m} \\ 0 & 0 & \frac{1}{J_l} \\ K_{sh} & -K_{sh} & 0 \end{bmatrix}$$

$$B = \begin{bmatrix} \frac{1}{J_m} \\ 0 \\ 0 \end{bmatrix}$$

$$C = \begin{bmatrix} 1 & 0 & 0 \end{bmatrix}$$

$$E = \begin{bmatrix} 0 \\ \frac{1}{J_l} \\ 0 \end{bmatrix}$$

$$U = T_m$$

To simplify the comparison of the dynamical performances of the drive systems of different powers, the mathematical model can be expressed in a per-unit system, using the following notation as new state variables:

$$\omega_1 = \frac{\Omega_1}{\Omega_N} \quad \omega_2 = \frac{\Omega_2}{\Omega_N}$$

$$m_e = \frac{M_e}{M_N} \quad m_s = \frac{M_s}{M_N} \quad m_L = \frac{M_L}{M_N}$$

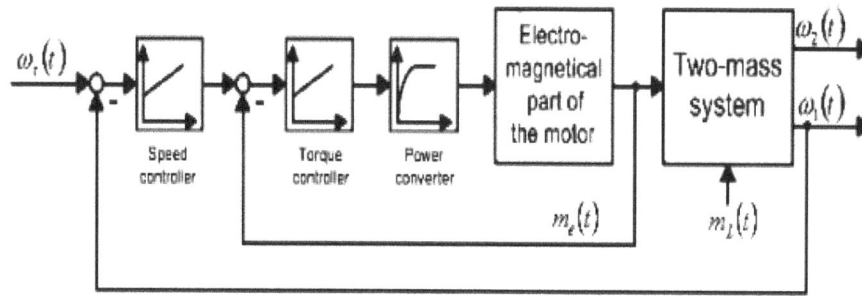

**Figure 2.** Classical control structure.

Where $\Omega_N$ is the nominal speed of the motor; $M_N$ is the nominal torque of the motor; $\omega_1$ and $\omega_2$ are the motor and load speeds, respectively; $m_e$, $m_s$, and $m_L$ are the electromagnetic, shaft, and load torques in the per-unit system, respectively. The mechanical time constant of the motor $T_1$ and the load machine $T_2$ are, thus, given as

$$T_1 = \frac{\Omega_N J_1}{M_N} \quad T_2 = \frac{\Omega_N J_2}{M_N}$$

The stiffness time constant $Tc$ and internal damping of the shaft $d$ can be calculated as follows:

$$T_c = \frac{M_N}{K_c \Omega_N} \quad d = \frac{\Omega_N D}{M_N}$$

Where,

D is the internal damping of the shaft.

The analyzed system is described by the following state equation (in per unit system):

$$\frac{d}{dt}\begin{bmatrix} \omega_1(t) \\ \omega_2(t) \\ \omega_2(t) \end{bmatrix} = \begin{bmatrix} 0 & 0 & -\frac{1}{T_1} \\ 0 & 0 & \frac{1}{T_2} \\ \frac{1}{T_c} & -\frac{1}{T_c} & 0 \end{bmatrix}\begin{bmatrix} \omega_1 \\ \omega_2 \\ \omega_2 \end{bmatrix} + \begin{bmatrix} \frac{1}{T_1} \\ 0 \\ 0 \end{bmatrix}\begin{bmatrix} T_m \end{bmatrix} + \begin{bmatrix} 0 \\ -\frac{1}{T_2} \\ 0 \end{bmatrix}\begin{bmatrix} T_l \end{bmatrix} \quad (5)$$

Where:

$\omega_1$: motor speed;
$\omega_2$: load speed;
$T_m$: motor torque;
$T_{sh}$: shaft (torsional) torque;
$T_l$: disturbance torque;
$T_1$: mechanical time constant of the motor;
$T_2$: mechanical time constant of the load machine;
$T_c$: stiffness time constant.

## PROPOSED CONTROL STRUCTURE

### General description

A typical electrical drive system is composed of a power-converter-fed motor coupled to a mechanical system; microprocessor-based speed and torque controllers, and current, speed, and/or position sensors used for feedback signals. The diagram of such system is presented in Figure 2.

The inner control loop performs motor torque regulation and consists of a power converter, the electromagnetic part of the motor and current sensor and respective current or torque controller. This control loop is designed to provide sufficiently fast torque control, so it can be approximated by an equivalent first-order term. The outer control loop consists of the mechanical part of the drive, speed sensor, and speed controller, and is cascaded to the inner torque control loop. It provides speed control according to its reference value.

Suitable oscillation damping of the two-mass system can be obtained using different additional feedbacks. The block diagram of the drive system with a simplified inner loop and additional feedbacks is presented in Figure 3. In a typical industrial drive internal damping coefficient $d$ of the shaft has a very small value and, therefore, will be neglected in the further analysis.

Three additional feedbacks $k2$, $k6$, and $k7$, which were not mentioned in the literature, were introduced, that is, the feedback from the derivative of the speed difference ($\omega1 - \omega2$) in group A, the feedback from the load speed in group B, and the feedback from the derivative of the shaft torque in group C. The control structures were divided into three different groups according to their dynamical characteristic (Szabat and Orlowska, 2006). The link between different feedbacks (in every group) can be found out from Figure 3. The relationship can be directly seen between feedbacks $k4$ and $k5$ in group B: The derivative of the shaft torque is simply the difference between the motor and load speeds multiplied by the stiffness coefficient. The same relationship exists between the feedbacks $k7$ and $k8$ in group C. The last feedback $k9$ is based on the motor and load speeds. The link between feedbacks $k1$ and $k2$ is not so clearly seen in group A. But, if the electromagnetic and load torques are neglected, the derivative of the difference speeds is the shaft torque multiplied by the following coefficient: $d(\omega1 - \omega2)/dt = -ms(1/T1 + 1/T2)$ (Szabat and Orlowska, 2007) .

The closed-loop transfer functions from the reference speed to the motor and load speeds, respectively, for the control structure demonstrated in Figure 3, are given by the equations (6) and (7) (Szabat and Orlowska, 2007), with the assumption that the optimized transfer function of the electromagnetic-torque control loop is equal to 1.The close loop transfer can be obtained by using signal flow graph (SFG) and Mason's gain formula.

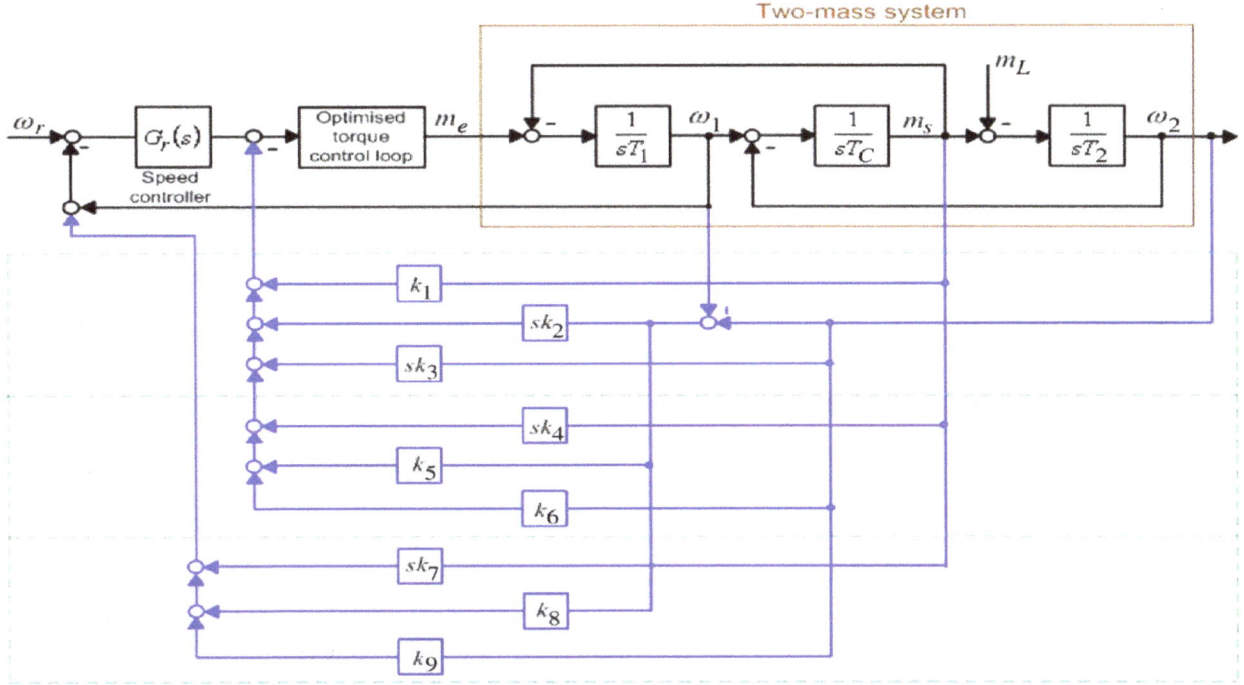

**Figure 3.** Control structure with different additional feedbacks.

$$G_{\omega_1}(s) = \frac{\omega_1(s)}{\omega_r(s)} = \frac{G_r(s)(s^2 T_2 T_c + 1)}{s^3 T_2 T_c (T_1 + K_2) + s^2 T_2 (G_r(s) T_c + G_r(s) K_7 + G_r(s) T_c K_8) + s(T_1 + T_2(1 + K_1 + sK_4 + sT_cK_5) + K_3) + G_r(s)(1 + K_9) + K_6}$$

$$G_{\omega_2}(s) = \frac{\omega_2(s)}{\omega_r(s)} = \frac{G_r(s)}{s^3 T_2 T_c (T_1 + K_2) + s^2 T_2 (G_r(s) T_c + G_r(s) K_7 + G_r(s) T_c K_8) + s(T_1 + T_2(1 + K_1 + sK_4 + sT_cK_5) + K_3) + G_r(s)(1 + K_9) + K_6}$$

(6) and (7)

Where:

$$G_r(s) = K_P + K_I \frac{1}{s} \qquad (8)$$

is the transfer function of the PI controller.

**Cascade control structure without additional feedbacks**

At first, the control structure without additional feedback as in Figure 4 was considered. The characteristic equation of the analyzed system is given by

$$s^4 + s^3 + s^2\left(\frac{K_I}{T_1} + \frac{1}{T_1 T_c} + \frac{1}{T_2 T_c}\right) + s\left(\frac{K_p}{T_1 T_2 T_c}\right) + \frac{K_I}{T_1 T_2 T} = 0 \qquad (9)$$

The desired polynomial of the system has the following form:

$$\left(s^2 + 2\xi\omega_o + \omega_o^2\right)\left(s^2 + 2\xi\omega_o s + \omega_o^2\right) = 0 \qquad (10)$$

Where $\xi$ is the damping coefficient and $\omega_o$ is the resonant frequency of the closed-loop system.

$$s^4 + s^3\left(4\xi\omega_o\right) + s^2\left(2\omega_o^2 + 4\xi^2\omega_o^2\right) + s\left(4\xi\omega_o^3\right) + \omega_o^4 = 0 \qquad (11)$$

Through the comparison of relationships (9) and (11), the set of four equations is created, that is

$$4\xi\omega_o = \frac{K_P}{T_1}$$

$$2\omega_o^2 + 4\xi^2\omega_o^2 = \frac{K_I}{T_1} + \frac{1}{T_1 T_c} + \frac{1}{T_2 T_c}$$

$$4\xi\omega_o^3 = \frac{K_P}{T_1 T_2 T_c} \qquad (12)$$

$$\omega_o^4 = \frac{K_I}{T_1 T_2 T_c}$$

Solving the equation set (12), the parameters of the system, that is, damping coefficient $\xi$ and resonant frequency $\omega_o$, as well as the controller parameters, that is, $K_P$ and $K_I$ are obtained as follows Equation 10 can be rewritten as follows:

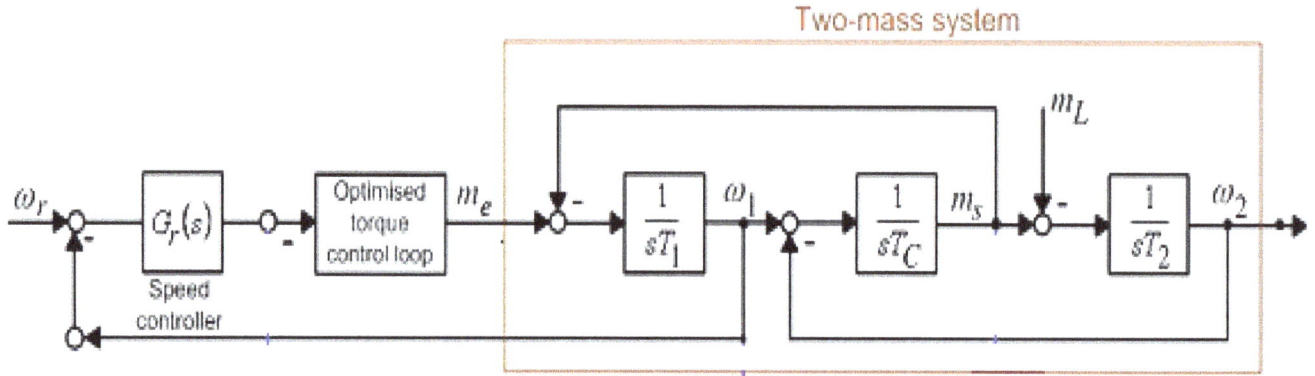

**Figure 4.** Control structure without additional feedback.

$$\xi = \frac{1}{2}\sqrt{\frac{T_1}{T_2}} \qquad \omega_o = \sqrt{\frac{T_1}{T_2 T_c}}$$

$$K_P = 2\sqrt{\frac{T_1}{T_c}} \qquad K_I = \frac{T_1}{T_2 T_c} \qquad (13)$$

### Control structures with additional feedbacks $K_1$, $K_2$ and $K_3$

This group includes the modified control structures with additional feedbacks from the shaft torque $k_1$, from the derivative of the difference between the motor and load speeds $k_2$, or from the derivative of the load speed $k_3$.

First, the control structure with additional feedback K1 was investigated. The damping coefficient and resonant frequency of this structure with the PI speed controller are the following:

$$\xi^{K_1} = \frac{1}{2}\sqrt{\frac{T_2(1+K_1)}{T_1}} \qquad \omega_o^{K_1} = \sqrt{\frac{1}{T_2 T_c}} \qquad (14)$$

Similarly the damping coefficient and resonant frequency of the second System, with additional feedback from the derivative of the difference between two speeds $k2$, are

$$\xi^{K_2} = \frac{1}{2}\sqrt{\frac{T_2 - K_2}{T_1 + K_2}} \qquad \omega_o^{K_2} = \sqrt{\frac{1}{T_2 T_c}} \qquad (15)$$

For the next control structure with additional feedback from the derivative of the load speed $k3$, the damping coefficient and resonant frequency are

$$\xi^{K_3} = \frac{1}{2}\sqrt{\frac{T_2 + K_3}{T_1}} \qquad \omega_o^{K_3} = \sqrt{\frac{1}{T_2 T_c}} \qquad (16)$$

The following equations allow setting the parameters of the feedback loop and the speed controller (Szabat and Orlowska, 2007).

$$K_1 = \frac{4\xi_r^2 T_1}{T_2} - 1 \qquad K_P^{K1} = 2\sqrt{\frac{T_1(1+K1)}{T_c}}$$

$$K_2 = \frac{T_2 - 4\xi_r^2 T_1}{4\xi_r^2 T + 1} \qquad K_I^{K1} = \frac{T_1}{T_2 T_2}$$

$$K_P^{K2} = 2\sqrt{\frac{(T_1 + K_2)(T_2 - K_2)}{T_2 T_c}} \qquad K_P^{K3} = 2\sqrt{\frac{T_1(T_2 - K_3)}{T_c}}$$

$$K_P^{K2} = \frac{T_1 + K_2}{T_2 T_c} \qquad\qquad K_I^{K3} = \frac{T_1}{T_2 T_2} \qquad (17)$$

$$K_3 = 4\xi_r^2 T_1 - T_2$$

In the three mentioned structures, the application of additional feedback ($k1$, $k2$, or $k3$) increases the damping coefficient of the drive system, yet the resonant frequency remains unchanged [see, e.g., (14), (15), and (16)].

### Control structures with additional feedbacks $K_4$, $K_5$ and $K_6$

Next, the control structures with additional feedbacks from the derivative of torsional torque $k_4$, the difference between motor and load speeds $k_5$, or the load speed $k_6$, inserted to the torque node, were tested one after another.

As in previous case, the damping coefficient and the resonant frequency of the system with additional feedback from the derivative of the shaft torque $k_4$ are:

$$\xi^{k_4} = \sqrt{\frac{T_c + x}{4T_1 T_c}(T_1 + T_2) + \frac{T_c}{4(T_c + x)} - \frac{1}{2}} \qquad \omega_o^{k_4} = \sqrt{\frac{1}{T_2(T_c + x)}}$$

Where:

$$x_{1,2} = \frac{-b \pm \sqrt{b^2 - 4ac}}{2a}$$

$$a = T_2^2(T_2 + T_2)$$

$$b = 2T_2^3 T_c - 4\xi_r^2 T_1 T_2^2 T_c$$

$$c = T_2^3 T_c - 4\xi_r^2 T_1 T_2^2 T_c$$

Then, the control structure with additional feedback from the difference between the motor and load speeds $k5$ was investigated. The damping coefficient and the resonant frequency of the analyzed system are:

$$\xi^{k5} = \sqrt{\frac{(T_c+x)(T_1+T_2)}{4T_1} + \frac{1}{4(T_c+x)^2} - \frac{1}{2}} \quad \omega_o^{k5} = \sqrt{\frac{1}{T_2 T_c (1+x)}}$$

The following equations allow setting the parameters of the feedback loop and the speed controller (Szabat and Orlowska, 2007).

$$K_4 = xK_P^{K4} \quad K_P^{K4} = \frac{4\xi_r \omega_0^{K4} T_1 T_c}{T_c+x} \quad K_I^{K4} = (\omega_o^{K4})^4 T_1 T_2 T_c$$

$$K_5 = xK_P^{K5} \quad K_P^{K5} = \frac{4\xi_r \omega_0^{K5} T_1}{1+x} \quad K_I^{K5} = (\omega_0^{K5})^4 T_1 T_2 T_c \qquad (18)$$

$$K_6 = xK_P^{K6} \quad K_P^{K6} = 4\xi_r \omega_0^{K6} T \quad K_I^{K6} = (\omega_o^{K6})^4 T_1 T_2 T_c$$

## Control structures with additional feedbacks: $K_7$, $K_8$ and $K_9$

In this case, control structure with additional feedbacks from the derivative of shaft torque $k_7$, the difference between the load and motor speeds $k_8$, or the load speed $k_9$. Unlike the previous two groups, the additional feedbacks are inserted to the speed node were investigated.

First, the control structure with additional feedback from the derivative of the torsional torque $k7$ was examined. The system damping coefficient and resonant frequency are

$$\xi^{k7} = \frac{1}{2}\sqrt{\frac{(T_c+x)(T_c+K_7)}{T_1 T_c} - 1} \quad \omega_o^{k7} = \sqrt{\frac{1}{T_2(T_c+K_7)}}$$

Next, the control structure with additional feedback from the difference between motor and load speed $k8$ was tested. The damping coefficient and resonant frequency are:

$$\xi^{k8} = \frac{1}{2}\sqrt{\frac{T_1 K_8 + T_2(1+K_8)}{T_1}} \quad \omega_o^{k8} = \sqrt{\frac{1}{(1+K8)T_2 T_c}}$$

Finally, the system with additional feedback from the load speed $k9$ was considered. The damping coefficient and resonant frequency of this system are defined as

$$\xi^{k9} = \frac{1}{2}\sqrt{\frac{T_1+T_2}{T_1(1+K_9)} - 1} \quad \omega_o^{k9} = \sqrt{\frac{1+K_9}{T_2 T_c}}$$

The following equations allow setting the parameters of the feedback loop and the speed controller (Szabat and Orlowska, 2007).

$$K_7 = \frac{(4\xi_r^2+1)T_1 T_c}{T_1+T_c} - T_c \quad K_8 = \frac{(\xi_r^2 4T_1)-T_2}{T_1+T_c} \quad K_9 = \frac{(T_1+T_2)}{T_1(4\xi_r^2+1)} - 1$$

$$K_P^{K7} = 4\xi_r(\omega_0^{k7})^3 T_1 T_2 T_c \quad K_P^{K8} = \frac{4\xi_r \omega_0^{k8} T_1}{1+K_8} \quad K_P^{K9} = 4\xi_r \omega_0^{k9} T_1 \qquad (19)$$

$$K_I^{K7} = (\omega_0^{k7})^4 T_1 T_2 T_c \quad K_I^{K8} = \frac{T_1}{(1+K_8)T_2 T_c} \quad K_I^{K9} = \frac{T_1(1+K_9)}{T_2 T_c}$$

## SIMULATION RESULTS AND DISCUSSION

The performance parameters or time domain transient response specifications are presented in details in Tables 1 and 2. For simulation purpose, the required damping co-efficient is taken as 0.7 that is $\xi r = 0.7$. A 5.5 KW Induction motor connected through a long shaft to a 6.4 KW servo induction motor which can induce disturbance torque given in Table 3 is taken for simulation purpose. The control structure is simulated in MATLAB/SIMULINK. Various control structures simulation results are presented in Figures 5 a - j.

Firstly, the control structure of a PI controller without any additional feedback was investigated. The load speed transient has a large overshoot and quite a long settling time. Next, the control structure using PI with various additional feedbacks $k_1$ to $k_9$ considering one feedback at a time was simulated. The detail effect of using various feedbacks can be seen from the simulated results presented in Tables 1 and 2. With additional feedbacks $k_1$, tremendous improvement in settling time is achieved on the load side transient. Overshoot is also reduced to 54.2 from 99.9%. Using feedback $K_2$ similar improvement is achieved; settling time is reduced to 4.7 s while overshoot is reduced to 27.9%.

In case where PI is supported by feedback $K_4$, we observed that both the load speed and motor speed transient exhibit critically damped response with very small overshoot of 0.22% while the settling time remains around 53 s. This is useful in application where overshoot is the main criteria for consideration like in dryer section of paper mills [Valenzuela et al., 2005] and robotic application. With additional feedback $K_5$, good performance on the motor is achieved but the load sides overshoot remains around 71.11%. When using K6, both motor side and load exhibit very good response. The detail performance parameters are given in Table 2. With $K_7$ load side transient is improved, but the motor side performance remains almost same as without feedback. But this can be further improved by re-tuning or changing $\xi r$. With feedback $K_8$ improvement on the load side transient is observed but the motor side remains unbounded similar to control structure using $K_2$.

Lastly, the control structure with $K_9$ was simulated. Both motor side and load side exhibit good transient response. The response time or the rise time of this control structure on both sides is very fast. It has rise time 0.4 s on the load side and 0.12 s on the motor side.

We observe that based on our requirement, we can choose different feedback. But the process operator will have to often do final tuning of the controller iteratively on the actual process to yield more satisfactory control.

## Graphical output

The following Figures 5a to j are obtained from the simulation. Blue line plot represent transient response on the load side while the red line plot represent transient response on the motor side. For simulation purpose the required damping co-efficient is taken as 0.7 that is, $\xi r = 0.7$.

**Table 1.** Load side transient response parameters.

|  | PI | PI + $K_1$ | PI + $K_2$ | PI + $K_3$ | PI + $K_4$ | PI + $K_5$ | PI + $K_6$ | PI + $K_7$ | PI + $K_8$ | PI + $K_9$ |
|---|---|---|---|---|---|---|---|---|---|---|
| Tp(s) | 2.6 | 1.41 | 2.03 | 1.33 | 141.08 | 1.44 | 1.66 | 1.99 | 1.19 | 1.19 |
| Tr (s) | 0.78 | 0.47 | 0.67 | 0.45 | 30.61 | 0.41 | 0.54 | 0.47 | 0.40 | 0.40 |
| Ts (s) | 1007.8 | 3.8 | 4.71 | 5.54 | 53.61 | 5.55 | 5.43 | 247.55 | 3.18 | 3.19 |
| %. O.S | 99.52 | 54.23 | 27.99 | 60.78 | 0.22 | 71.11 | 39.08 | 109.69 | 54.32 | 54.31 |

**Table 2.** Motor side transient response parameters.

|  | PI | PI + $K_1$ | PI + $K_2$ | PI + $K_3$ | PI + $K_4$ | PI + $K_5$ | PI + $K_6$ | PI + $K_7$ | PI + $K_8$ | PI + $K_9$ |
|---|---|---|---|---|---|---|---|---|---|---|
| Tp(s) | 1.83 | 1.83 | Unbound response | 1.93 | 137.57 | 1.90 | 1.90 | 5.8 | Unbound response | 1.83 |
| Tr (s) | 1.13 | 1.05 | Unbound response | 1.17 | 29.99 | 0.98 | 0.98 | 0.87 | Unbound response | 0.12 |
| Ts (s) | 3.42 | 3.97 | Unbound response | 6.92 | 52.69 | 5.96 | 5.95 | 219.83 | Unbound response | 6.70 |
| %. O.S | 21.54 | 32.71 | Unbound response | 13.13 | 0.22 | 36.36 | 36.37 | 83.17 | Unbound response | 24.41 |

**Table 3.** System parameter.

| Motor | Load | Mechanics |
|---|---|---|
| **Power** | **Power** | **Inertia of motor** |
| 5.5 KW | 6.4 KW | 0.037 Kgm$^2$ |
| **Torque** | **Torque** | **Inertia of load side** |
| 36 Nm | 39 Nm | 0.125 Kgm$^2$ |
| **Speed** | **Speed** | **Shaft stiffness** |
| 1455 min$^{-1}$ | 2490 min$^{-1}$ | 2070 Nm/rad |

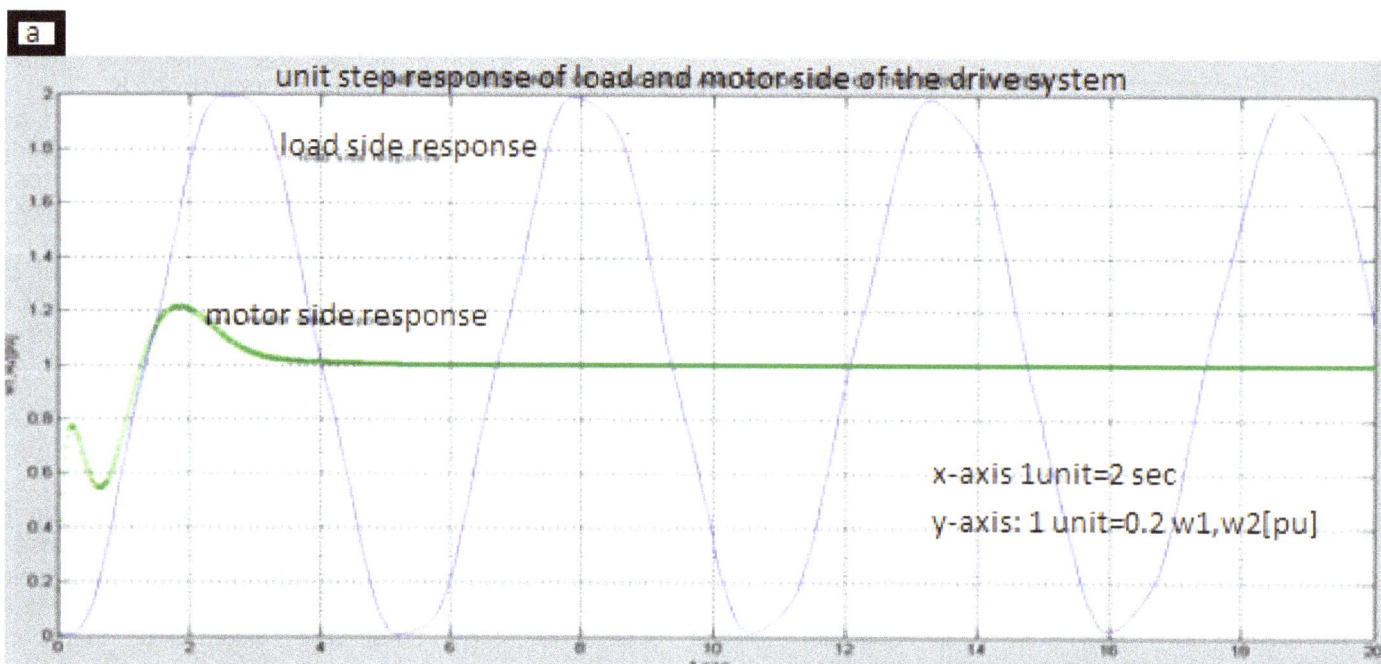

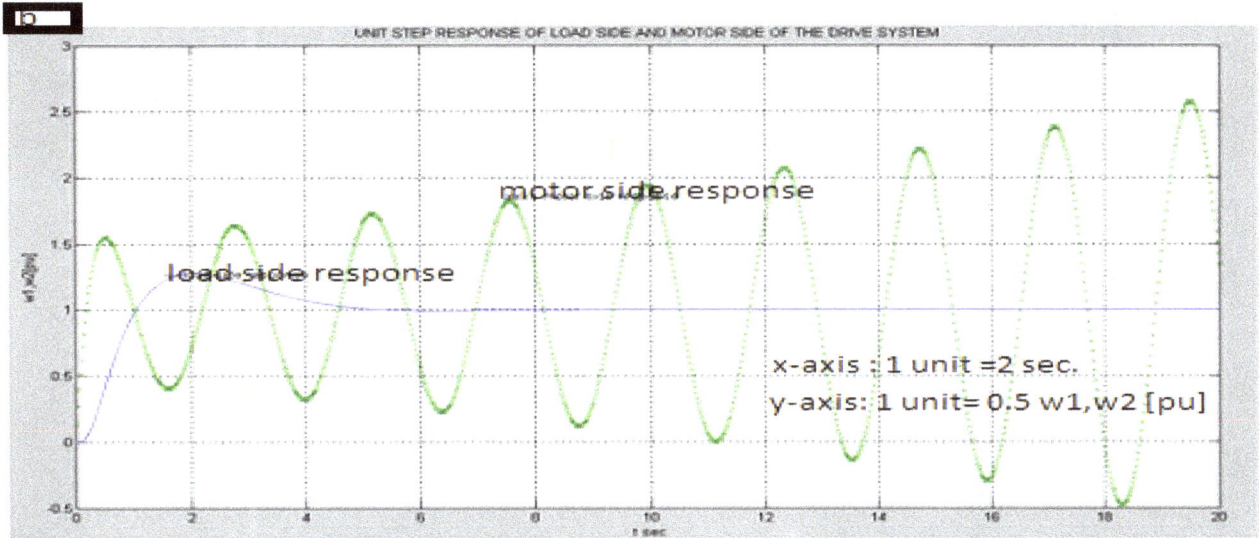

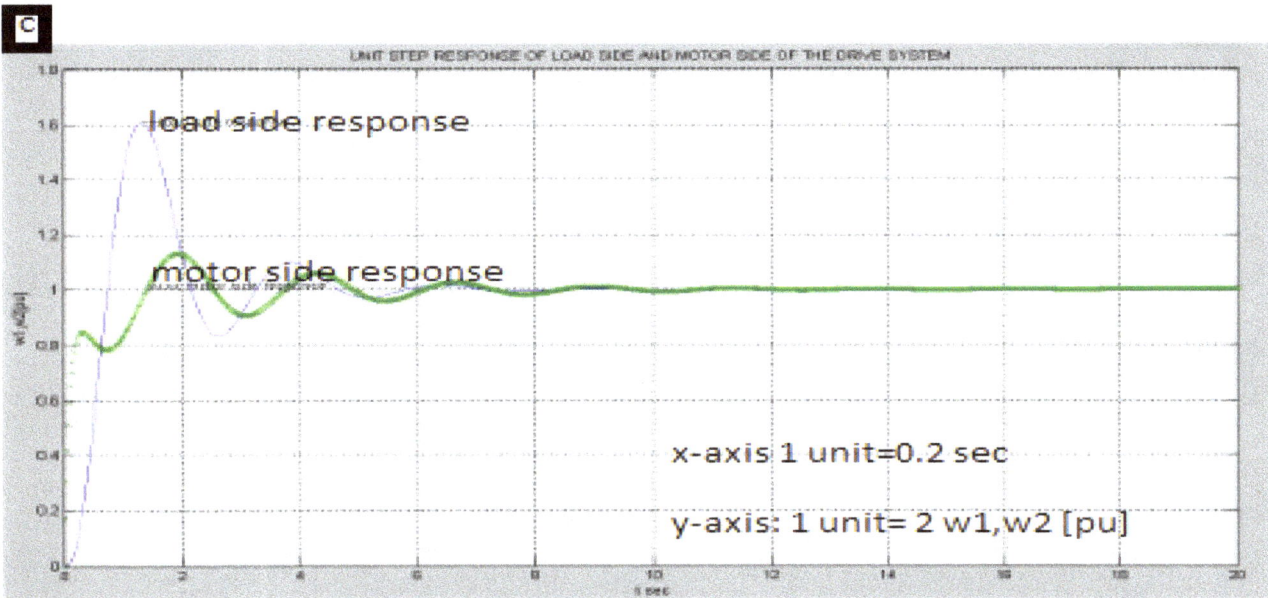

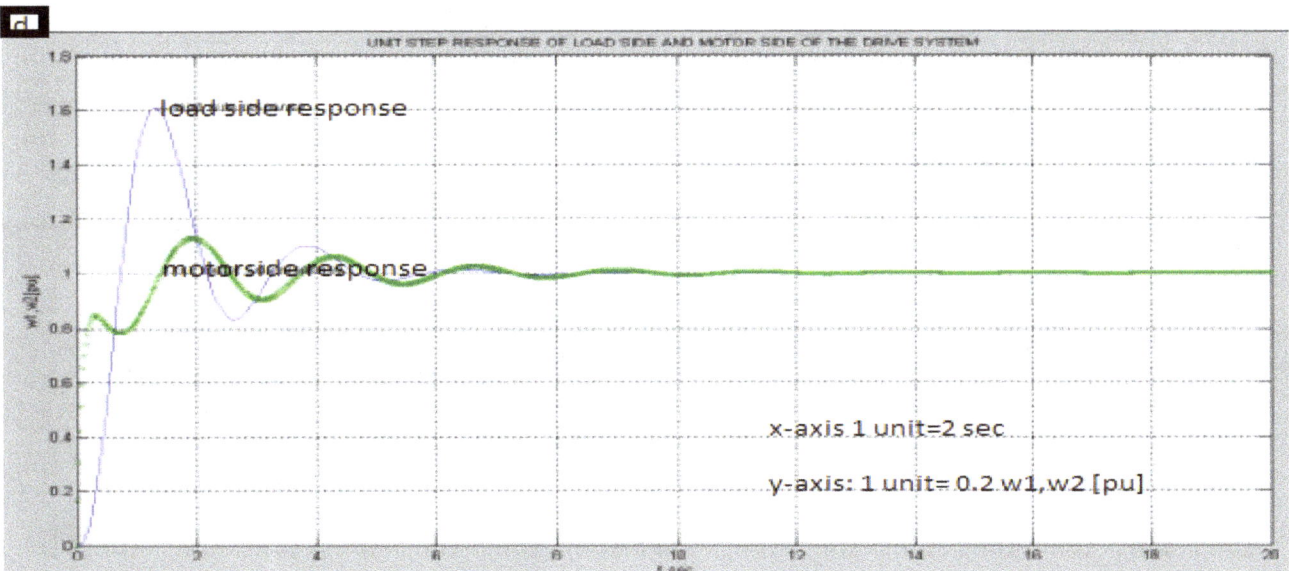

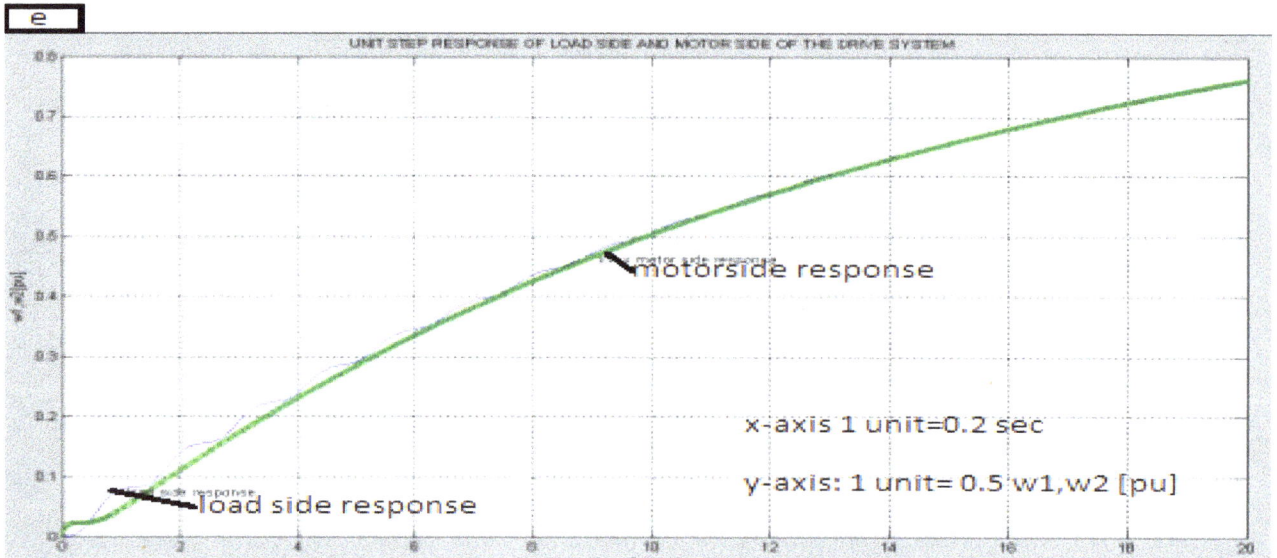

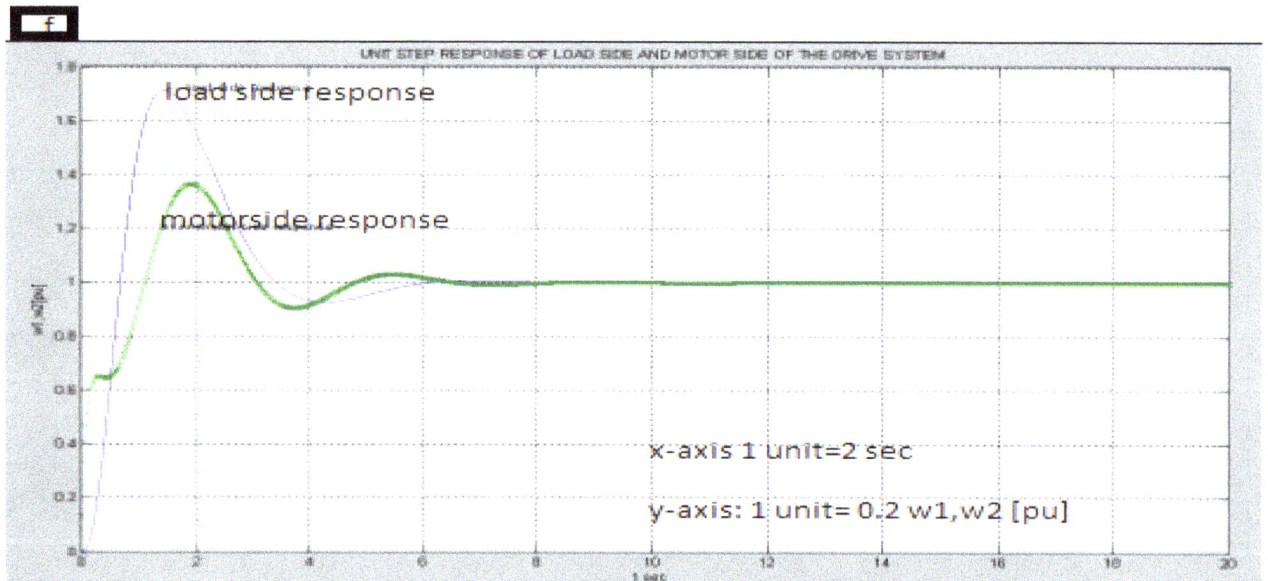

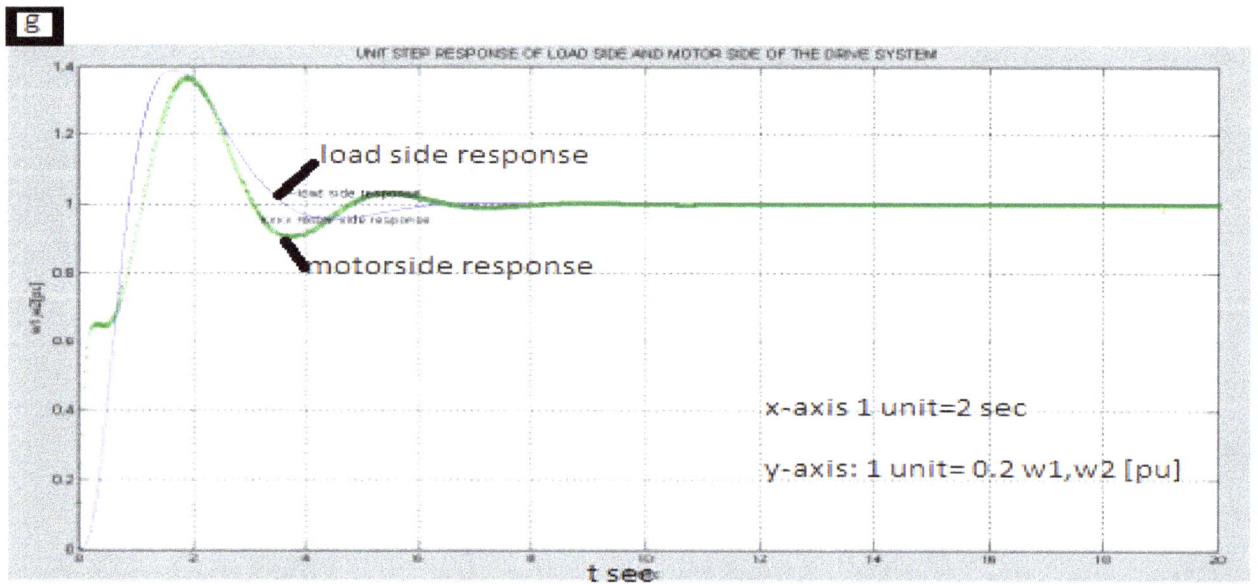

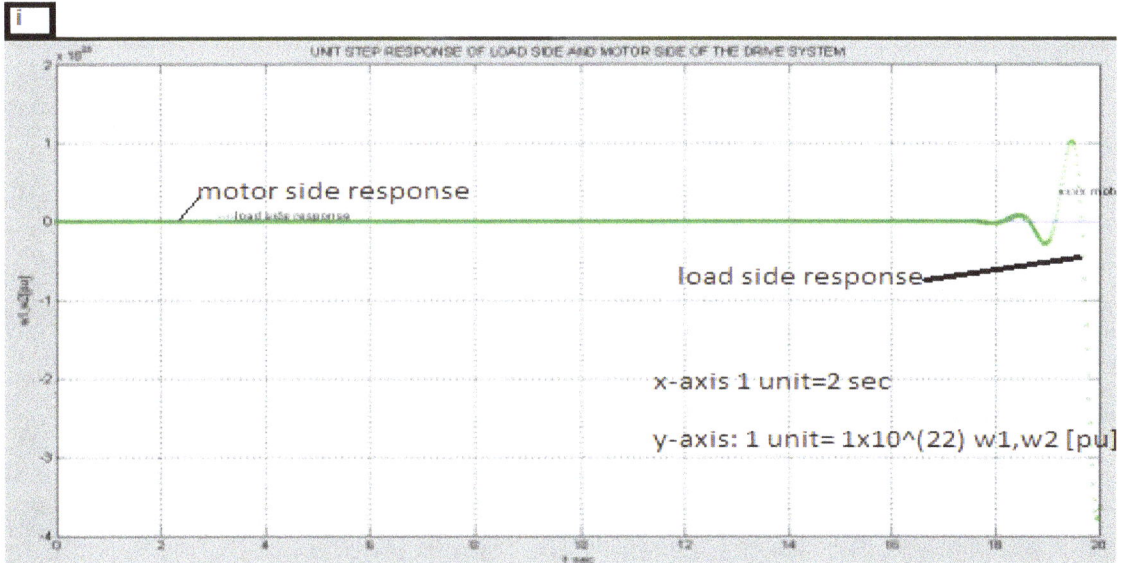

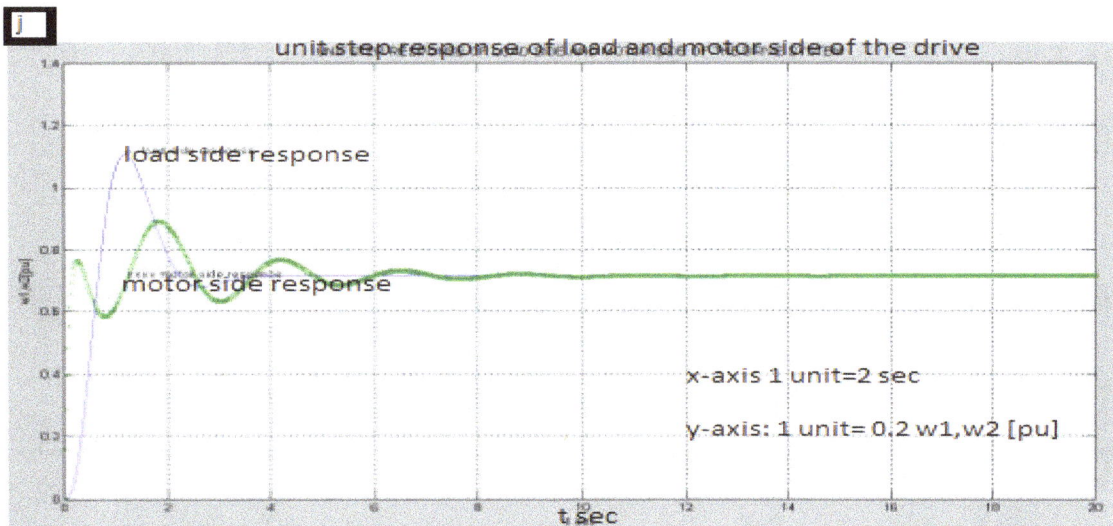

**Figure 5.** a.transient response when only PI present; b. transient response (PI + KI); c. transient response (PI + K$_2$); d. transient response (PI + K$_3$); e. transient response (PI + K$_4$); f. transient response (PI + K5); g. transient response (PI + K6); h. transient response (PI + K7); i. transient responsje (PI + K8), and j. transient response (PI + K9).

## Performance specification (transient response parameter) results

Tables 1 – 3 give details of the transient response parameter obtained from the simulation. The parameters obtained are similar to the response curve obtained. Table 1 gives the load side transient response parameters while Table 2 gives the motor side transient response parameters.

## Conclusion

Different cascade control structure using PI supported by various additional feedbacks for industrial drive systems with elasticity were investigated. Classical pole placement method was implemented to calculate the control system controller parameters. The performance of the control structure without additional feedbacks depends only on the mechanical parameters of the drive system and is rather poor. The fact is that the system is of fourth order and only two parameters $K_P$ and $K_I$ available which makes it difficult to achieve desired performance. We observed that in order to damp torsional vibration effectively, application of additional feedbacks is necessary. Resulting from the review of the literature, the application of different feedbacks is possible. The structures with one additional feedback ensure setting the desired value of the damping coefficient, yet the required value of the resonant frequency cannot be adjusted at the same time.

If the design specifications require the free setting of the damping coefficient and resonant frequency simultaneously, then the application of two additional feedbacks is necessary. This work could be carried out in future analysis because with this structure the closed-loop poles can be placed in every desired position. Particular feedbacks can be selected according to our requirement. Conventional control structure using PI alone are rather poor, but with various feedbacks torsional oscillation are being damped. Thus, this methodology acts as sophisticated methods available to develop a controller that will meet transient response specification and improves performance of various industrial drives. The controller parameters setting according to this methodology will provide acceptable response for many drive systems. But the process operator will have to often do final tuning of the controller iteratively on the actual process to yield more satisfactory control.

### REFERENCES

Ji JK, Sul SK (1995). Kalman filter and LQ based speed controller for torsional vibration suppression in a 2-mass motor drive system", IEEE Trans. Ind. Electron., 42: 564–571.

O'Sullivan T, Bingham CC, Schofield N (2006). High-performance control of dual-inertia servo-drive systems using low-cost integrated SAW torque transducers, IEEE Trans. Ind. Electron, 53(4): 1226–1237.

Rached D, Kenji K, Masahiro T (1994). Analysis and Compensation of Speed Drive Systems with Torsional Loads, IEEE Trans. on Ind. Appl., 30(3): 760-766.

Sugiura K, Hori Y (1996). Vibration suppression in 2- and 3-mass system based on the feedback of imperfect derivative of the estimated torsional torque, IEEE Trans. Ind. Electron., 43(1): 56–64.

Szabat K, Orlowska KT (2007). Vibration suppression in a two-mass drive system using PI speed controller and additional feedbacks comparative study", IEEE Trans. on Ind. Electron., 54(2): 1193-1206,.

Szabat K, Orlowska KT (2008). Performance Improvement of the Industrial Drives with Mechanical Elasticity using Nonlinear Adaptive Kalman Filter", IEEE Trans. on Ind. Electron., 55(3): 1075-1084.

Szabat K, Orlowska-Kowalska T, Dyrcz K (2006). Extended Kalman filters in the control structure of two-mass drive system", Bull. Pol. Acad. Sci. Tech. Sci., 54(3): 315–325.

Valenzuela MA, Bentley JM, Lorenz RD (2005). Evaluation of torsional oscillationss in paper machine sections", IEEE Trans. Ind. Appl., 41(2): 493–501.

Zhang G (1999). Comparison of control schemes for two-inertia system", in Proc. Int. Conf. PEDS, Hong Kong, pp. 573–578.

Zhang G, Furusho J (2000). Speed control of two-inertia system by PI/PID Control", IEEE Trans. Ind. Electron, 47(3): 603–609.

Zhang R, Chen Z, Yang Y, Tong C (2007). Torsional Vibration Suppression Control in the Main Drive System of Rolling Mill by State Feedback Speed Controller Based on Extended State Observer", IEEE International Conference on Control and Automation.

# 9

# Radio access network (RAN) architecture development for a 4G Mobile system with reduced interference

**Jalal J. Hamad Ameen\* and Widad Binti Ismail**

School of Electrical and Electronics Engineering, University Sains Malaysia (USM), Malaysia.

The fourth-generation wireless network communications technology standard is called 4G. Users of 4G devices will have the ability to access applications ranging from basic voice communication to seamless real-time video streaming. This paper proposes new techniques for the 4G mobile system, which include base station and cell distribution, channel allocation, and carrier frequencies. In addition, the connection between cells with a new topology using the fiber optic system instead of the microwave link in the air interface is proposed. The results show that there is very good improvement in terms of signal strength, data rate, and quality of service by using the proposed radio access network, the fiber length depends on the cluster size, also, co-channel and adjacent channel interferences has been further reduced. The fourth generation system requires accessing and air interface at higher data rates, which can be achieved by these new proposed techniques as shown in the results.

**Key words:** 2G, 3G universal mobile telecommunications system (UMTS), long term evolution (LTE), 4G, cell planning, frequency, channel allocation.

## INTRODUCTION

Frequency assignment and channel allocation are the main procedures in the planning process for mobile systems. Each mobile system has a frequency band assigned by the International Telecommunications Union (ITU) and the Federal Communications Committee (FCC). The initial channel allocation for GSM provides 124 carriers with frequency division duplex uplink and downlink sub-bands with a width of 25 MHz, duplex spacing of 45 MHz, and frequency spacing between carriers of 200 kHz. One carrier is used for the guard-bands, giving the following (Jalal, 2008):

Total number of carriers Absolute Radio Frequency Channel Number (ARFCN) = (25 − 0.2)/0.2 = 124

$$fu(n) = 890 + 0.2n \quad (1)$$

Where, (1 ≤ n ≤ 124)

$$fd(n) = fu(n) + 45 \quad (2)$$

Where $fu(n)$ and $fd(n)$ are the up and down carrier frequencies, respectively, and ARFCN stands for absolute radio frequency carrier number (Jalal, 2008).

For other mobile systems, the above parameters are calculated as shown in Table 1. In second-generation mobile systems, detailed planning is concentrated strongly on the coverage planning. On the other hand, in third-generation systems, more detailed interference planning and capacity analysis, rather than a simple coverage optimization, are required. The planning tool should aid the planner in optimizing the base station configurations, antenna selections, antenna directions, and even site locations such as real networks. The uplink and downlink coverage probability is determined for a specified service by testing the service availability in each location of the plan. The coverage of a sample geographic area according to 3G UMTS mobile is shown in Figures 1 and 2, which is covered by 13 base stations (CELCOM mobile communication system in Malaysia,

\*Corresponding author. E-mail: jalal3120002000@yahoo.com.

**Table 1**. Frequency Related Specifications of Mobile Systems

| System | Uplink frequency band (MHz) | Downlink frequency band (MHz) | Bandwidth (MHz) | Channel spacing (kHz) | No. of channels | Duplex distance (MHz) |
|--------|------------------------------|-------------------------------|-----------------|------------------------|-----------------|------------------------|
| P-GSM900 | 890 to 915 | 935 to 960 | 25 | 200 | 125 | 45 |
| E-GSM900 | 880 to 915 | 925 to 960 | 35 | 200 | 175 | 45 |
| GSM1800 | 1710 to 1785 | 1805 to 1880 | 75 | 200 | 375 | 95 |
| GSM1900 | 1850 to 1910 | 1930 to 1990 | 60 | 200 | 300 | 80 |
| UMTS | 1920 to 1980 | 2110 to 2170 | 60 | 5000 | 12 | 90 |
| LTE | 2110 to 2170 | 2300 to 2360 | 60 | 5000 | 12 | 190 |

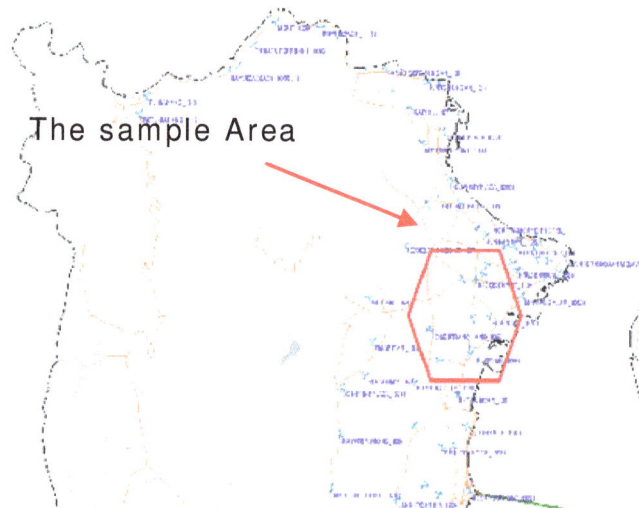

**Figure 1.** Sample area.

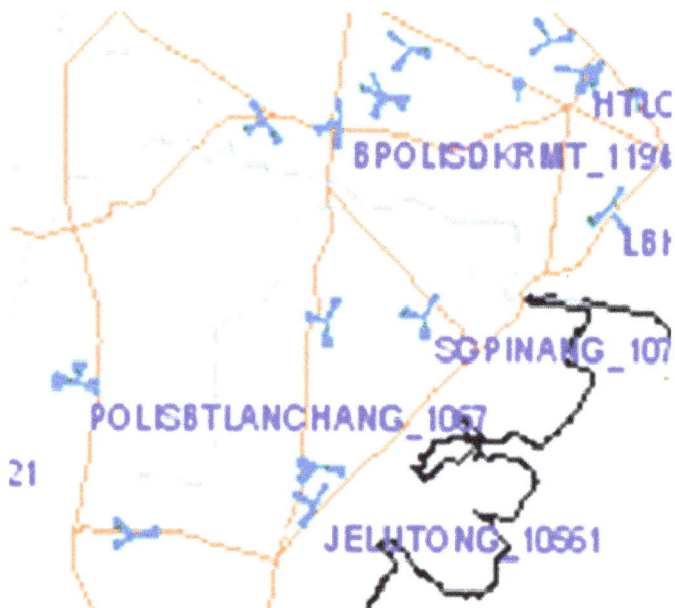

**Figure 2.** Maximized view of the sample area.

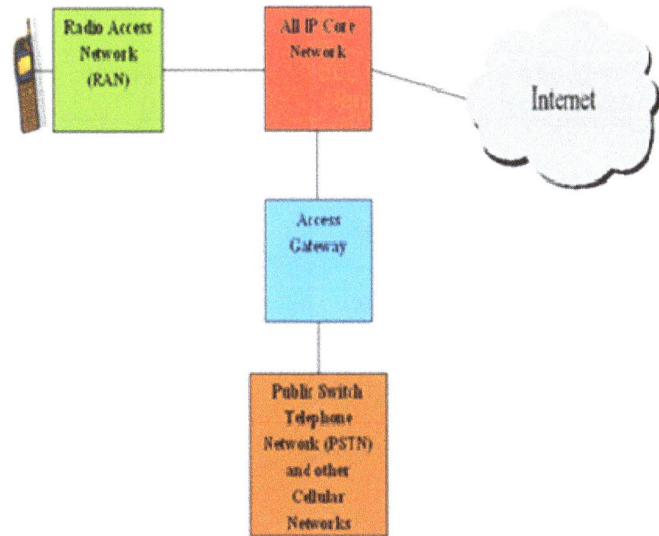

**Figure 3.** Simplified 4G RAN architecture.

Penang section).

## CO-CHANNEL AND ADJACENT CHANNEL INTERFERENCE

Co-channel interference (CCI) is one of the major limitations in mobile systems, and it is caused mainly by frequency reuse spectrum allocations, especially when frequency planning has not been optimized. This type of interference decreases the ratio of carrier-to-interference (C/I) powers at the periphery of the cells, causing a diminished system capacity, more frequent band-offs, and dropped calls. Adjacent channel interference (ACI) is another type of interference that affects a mobile system's signal strength and quality. This type of interference occurs when two channels are assigned with two adjacent carriers. For example, in a GSM system, two channels are separated by 200 kHz; thus, without cell planning, ACI will affect the signal quality in this case. CCI can be expressed as follows (Arango, 2001):

$$C/I = 10 \log (P_C/P_I) \quad \text{(dB)} \quad (3)$$

Where $P_C$ = wanted signal power
$P_I$ = signal power from the interfering transmitter

i) The GSM specification recommends that the C/I ratio should be greater than 9 dB.
ii) Ericsson recommends that 12 dB should be used as the planning criterion.
iii) Operators generally plan for higher C/I values.

ACI can be expressed as follows:

$$C/A = 10 \log (P_C/P_A) \quad \text{(dB)} \quad (4)$$

Where, $P_C$ = power from a wanted signal
$P_A$ = power from an adjacent channel

iv) The GSM specification states that the carrier-to-adjacent (C/A) ratio must be greater than –9 dB.
v) Ericsson recommends that a C/A ratio higher than 3 dB should be used as the planning criterion.

## 4G MOBILE SYSTEM

A 4G mobile system is the future mobile system beyond 3G, which may have the following specifications (Arango, 2001):

i) High-speed transmission (peak of 50 to 100 Mb/s and average of 200 Mb/s).
ii) Larger capacity (~10 times greater than 3G systems).
iii) Next-generation Internet support (IPv6, QoS).
iv) Seamless services.
v) Flexible network architecture.
vi) Use of microwave band (2 to 8 GHz).
vii) Low system costs (1/10~1/100 of 3G systems).

A simplified radio access network (RAN) architecture for 4G mobile is shown in Figure 3. The proposed frequency band for this type is 2 to 8 GHz. Owing to the higher frequency carriers, the coverage of the base stations will be smaller than that of 3G. To cover large geographic areas, the number of base stations will be increased, which causes a larger interference (both CCI and ACI).

## THE PROPOSED 4G CHANNEL ALLOCATIONS AND CARRIER FREQUENCY ASSIGNMNET

In this paper, a proposed channel allocation scenario and carrier frequencies have been presented, as shown in Table 2. The proposed allocation shows that after cell planning using the proposed channel allocation, when the C/I ratio increases, the signal strength also increases. The coverage of a certain sample area is obtained, and CCI and ACI will be reduced as shown in Figure 4 and Figure 5.

### RELATED WORK

After proposing the channel allocation and carrier frequencies assignment as shown in Table 2, the next

**Table 2.** The proposed frequency and channel allocation for 4 g mobile system.

| Sub-band no. | Uplink frequency band (GHz) | Downlink frequency band (GHz) | Sub-band band Width (MHz) | No. of sub-channels | Channel spacing (MHz) |
|---|---|---|---|---|---|
| 1 | 2.0 to 2.2 | 2.3 to 2.5 | 200 | 40 | 5 |
| 2 | 2.6 to 2.8 | 2.9 to 3.1 | 200 | 40 | 5 |
| 3 | 3.2 to 3.4 | 3.5 to 3.7 | 200 | 40 | 5 |
| 4 | 3.8 to 4.0 | 4.1 to 4.3 | 200 | 40 | 5 |
| 5 | 4.4 to 4.6 | 4.7 to 4.9 | 200 | 40 | 5 |
| 6 | 5.0 to 5.2 | 5.3 to 5.5 | 200 | 40 | 5 |
| 7 | 5.6 to 5.8 | 5.9 to 6.1 | 200 | 40 | 5 |
| 8 | 6.2 to 6.4 | 6.5 to 6.7 | 200 | 40 | 5 |
| 9 | 6.8 to 7.0 | 7.1 to 7.3 | 200 | 40 | 5 |
| 10 | 7.4 to 7.6 | 7.7 to 7.9 | 200 | 40 | 5 |

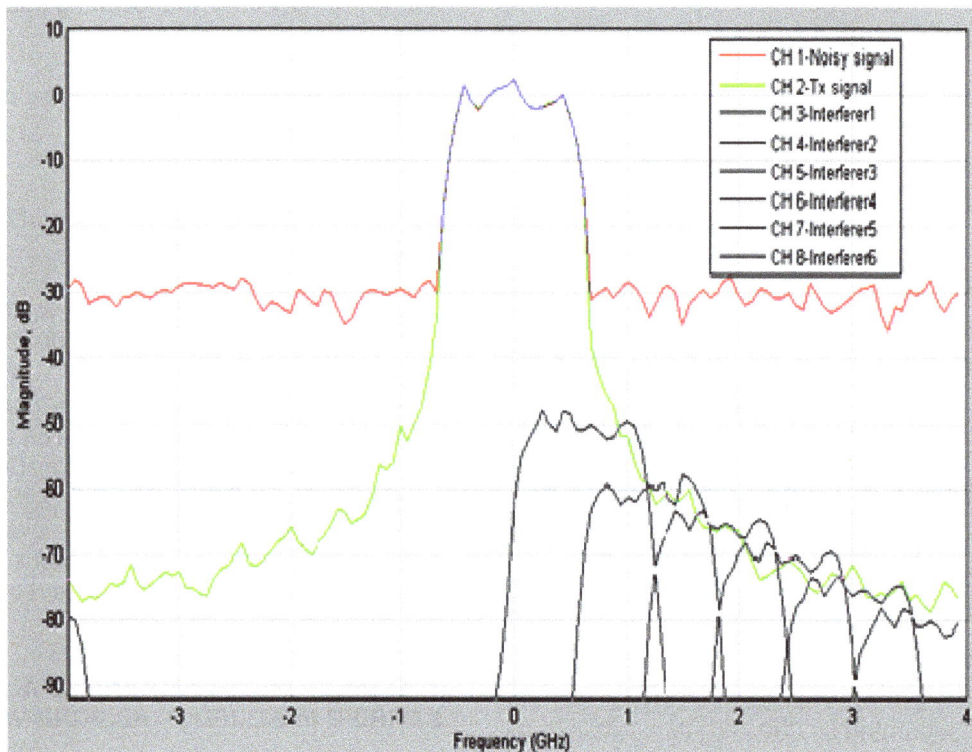

**Figure 4.** ACI effect on the transmitted signal power spectrum (8PSK) modulation.

step is to develop the base stations topology with a new connection model. As a survey for the work, in (Kwansoo, 2005), radio over fiber has been shown between the radio access unit with the antenna and the switching unit for the base station. Also, (Yasushi et al., 2000) has proposed cluster-cellular radio access network but also using microwave link not fiber optic link. The same for (Halim et al., 2008) and (Seungwan et al., 2004). In (Noureddine et al., 2008), ring topology has been proposed but cell with Wireless Local Area Network (WLAN). Whereas, (Toru et al., 2001; Gary et al., 2009;

Afaq et al., 2009) are for the cell radius versus frequency relationship, the interference effect and some 4G expected specifications.

Finally, (Istivan, 2004) has proposed a double-multiple access (MA) wireless network and a two-service wave division multiplexing (WDM) optical transfer network using fiber optic link for cellular wireless networks realized via radio over fiber composing two-sub layers : conventional wireless layer and below it the optical layer. In this new developed model of connection, there will be central base stations (CBS), such as radio network

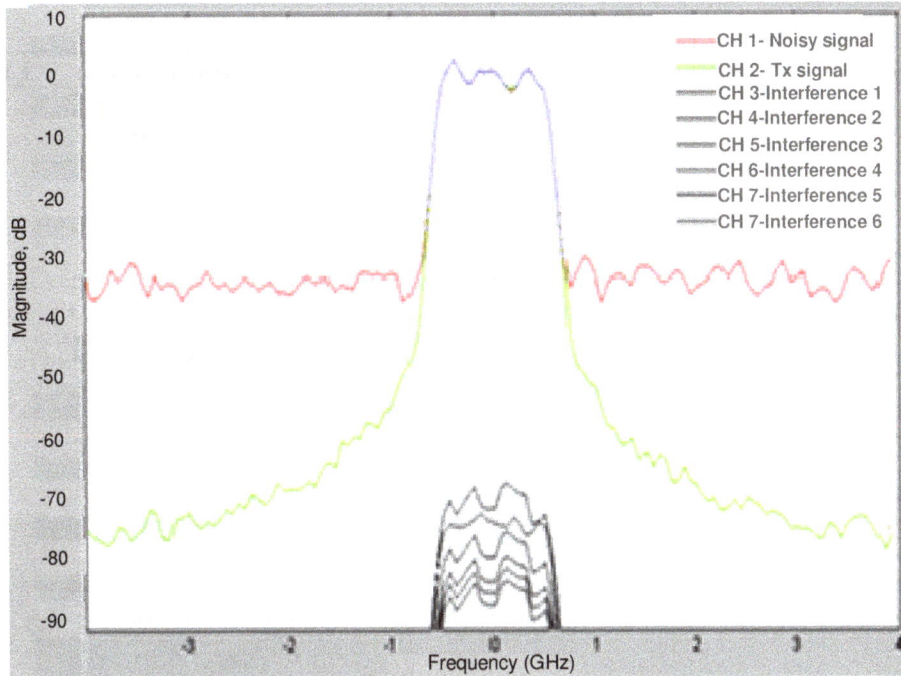

**Figure 5.** CCI effect on the transmitted signal power spectrum (8PSK) modulation.

controllers (RNCs) in 3G mobile. Each CBS represents a center for a group of base stations connected to it. Each group of these base stations connections will pass through fiber optic links instead of microwave links. In this case, there will be no free space losses because of the use of the fiber optic channel. Although the installation cost is high, but the running cost will be low. Also, all the base stations are grouped according to the cluster size, then the topology is tree unlike Toru et al. (2001) from which ring topology has been proposed, in ring topology, any fault between two adjacent base stations causes overall disconnection, also there will be no central base station to control it's base stations group, in our proposed technique in this paper is tree topology, fault between base stations link will not affect the others, also there will be central base stations as a controller for the base stations group connected with it.

## RESULTS

Figures 4 and 5 show the CCI and ACI effects without using the proposed technique, while Figures 6 and 7 shows the effects of using the new technique. This indicates that there will be no interference when CCI and ACI are removed, so the signal-to-interference ratio (SIR) will be increased. Also, the length of the fiber optic depends on the cluster size, as shown in Table 3. The fiber length increases if the cluster size increases, which in turn raises the total dispersion and data rate. After

optimization, the cluster size of N = 3 and 7 will have the best cases, as shown in Figures 8 and 9. Figures 10 and 11 show the practical view of connection for the proposed topology.

This new proposed topology for the base station connection results in reduced free space losses and an improvement in signal strength, although, the installation cost is high using this proposed technique, but the running cost is very low because fiber optic system running cost is low. Whereas, microwave link system has very high installation and running cost. Also the atmosphere effects on microwave link, but fiber optic link is under ground, so the atmosphere effect is negligible. Meanwhile, the red color refers to the fiber optic, the black nodes refer to the BSs and blue colored node for central base station (CBs). It is shown that for the same sample area, the number of BSs for 3G is 13, while that for 4G is 36 (that is approximately three times that of 3G). This is because of the higher frequency bandwidth for 4G mobile, resulting in a smaller cell radius.

## CONCLUSIONS

This paper describes 4G mobile base station distribution, connection, and channel allocation. A new channel allocation and carrier frequencies for the system cells with new cell connection and distribution are also introduced, yielding some measured results. It is shown that the new allocation and connection increases the SIR

**Table 3**. Summary of coverage planning and fuber length

| Cluster size (N) | No. of tiers | No. of base stations (BSs) | Coverage area (km$^2$) | L (fiber length, km) |
|---|---|---|---|---|
| 1 | 6 | 127 | (2.6*R^2)*127 | >2R |
| 3 | 6 | 381 | (2.6*R^2)*381 | > 3R |
| 4 | 6 | 508 | (2.6*R^2)*508 | > 3.4R |
| 7 | 6 | 889 | (2.6*R^2)*889 | >4.5R |
| 12 | 6 | 1,524 | (2.6*R^2)*1524 | > 6R |
| 13 | 6 | 1,651 | (2.6*R^2)*1651 | >6.2R |

Where, R is the cell radius in km, and BSs are the base stations. If N = 3, the number of BSs = 381, R = 2 km, and the overall coverage area = 3,962.4 km$^2$ , the fiber length must be greater than 6 km between each two base stations; if the overall fiber length (126 part of fibers) = 756 km for each group and the number of groups = 3, then the fiber length equals 2,268 km for all the three groups. On the other hand, if  N = 7 and the total coverage area = 9,245.6 km$^2$, the fiber length between each two base stations must be greater than 9.165 km; if the total fiber parts for each group (126 part of fibers) = 1,154 km, the fiber length for the three groups is 8,083.66 km.

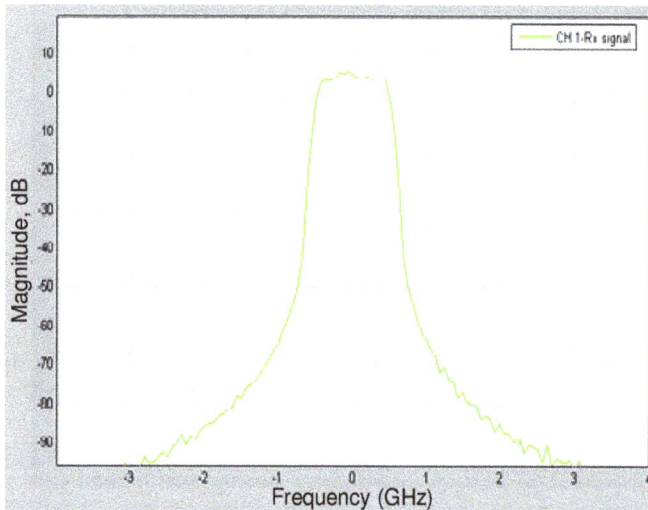

**Figure 6.**  Received signal power spectrum (8PSK) using the new proposed technique.

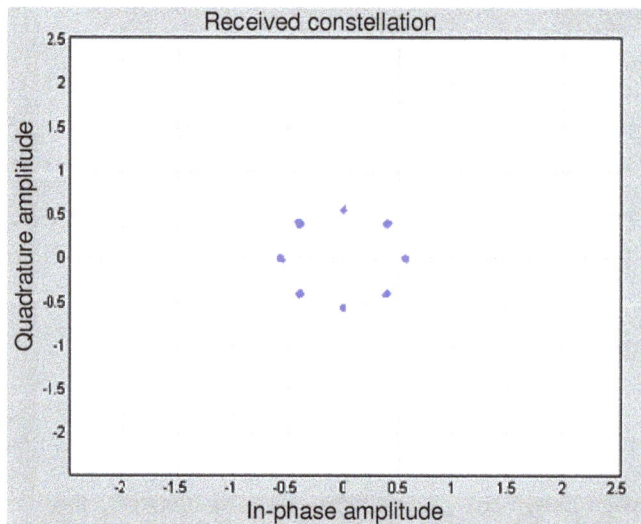

**Figure 8.** The new proposed topology (N = 3) for the same sample area shown in Figure (2). There will be the same step and connection for cell groups B and C because N = 3; the yellow-colored cell is the CBS for this cell group.

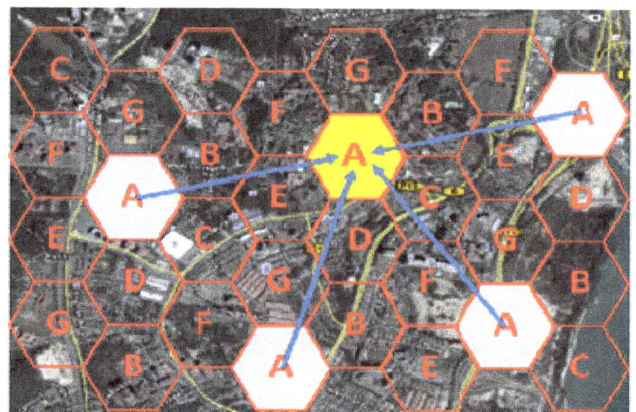

**Figure 7.**  Received signal constellation (8PSK).

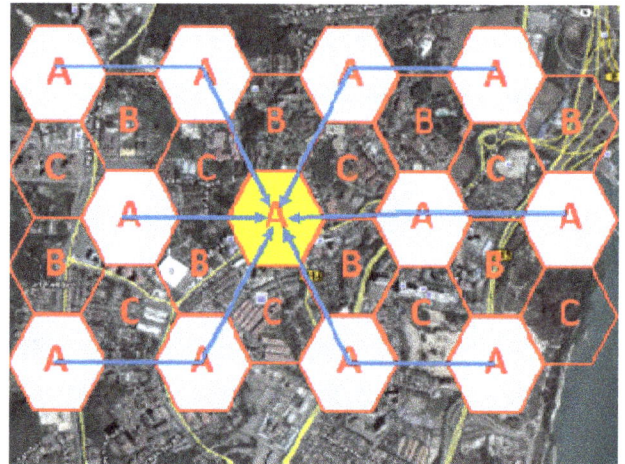

**Figure 9.**   The new proposed topology (N = 7) for the same sample area as shown in Figure (2).There will be the same step and connection for cell groups B, C, D, E, F, and G because N = 7; the yellow-colored cell is the CBS for this cell group.

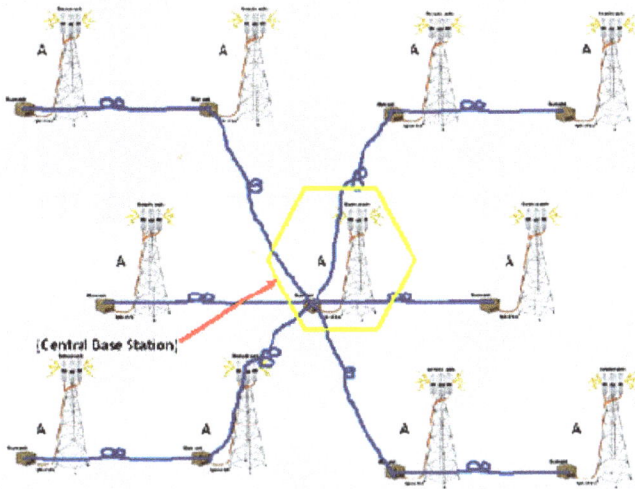

**Figure 10.** The new proposed topology (N = 3), practical view.

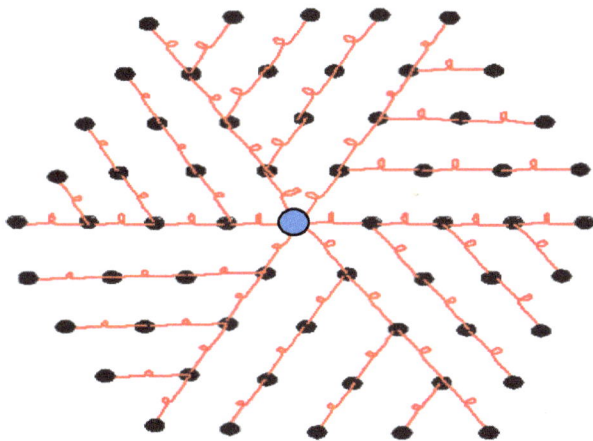

**Figure 11.** The new proposed topology (N = 3), practical view for each group.

and reduces the interference because the carrier frequencies are allocated with minimum interference. Moreover, the connection between the cells with the newly introduced technique improves the data rate and QoS. The signal strength also improves because there are no free space losses, and the fiber optic has very small losses of approximately 0.02 dB/km.

The proposed new techniques improve the quality of 4G services and the signal quality as given in the results, the results shows the proposed radio access network gives more suitable network for 4G mobile system by increase in data rate to about more than ten times that for 3G mobile system, although the installation cost is

higher. Nevertheless, the running cost is very small because of the use of the fiber system.

## ACKNOWLEDGMENTS

The authors would like to thank the School of Electrical and Electronic Engineering, USM, and the Secretariat of the Ministry of Science, Technology, and Innovation of Malaysia (MOSTI) e-Science fund: 01-01-05-SF0239 for sponsoring this work. Also, our gratitude goes to Celcom Mobile Communication System in Malaysia, Penang section, for their assistance.

## REFERENCES

Afaq HK, Mohammed AQ, Juned AA, Sariya W (2009). 4G as a Next Generation Wireless Network, IEEE computer Society International Conference on Future Computer and Communication, ICFCC2009, 3-5 April 2009, pp. 334-338.

Arango M (2001). Architecture Issues in 4G Networks, Sun Microsystems, VON, 3/23/01.

Gary B, John P, Ning G, Rui C, Neng W, Sophie V (2009). Interference Coordination and Cancellation for 4G Networks, IEEE Communications Magazine, April 2009, pp. 74-78.

Halim Y, Jietao T, Ony ZG (2008). Beyond-4G Cellular Networks : Advanced Radio Access Network (RAN) Architectures, Advanced Radio Resource Management (RRM) Techniques, and Other Enabling Technologies, WWRF Meeting, 13-15 October 2008, Stockholm, Sweeden, pp.1-5.

Istivan F (2004). Radio Over Fiber: Application, Basic Design and Impact on Resource Management, Budapest Univrsity of Technology and Economics, Mobile, Radio Network Management '04 (RNM '04) and Mobile Location Workshop '04 (MLW '04) respectively, 27-28 May, 2004, Athens, Greece : http://www.telecom.ntua.gr/mobilevenue04/Papers.

Jalal JHA (2008). Cell planning in GSM mobiles, WSEAS Transactions on Communications, 7 (5): 393-398.

Kwansoo L (2005). Radio over Fiber for Beyond 3G, Microwaqve Photonics (MWP), International Topical Meeting, 12-14 Oct. 2005, pp. 9-10.

Noureddine B, Mohammad SO, Faouzi Z (2008). Intelligent Network Functionalities in Wireless 4G Networks: Integration Scheme and Simulation Analysis, ELSEVIER, Computer Communications, 31: 3752-3759.

Seungwan R, Donsung O, Gyungchul S, Daesik K, Kichul H (2004). The Next Generation Mobile Services and a Proposed Network Architecture, Vehicular Technology Conference, 2004, VTC 2004, IEEE 60th, 26-29 Sept. 2004, pp. 3306-3309.

Toru O, Ichiro O, Narumi U, Yasushi Y (2000). Network Architecture for Mobile Communications Systems Beyond IMT-2000, IEEE Personal Communications, October 2001, 8(5): 31-37.

Yasushi Y, Hirohito S, Narumi U, Nobuo N (2000). Radio Access Network Design Concept for the Fourth Generation Mobile Communication System, Vehicular Technology Conference Proceedings, 2000. VTC 2000-Spring Tokyo, IEEE 51st, 15-18 May 2000, 3: 2285-2289

# 10

# Effect of work related variables on human errors in measurement

**Vinodkumar Jacob[1]\*, M. Bhasi[2] and R. Gopikakumari[3]**

[1]Department of Electronics and Communication Engineering, MACE, Kothamangalam, 686666, Kerala, Ph. 91 98461 21223, India.
[2]School of Management Studies, Cochin University of Science and Technology, Kochi, Kerala, India.
[3]School of Engineering, Cochin University of Science and Technology, Kochi, Kerala, India.

**What one cannot measure one cannot control. Measurement plays a key role in science, technology and industry. Where there are measurements there is associated errors. Study of measurement errors has a long history. Attempts have been made to classify and understand the factors that contribute to errors in measurement. The understanding of this is useful for error reduction and also providing the margin for errors and reducing damage caused due to errors. For the purpose of classification and study, measurement errors have been divided into instrument error, method error and human error. The former two are easier to study and correct, but the later is less understood. In this study an attempt has been made to study the effect of selected work related variables on human errors in observing and noting measurements which contribute to measurement errors. In a measuring system, though some of the effects of variables on measurement errors can be guessed, only an experimental study will be able to isolate, quantify and present the effect of each variable separately. Hence an experimental study was designed and conducted to quantify and present the effect of selected work related variables of two sets of human subjects used in the experiments. Analysis of the results revealed that the variables identified and studied have significant effect on measurement errors, and their effects were also separately quantified. This will be of use to professionals trying to reduce measurement errors, especially in industrial environments, where knowing the variables and the extent of error they induce, appropriate work related settings can be adopted to keep human errors within the tolerable limits.**

**Key words:** Measurement error, test type differences, instrument differences, time of work, time pressure, environment.

## INTRODUCTION

Measurement is essential for technological investigations. It is so fundamental and important to science and engineering that the whole science can be said to be dependent on it (Blanchard, 1973). Instruments are developed for measuring and displaying physical variables. Every act of measurement has to deal with errors (Carmen, 2005). Errors can result in negative consequences [example:

loss of time, faulty products] as well as positive ones [example: learning, innovation]. The large negative consequences for example, accidents such as the chernobyl or challenger disasters tend to be widely observed (Meijman and Mulder, 1998) and have been of high interest to scholars and laypeople alike (Reason, 1990). The scientific understanding of the negative effects of errors is much better developed than that of the potential positive effects of errors (Carter, (1986). One way to contain the negative and to promote the positive consequences of errors is to use error management (Cannon and Edmondson (2001). This approach assumes that

---

\*Corresponding author. E-mail: vinodkumar_jacob@rediffmail.com.

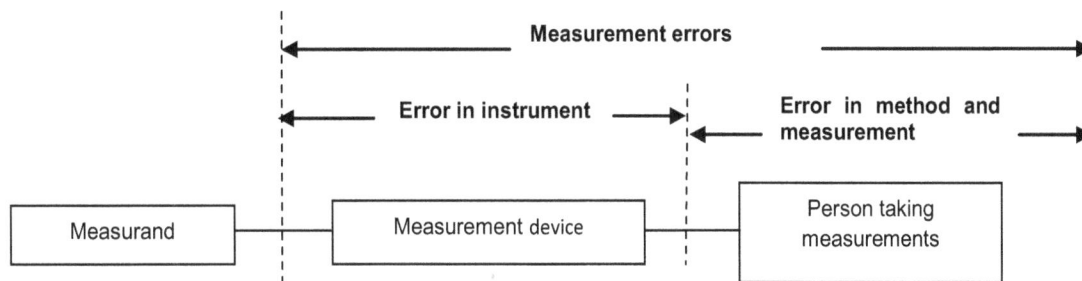

**Figure 1.** Measurement system.

human errors per se can never be completely prevented, and, therefore, it is necessary to ask the question of what can be done after an error has occurred (Frese, 1991; Frese, 1995). Errors are not easily defined (Cathy et al., 2005). Errors may be unintended deviation from goals, standards, and a code of behavior, the truth, or from some true value (Carmen, 2005). A measurement system comprises generally three parts as shown in Figure1.

Measurand is the physical parameter being measured. The measuring device can be of different types such as analog, digital, electronic, mechanical etc. Measurement errors may be due to the measuring device and/or the method and the person involved in measurements Chesher, (1991). Measurement error is defined as the difference between the output of the measurement system and the reference [known, actual, true, master, and standard] value (Parasuraman et al., 2000). Now, the measurement system could be defined as only the measuring instrument or as comprising of the measuring instrument and the person taking or doing the measurement and reporting the measured output Douglas and Esa, (2002). The later definition of a measuring system is more in tune with the practice of measurement. There is no general consensus in the literature about the terminology used to categorize and classify errors (Cathy et al., 2005). Several different taxonomies of human error exist with varying degrees of overlap Gawron et al., (1989). For example, studies on "decision errors" may classify errors in terms of knowledge based mistakes (Zapt et al., 1992). The other errors are diagnostic errors and planning errors Wiegmann and Shappell, (1997).

Production and quality control engineers who deploy human resource to take readings from instruments need

**Abbreviations: VAP;** Voltage analog parameter, **VDP;** voltage digital parameter, **RAP;** resistance analog parameter, **RDP;** resistance digital parameter, **ET;** experienced technicians, **IE;** inexperienced subjects, **IQ1;** above average intelligent quotient, **IQ2;** average intelligent quotient, **IQ3;** below average intelligent quotient, **B. Tech.;** Bachelor of Technology (four year engineering degree program in India).

to understand the effect of various factors on the errors induced by the human resource Helmreich and Merritt, (2000). Errors induced by humans during measurement can be further split to arise from two sources (a) from the work related factors (b) from the human related factors Mark and Brian, (2000); Nordstrom et al., (1998). In this study the focus is on the work related variables and their influence on human induced errors in measurement. To study the effect of only work related variables using experiments, it was necessary to remove the effect of measurement error due to human factors and the method of measurement (David, 1996). For doing this and isolating the effect of work related variables only on measurement errors, the same persons were asked, to use the same set of instruments and standardized methods, for making measurements with only one work related variable changed at a time over two different experimental setups. Thus reducing the error involved to almost only the human error induced due to change in work related variables.

In order to identify the variables to be taken in the study, a review of literature was done to identify some work related variables influencing human induced errors in measurement. A survey was also carried out among experts supervising production and quality control in different production environments to generate a list of possible variables usually found in industry, the effect of which would be useful to explore. These two sources were used to make the list of variables for the study. These variables were then classified into stable and transient (Senders and Moray (1991). Stable variables were those work related variables that Drury et al., (1989) would remain the same over time for the experimental setup (voltage, resistance, analog, digital, A/c, non A/c, forenoon, afternoon etc). Transient variables would change over time for a given experimental setup (instrument temperature, input values, aging etc). The study concentrated on the effect of stable work related variables on human errors in measurement. Now when human errors due to observing and noting become the area of study, the major work related factors influencing these are test type variables, time of day, time pressure and environment. The test type variables studied are

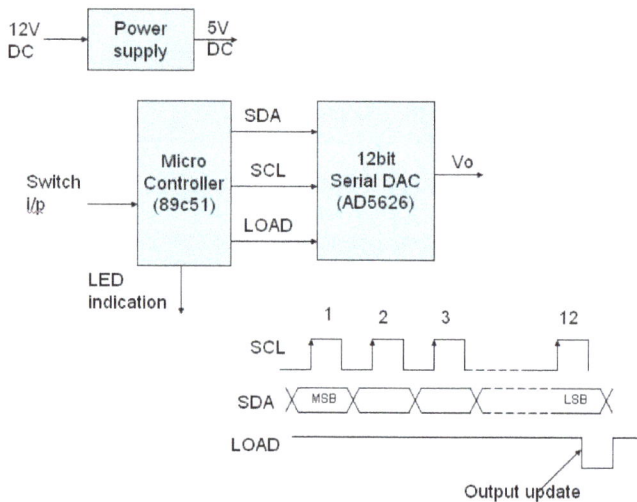

Figure 2 . Test pattern generator.

task type (voltage and resistance measurement) and technological difference in the measuring instrument (analog and digital). The effect of time of day was studied by doing the measurement during forenoon and afternoon. To study effect of time pressure, work was carried out in a set of experiments without any upper time limit and in the second set of experiments the time allowed was limited to the normal time required (as per work study). Experiments were carried out under normal laboratory environment and in an air-conditioned laboratory. All the subjects were given training Dormann and Frese (1994), before doing the experiments. Two sets of human resource, one Diploma holders with work experience and the second, B. Tech. (Bachelor of Technology) and Diploma Holders without work experience were used in the study.

## METHODOLOGY

### Procedure for testing work related variables

#### *Voltage*

There was need to isolate the effect of work related variable, therefore in the experiments one variable was changed at a time. In this case, the work related variable, type of work was set to Voltage measurement. A setup was made to generate predetermined set of 50 voltages, one after the other to be provided as an input for the subject doing the experiment to measure voltage using two types of voltmeters one analog and the other digital. The test voltage generator was to ensure that all subjects were given the same set of values. For doing this a microcontroller based test pattern generator which gives 50 different voltage outputs has been designed and used for the study. This setup kept errors due to system being measured out of the experiment and the focus could be maintained on error due to observation and noting. Different experiments were carried out using the microcontroller based test pattern generator

where analog and digital meters were used for measurement and the other variables such as time of day, time pressure, environment and experience of human resource were varied one at a time.

A block schematic of digitally controlled analog test pattern generator is shown in Figure 2. This generator gave different prefixed voltages for each hit of a switch. Display devices such as analog multi-meter (Make: SUNWA, Model: YX-3600TREB) and digital multi-meter (Make: CLASSIC, Model: 333) were used for testing VAP and VDP. In an experiment a subject had to push the switch for the next reading and make note of fifty such consecutive readings.

### Resistance

A set of fifty different valued [covered] resistances were used for these set of experiments to study the effect of work type: measurement of resistance, on human errors in measurement. These covered resistances numbered from 1 to 50 were given to subjects in experiments where they used digital and analog multimeters to make the measurements and note the same. In different experiments the other variables such as time of day, Time pressure, environment and experience of human resource were varied one at a time. Though the method involved in this case is very simple all subjects were also trained in it.

### Subjects for the experiments

Experiments were conducted using different subjects, the key differentiating factors of the subject groups are given as follows:

1) Experienced technicians (ET) in the age group of 31 to 40 years and 41 to 50 years. Their IQ test showed that all were in the below average (IQ3) category.
2) Inexperienced (IE) B. Tech. and Diploma holders in the age group of 21 to 30 years and with different IQ levels Above Average (IQ1), Average (IQ2) and Below Average (IQ3).
3) 20 subjects from each of the different categories participated in the experiment. This was found to be statistically sufficient for the mean error measurement which has been taken for analysis in this study (There is no significant change in the measure of mean and standard deviation of the error when sample was increased from 15 to 20).

Note: It was very difficult to get sufficient number of experienced technicians in IQ1 and IQ2 category. A summary of the subjects and the experiments is given in the Table 1.

### Experiments

The impact of instrument differences were studied by comparing the errors occurred when using analog and digital readouts. To check the effect of task differences, subjects were asked to measure resistances and voltages using analog and digital measuring devices. The subjects were made to do the measurements during both forenoon and afternoon sessions. The following procedure was adopted for the experiment:

1) Twenty subjects of each category were selected.
2) On a given day, one category subjects were made to take one set of measurement (say only resistance measurement using analog device) both in the forenoon and afternoon. This was repeated

on different days till all subjects had done all types of measurement experiments.

3) No feedback on their performance was given to them.

4) The measurements were conducted within a time frame as given as follows:

a) 30 min to make fifty measurements when using analog device for measuring both resistance and voltage. (This being the standard allowed time for such work in India as based on work study).

b) 20 min to make fifty measurements when using digital device for measuring both resistance and voltage (This being the Standard allowed time for such work in India as based on work study).

c) Inexperienced subjects were also allowed to do the measurement in a relaxed environment without time limit to complete the fifty measurements.

d) The experiments were done in a normal laboratory environment and air conditioned environment.

e) Training was given to all subjects for taking the measurements before start of the experiments.

## RESULTS AND DISCUSSION

The results from the experiments were analyzed and descriptive statistics such as mean, standard deviation and coefficient of variance of error were calculated. The mean percentage error for each category was used for further analysis to remove the effect of individual characteristics. Parametric analysis such as one sample T-test, paired sample T-test and ANOVA were carried out to understand the effect of different factors on the measurement errors observed Holland, (1986). The results of the analysis are discussed variable-wise in the next section.

### Test type differences

#### Instrument differences

The effect of technology on work reduction and simplification is well known Wickens et al. (1998). In this study the effect of two technologies in measuring instruments that is analog and digital on human error in measurement has been examined. For this, subjects with experience and without experience and having different IQ levels were asked to do measurements of voltages and resistances using analog and digital devices and to note their readings using pen and paper. A graph showing changes in the mean percentage error with change in category of the subjects for both analog and digital measurement of voltages are shown in Figure 3.

The errors are more when using analog technology. The error was seen to change with IQ level of the subjects and a significant jump in error occurs when the below average (IQ3) category was given analog instruments for measurement. It can be observed from Tables 2 and 3 that the errors were always more in the afternoon when compared with forenoon. The magnitude

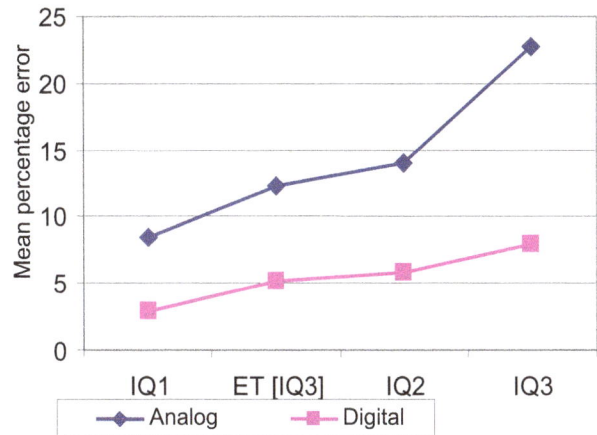

**Figure 3.** Instrument differences.

of increase of error between forenoon and afternoon was more when using analog devices.

The measurement error occurring in the afternoon session for the IQ1 category when using digital instruments for measurement was even less than the error they make in the forenoon when using analog instruments as is instruments as is evident from Tables 3 and 6.

The experienced technicians were seen to make 7.3% more error when measuring with analog devices than with digital devices. Figure 3 also shows that when using analog devices 9.5% more error occurs than when using digital devices on an average for inexperienced subjects. It can be therefore said that digital devices when used for measurement will result in only one by ninth Human error in measurement when compared to analog.

ANOVA test was carried out to check the effect of different variables on human error. It can be inferred that all variables studied have significant (at 0.05 level) impact on human errors in measurement. The sum of squares also shows that Digital measurement is superior to Analog Measurement.

#### Task differences

The other work related variable is the difference in parameters to be measured. In this study the parameters measured were voltage and resistance using both analog and digital devices thus there were two tasks.

It can be noticed that irrespective of the subject's experience or IQ, the errors when measuring voltage were more than when measuring resistances. The IQ1 category makes least error for both resistances and voltage measurements followed by experienced technicians with IQ level IQ3 (Figure 4). The results of ANOVA test shown in Table 3 and paired sample test shown in Table 4, shows that there was significant change in error

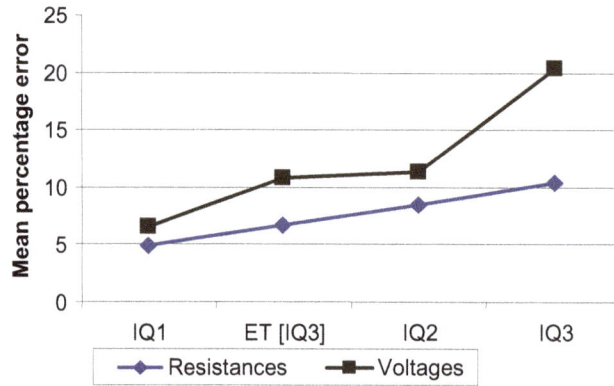

**Figure 4.** Test type differences.

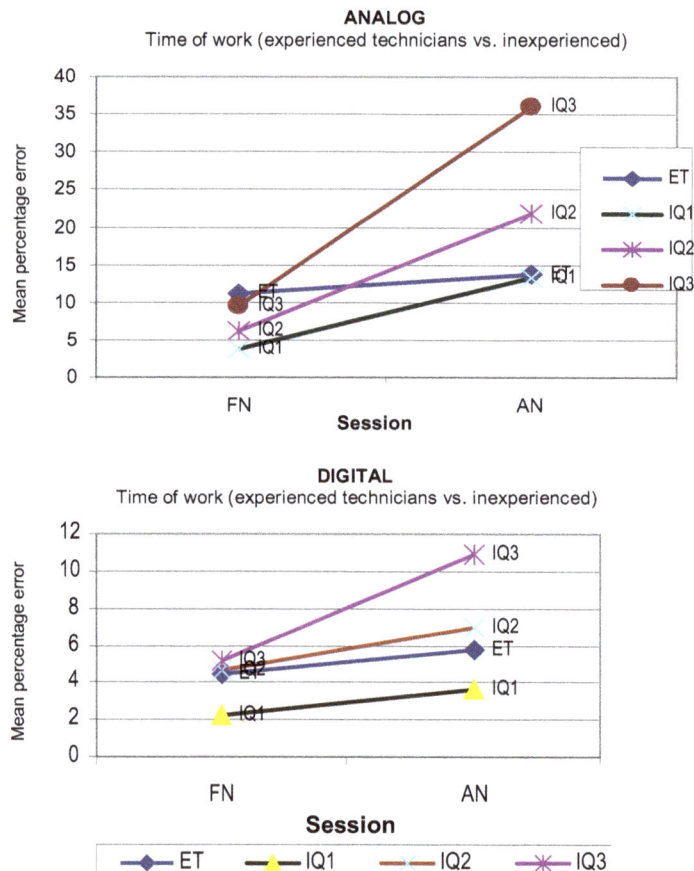

when doing both resistance and voltage measurement using digital and analog devices except for the pair RDP-VDP (Resistance measurement using digital device voltage measurement using digital device. An interesting observation in this experimental work is that though the voltage measurement work is simpler compared to resistance measurement, more errors were seen in voltage measurement. A possible explanation could be

that when a simple task is given subjects may pay less attention to the work and thus human errors could become higher.

**Time of work**

The subjects were seen to make more errors in the

## a. Inexperienced subjects

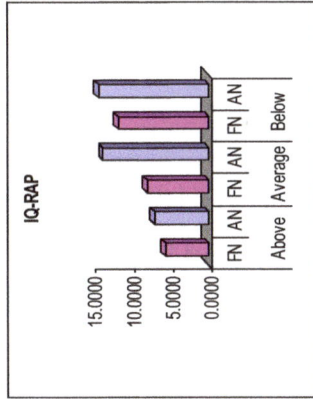

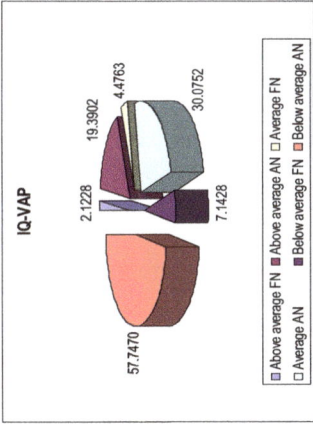

## b. Experienced technicians

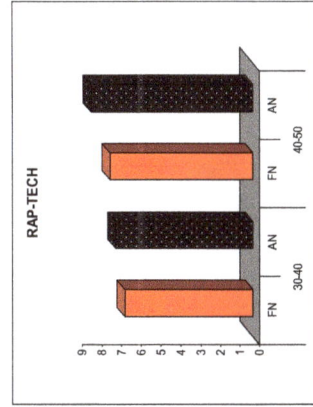

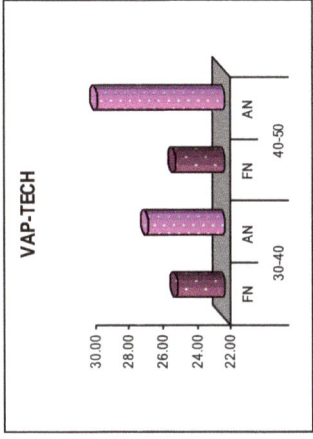

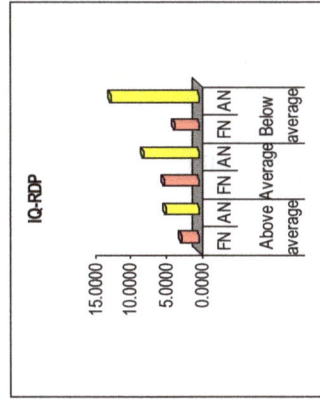

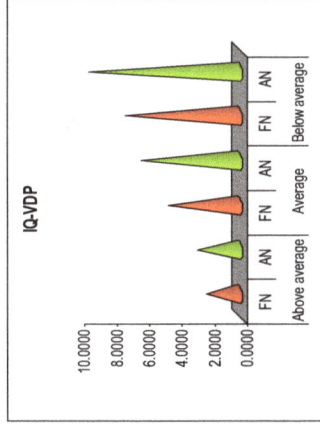

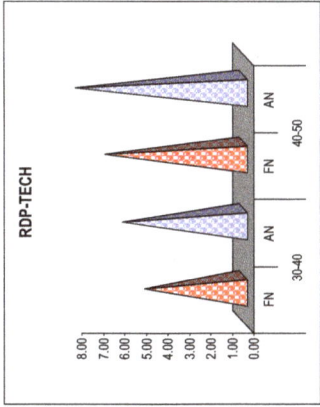

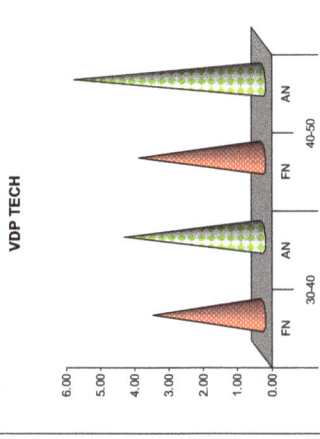

**Figure 6.** Inexperienced vs. experienced subjects.

afternoon irrespective of their experience, intelligent quotients (IQ), instrument differences and type of measurement they were carrying out.

It can be observed from Figure 5 that in the case of experienced technicians, the error increase for analog devices in the afternoon compared to the forenoon is about 2.6% and for digital devices it is 1.3%. But for the inexperienced subjects with IQ levels IQ1, IQ2 and IQ3, the error increase is in the order of 9.4, 15.7 and 26.6% respectively with analog devices. For digital devices it is 1.4, 2.3 and 5.7% respectively. It can also

be noticed that even though inexperienced B. Tech. and Diploma Holders with IQ above average [IQ1] were making less error compared to the experienced subjects, but the error growth in the afternoon was very high with inexperienced subjects. This may be because of, the experienced technicians are more tuned to long and tiring working hours and the increased room temperature in tropical climate afternoons and therefore make fewer errors in adverse working conditions when compared to inexperienced subjects.

Figure 6 shows that the error is less with digital measurements compared to analog measurements. The error is more with voltage measurements than resistance measurements. It can also be observed that the error is more in the afternoon than forenoon. The t-test given in Table 7 shows the significance in error difference between forenoon and afternoon.

**Time pressure**

Inexperienced subjects in the IQ1, IQ2 and IQ3

## Environment analog

## Environment digital

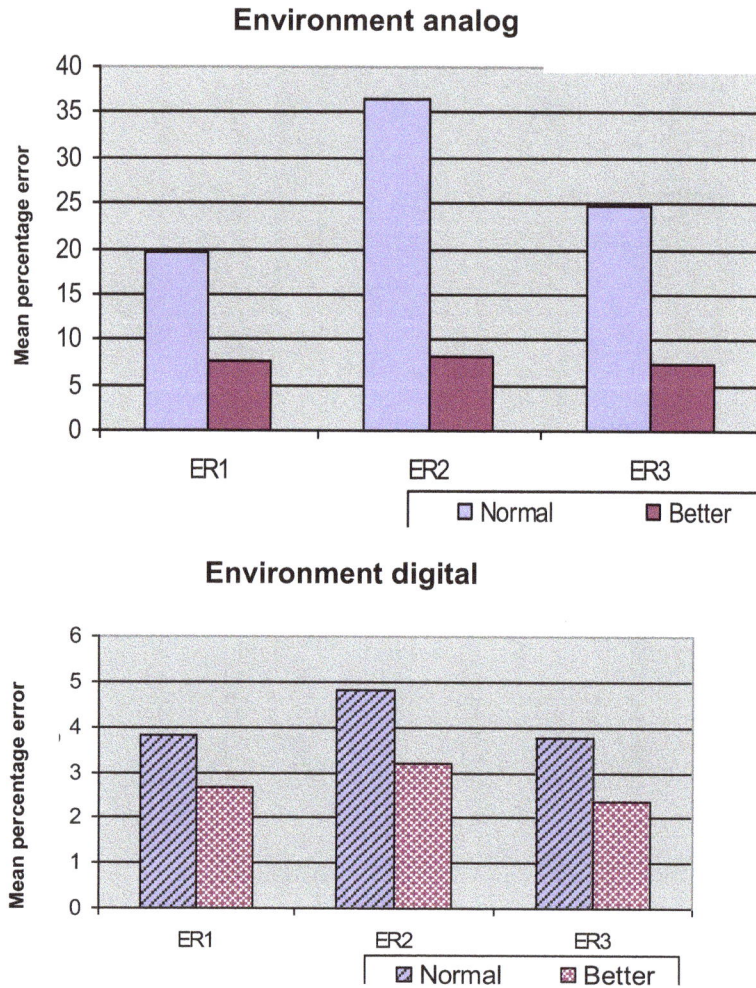

**Figure 7.** Environment.

categories were also asked to do the resistance measurement using analog and digital multimeter without any upper limit of time for completing the work. Table 8 shows that with relaxed time the analog measurement errors reduced by 13.4%, 18.6% and 21.3% respectively for IQ1, IQ2 and IQ3 categories. The corresponding reduction in errors when digital multimeter was used was 1.4%, 0.6% and 0.8%.

The t-test helps us to conclude that there is significant reduction in error when the time limits for doing the work are relaxed, in the case of analog measurement of resistance. But in the case of digital measurement, there is a significant error reduction only in the IQ1 category. Similar results were observed in the case of forenoon and afternoon sessions. This gives that relaxed time gives better result especially when a keen observation is involved in measurements (as in analog meter) rather than just read out (as in digital meter).

## Environment

The IQ1, IQ2 and IQ3 categories were done the measurements of resistance both in normal and better conditions.

Figure 7 gives that air conditioned environment provides error reduction of 12, 28.2 and 17.3% respectively for IQ1, IQ2 and IQ3 categories in analog measurements. In the case of digital measurements it is 1.1, 1.6 and 1.4% respectively.

## Conclusion

The outcome of a manual measuring system includes the results produced by the measuring instrument and what was observed and noted by the human subject involved in the manual measurement exercise. This paper focused

**Table 1.** Subjects for the experiments.

| Measurand subject | Voltage | | Resistance | |
|---|---|---|---|---|
| | Analog | Digital | Analog | Digital |
| ET - IQ3  (Age=31 to 40 yrs) [ET = Experienced technicians] | 20 | 20 | 20 | 20 |
| ET - IQ3 (Age=41 to 50 yrs) | 20 | 20 | 20 | 20 |
| IE  - IQ1 (Age=21 to 30 yrs) [IE = Inexperienced subjects ] | 20 | 20 | 20 | 20 |
| IE - IQ2  (Age=21 to 30 yrs) | 20 | 20 | 20 | 20 |
| IE – IQ3 (Age=21 to 30 yrs) | 20 | 20 | 20 | 20 |

**Table 2.** Experienced technicians total data.

| Subject | | RAP | RDP | VAP | VDP |
|---|---|---|---|---|---|
| | Mean | 6.92 | 5.51 | 15.22 | 3.39 |
| FN | Standard deviation | 0.73 | 1.62 | 4.71 | 0.36 |
| | CV | 10.58 | 29.33 | 18.93 | 10.67 |
| | Mean | 7.63 | 6.71 | 19.68 | 4.76 |
| AN | Standard deviation | 1.92 | 2.11 | 4.98 | 1.30 |
| | CV | 25.15 | 31.46 | 17.75 | 27.28 |
| Total | Mean | 7.28 | 6.11 | 26.45 | 4.08 |
| | Standard deviation | 1.46 | 1.93 | 4.99 | 1.16 |

**Table 3.** Inexperienced subjects total data.

| Subject | | N | FN  - Mean | AN – Mean |
|---|---|---|---|---|
| | Above average [IQ1] | 20 | 5.48 | 6.92 |
| RAP | Average [IQ2] | 20 | 7.87 | 13.61 |
| | Below average [IQ3] | 20 | 11.60 | 14.25 |
| | Total | 60 | 8.32 | 11.60 |
| | Above average [IQ1] | 20 | 2.34 | 4.68 |
| RDP | Average [IQ2] | 20 | 4.78 | 7.70 |
| | Below average [IQ3] | 20 | 3.27 | 12.40 |
| | Total | 60 | 3.47 | 8.26 |
| | Above average [IQ1] | 20 | 2.12 | 19.39 |
| VAP | Average [IQ2] | 20 | 4.48 | 30.08 |
| | Below average [IQ3] | 20 | 7.14 | 57.75 |
| | Total | 60 | 4.58 | 35.74 |
| | Above average [IQ1] | 20 | 2.10 | 2.63 |
| VDP | Average [IQ2] | 20 | 4.44 | 6.12 |
| | Below average [IQ3] | 20 | 7.01 | 9.33 |
| | Total | 60 | 4.52 | 6.03 |

**Table 4.** Inexperienced subjects ANOVA.

| Subject | | Sum of squares | Df | Mean square | F | Sig. |
|---|---|---|---|---|---|---|
| RAP | Between groups | 470.441 | 2 | 235.221 | 8.499 | 0.001 |
| | Within groups | 1577.613 | 57 | 27.677 | | |
| | Total | 2048.054 | 59 | | | |
| RDP | Between groups | 191.349 | 2 | 95.675 | 4.397 | 0.017 |
| | Within groups | 1240.304 | 57 | 21.760 | | |
| | Total | 1431.654 | 59 | | | |
| VAP | Between groups | 4953.264 | 2 | 2476.632 | 7.165 | 0.002 |
| | Within groups | 19702.204 | 57 | 345.653 | | |
| | Total | 24655.468 | 59 | | | |
| VDP | Between groups | 336.983 | 2 | 168.491 | 96.087 | <0.001 |
| | Within groups | 99.951 | 57 | 1.754 | | |
| | Total | 436.934 | 59 | | | |

**Table 5.** Paired sample test type differences.

| Paired subject | | t | df | Sig. (2-tailed) |
|---|---|---|---|---|
| Pair 1 | RAP - RDP | 5.087 | 59 | <0.001 |
| Pair 2 | RAP - VAP | -4.224 | 59 | <0.001 |
| Pair 3 | RAP - VDP | 6.863 | 59 | <0.001 |
| Pair 8 | RDP - VAP | -6.268 | 59 | <0.001 |
| Pair 9 | RDP - VDP | 1.048 | 59 | 0.299 |
| Pair 14 | VAP - VDP | 6.155 | 59 | <0.001 |

**Table 6.** Time of work experienced vs. inexperienced.

| Experienced | | | | | Inexperienced | | | |
|---|---|---|---|---|---|---|---|---|
| Analog FN | Analog AN | Digital FN | Digital AN | | Analog FN | Analog AN | Digital FN | Digital AN |
| 11.07 | 13.66 | 4.45 | 5.74 | IQ1 | 3.8 | 13.16 | 2.22 | 3.66 |
| | | | | IQ2 | 6.18 | 21.85 | 4.61 | 6.91 |
| | | | | IQ3 | 9.37 | 36 | 5.14 | 10.87 |

**Table 7.** Time of work t-test.

| Subject | Session | Mean | Standard deviation | CV | t | df | Significance (2-tailed) |
|---|---|---|---|---|---|---|---|
| RAP | FN | 21.51 | 10.24 | 47.62 | -2.977 | 58 | 0.004 |
| | AN | 30.62 | 13.27 | 43.33 | | | |
| RDP | FN | 2.04 | 0.71 | 34.55 | -5.887 | 58 | 0.001 |
| | AN | 3.79 | 1.46 | 38.64 | | | |
| VAP | FN | 41.10 | 21.02 | 51.15 | -2.132 | 58 | 0.037 |
| | AN | 52.32 | 19.72 | 37.69 | | | |
| VDP | FN | 5.77 | 2.49 | 43.11 | -3.760 | 58 | 0.001 |
| | AN | 9.01 | 4.00 | 44.44 | | | |

The t-test results between errors in forenoon and afternoon of all categories show that there is a significant increase in error in the afternoon.

**Table 8.** Time pressure t-test.

| Category | Subject | Time | Mean | t | Sig. (2-tailed) |
|---|---|---|---|---|---|
| IQ1 | RAP | Normal | 20.4850 | 5.802 | <0.001 |
| | | Relaxed time | 7.1237 | | |
| | RDP | Normal | 3.7471 | -2.775 | 0.012 |
| | | Relaxed time | 2.3405 | | |
| IQ2 | RAP | Normal | 25.8860 | 11.012 | <0.001 |
| | | Relaxed time | 7.3075 | | |
| | RDP | Normal | 3.6338 | -0.922 | 0.369 |
| | | Relaxed time | 3.0521 | | |
| IQ3 | RAP | Normal | 29.3066 | 4.113 | 0.001 |
| | | Relaxed time | 8.0431 | | |
| | RDP | Normal | 4.6823 | -1.191 | 0.249 |
| | | Relaxed time | 3.8865 | | |

on the effect of a few variables such as task assigned, type of instruments used, time of day, time pressure and environmental difference when experiment was done on measurement error. Using a standardized setup for the experiments, and only changing one parameter at a time, the measurement error in this study has been reduced to human errors in observation and noting. It was seen from the experiments conducted that, humans produce significant errors in measurement. The human errors due to observation and noting ranged from a minimum of 0.6 % to a maximum of 28.2%. It was necessary to study and bring out these values, since knowing the values would help in dealing with the errors when they are significant in the system. This study brings out values of human errors under different work related variable combinations; these may be taken as expected error values under those conditions for work system design.

The study has also been able to examine and quantify the effects of different work related variables on human errors induced in different types of subjects working under normal and air-conditioned environment. It was seen that, experienced technicians with below average IQ, were able to reduce the measurement error by 7.3% when using digital display devices instead of analog. A large segment of the workforce used in Industry for production and quality control measurements belong to this group. Digital measuring devices should only be used by them especially in areas where errors would be disastrous. Reduction in error with change over from ana-log to digital for the inexperienced category of subjects is also significant and stands at 9.5% on an average. Moving from use of analog to digital will therefore reduce human errors at least by 7.3%. The simpler task of voltage measurement, in all categories of subjects, unexpectedly, produced maximum error. A possible

explanation being that, voltage measurement being a relatively simple task, having this in mind and not paying enough attention to the simple work at hand could have induced more human errors. It can also be noted that, persons with high IQ make less error. This is substantiated by the observation that inexperienced B. Technicians and Diploma holders with IQ above average [IQ1] were making less error when compared to the experienced subjects with lesser IQ. The increase of errors between forenoon and afternoon was least for experienced technicians. This could be because of their being used to working in the forenoon and afternoon regularly. The effect of doing the measurement without time limits for completion, on inexperienced subjects, is a reduction of errors on an average by 18% and 1% in analog and digital measurements respectively. The inexperienced subjects reduce error by 19 and 1.5% on an average in analog and digital measurements respec-tively, when they have performed the measurement under air-conditioned conditions.

The study has identified some variables that contribute to human errors in measurement. In the decreasing order of influence on human errors the parameters are: Instrument differences (digital-analog), working environ-ment, time pressure, time of day and type of work. Identification of these parameters and an assessment of the extent of errors they produce will be of use to practitioners who rely on measurement for research, control and production, and quality control. This work has also demonstrated a simple methodology that can be used for such work. This work was started small with only a few parameters, since taking too many parameters all at a time would make the experiment very difficult to conduct and control. It is hoped that this work will help users of measurement in practice to better understand

and manage the phenomena of Human errors in measurement. In Future, studies need to be done to examine the effect of other type of variables on human errors in measurement. There is also need to look at the individual and interaction effects among different variable types.

## REFERENCES

Blanchard RT (1973). "Requirements, concept, and specification for a navy human factor performance data store". Int. J. Man Mach. Stud. 31:643–672.

Cannon MD, Amy CE (2001). "Confronting failure: Antecedents and consequences of shared beliefs about failure in organizational work groups". J. Org. Behav. 22:161–177.

Carmen S (2005). "The Study of Measurement Equipment Bias." Faculty of Engineering, University "Lucian Blaga" from Sibiu, "Hermann Oberth" Romania.

Carter JA Jr. (1986). "A taxonomy of user – oriented functions". Int. J. Man Mach. Stud. 24:270–286.

Cathy VD, Michael F, Markus B, Sabine S (2005). "Organizational error management culture and its impact on performance: A two- study replication". J. Appl. Psychol. 90(6):1228–1240.

Chesher A (1991). "The effect of measurement error". Biometrika 78(3):451–462.

David E (1996). "Understanding Human Behavior and Error". Human Reliability Associates. Wigan, Lancashire, WN8 7RP.

Dormann T, Frese M (1994). "Error training: Replication and the function of exploratory behavior". Int. J. Hum. Comp. Interact. 6:365-372.

Douglas AW, Esa R (2002). "Defining the Relationship Between Human Error Classes and Technology Intervention Strategies". Aviation Research Laboratory, Institute of Aviation, University of Illinois at Urbana – Champaign, Savoy, IL 61874.

Frese M (1991). "Error Management or error prevention: two strategies to deal with error in software design". In: Bullinger HJ (Ed.), Human Aspects in Computing, Design and use of Interactive Systems and Work with Terminals. Elsevier, Amsterdam. pp. 776 – 782,

Frese M (1995). "Error Management in training: Conceptual and Empirical Results." In: Zuccermaglio C, Bagnara S and Stucky S (Eds.), Organizational learning and technological change. Springer – Verlag, Berlin, Germany. pp. 112 - 124.

Gawron VJ, Drury C, Wikiris DM (1989). "A taxonomy of independent variables affecting human performance." Int. J. Man – Machine Stud. 31:643-672.

Helmreich RL, Merritt AC (2000). "Safety and Error Management: the Role of Crew Resource Management". Ashgate Publishing, Aldershot, England. pp. 107 – 119.

Holland P (1986). "Statistics and causal interface". J. Am. Stat. Assoc. 81(260):663–685.

Mark AH, Brian EB (2000). "Comment on measurement error in research on human Resources and firm performance: How much error is there and how does it influence effect size estimates? by Gerhart, Wright, Mc Mahan and Sneha", Personal Psychol., Inc. 53

Meijman TF, Mulder G (1998). Psychological aspects of workload. In: P Drenth, Thierry H and De Wolff C (Eds.). Hand book of work and organizational psychology. Psychology Press, 2nd ed, London. pp. 5-33.

Nordstrom CR, Wendland D, Williams KB (1998). "To err is Human. An Examination of the Effectiveness of Error Management Training". J. Bus. Psychol. 12:269–282.

Parasuraman R, Sheridan TB, Wickens CD (2000). "A model for types and levels of human interaction with automation." IEEE Trans. Syst. Man Cybernet. A30(3):286-295.

Reason J (1990). "Human Error." Cambridge University Press. Cambridge, England.

Senders JW, Moray NP (1991). "Human error: Case, Prediction and Reduction". Hillsdale NJ, Erlbaum.

Drury VJ, Czaja CGSJ, Wilkins DM (1989). "A taxonomy of independent variables affecting human performance." Int. J. Man- Machine Stud. 31:643-672.

Wickens CD, Gordon SG, Liu Y (1998). "An introduction to human factors engineering. Addison Wesley Longmann, New York.

Wiegmann D, Shappell S (1997). "Human factors analysis of post-accident data : Applying theoretical taxonomies of human error". Int. J. Aviation Psychol. 7:67-81.

Zapt D, Brodbeck FC, Frese M, Peters H, Prumper J (1992). "Errors in working with computers: A First Validation of Taxonomy for Observed Errors in a Field Setting". Int. J. Hum Comput. Interact. 4:311–339.

# 11

# A feed antenna for dielectric spheres lens in the Ka band

Yinfang Xu*, Yongjun Xie, Zhenya Lei and Chao Deng

National Key Laboratory of Antennas and Microwave Technology, Xidian University, Xi'an 710071, China.

In this paper, a feed antenna for dielectric spheres lens in the Ka band, which is a circularly polarized short dielectric rod antenna, is presented. The structure combines with circular polarizer and short dielectric rod antenna skillfully. The circular polarizer with dielectric septum realizes the circular polarization. The short dielectric rod antenna achieves the high gain. Compared with the long dielectric rod antenna, the short dielectric rod antenna in the design has realized miniaturization. The antenna is characterized by simple structure, manufacture and test easily. In addition, the performance of the antenna as a feed for the dielectric sphere lens antenna is also studied, whose simulated gain with a quarter wavelength choke is 20.369 dB and beam width is 16°, which show that the antenna is an excellent feed antenna.

Key words: Ka band, dielectric spheres lens, circularly polarized, short dielectric rod antenna, a quarter wavelength choke.

## INTRODUCTION

In order to make full use of the limited frequency in modern wireless communications, multi-beam antenna, whose signal could cover wider range of area with high gain, and whose beam configurations can be adaptively adjusted according to the need, has been thoroughly studied and widely used. As a multi-beam antenna, dielectric spheres lens multi-beam antenna mainly be used in millimeter wave and sub-millimeter wave band. And the design of the feed antenna for dielectric spheres lens hold the balance in the whole design.

At present, the main feed antenna for dielectric spheres lens are: planar yagi antenna (Phillip et al., 2004), tapered slot antenna (TSA) (Bernhard et al., 2002), circular waveguide (Bo et al., 2009). In this paper a new type of feed antenna for dielectric spheres lens is presented: circular polarizer combined with dielectric rod antenna, which not only meet the performance but also realize the circular polarization. Circular polarized antennas play an important role in the communication system. Circular polarizer is an important part of circular polarized antenna waveguide feed system, whose

performance have directly influence on the axis. In recently year, circular polarizer has been widely studied and discussed (Naofumi et al., 2000; Shih-Wei et al., 2004; Tian-ling and Ze-hong, 2006; Zhiru and Jin, 2009).

The dielectric rod antenna consists of a dielectric cylinder excited by a hollow waveguide. In order to reduce the reflection, the end part is generally conical. Dielectric rod antenna is usually traveling wave antennas. Its dielectric rod body part has much larger length than a wavelength and usually consists of 3 parts: feed gradient, rod body and end gradient. So, the long dielectric rod antenna is of great length. On the other hand, the short dielectric rod antenna, which refers to the length of the rod is little shorter than a wavelength, can satisfy the requirement of small size. The long dielectric rod antenna have been widely studied (Pramendra et al., 2008; Qiu and Wang, 2009), while the short dielectric rod antenna is mostly discussed before the 1980s (Watson and Horton, 1948; Kishk, and Lotfollah, 987). In recent years, few designs are reported in public papers. Moreover, the short dielectric rod antenna, which has the advantages of small size, highly-directivity, and adjustable direction diagram, has broad application prospect in microwave and millimeter waves (MMW) fuze. Therefore, it is necessary to study the short dielectric rod antenna.

---

*Corresponding author. E-mal: xyf074@yahoo.com.cn.

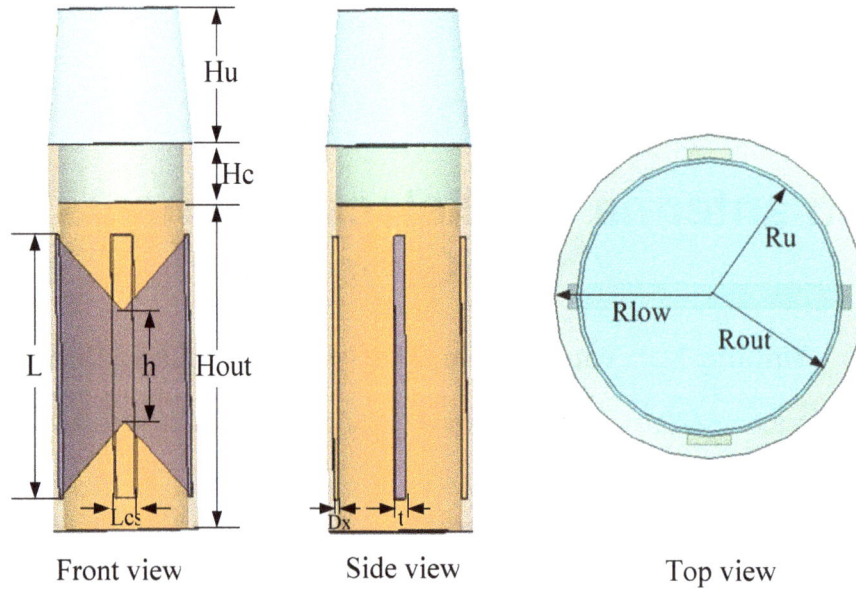

**Figure 1.** Schematic view of the circular polarizer with short dielectric rod antenna.

**Table 1.** Dimensions of the proposed antenna (unit: mm).

| Parameters | Ru | Rout | Rlow | t | Hu | Hc | Hout | L | h | Lcs | Dx |
|---|---|---|---|---|---|---|---|---|---|---|---|
| Values | 2.9 | 3.0 | 3.5 | 0.5 | 6.2 | 2.7 | 15.0 | 12.1 | 5.1 | 1.0 | 0.2 |

This paper skillfully assembles the circular polarizer and the short dielectric rod together to design a new type of antenna. Circular waveguide polarizer can achieve the requirements of circular polarization. The part of short rod antenna inserted into the waveguide medium is cylinder, plays the role of fixing dielectric rod antenna into the metal waveguide. While the exposed parts outside the metal waveguide, whose terminal is smaller and smaller, plays the role of guiding electromagnetic wave to radiate. Meanwhile, for being able to support the other antenna (such as dielectric ball), dielectric rod antenna terminal is of cone shape. The center frequency of the antenna is 35.4 GHz. The simulation results show that the combined antenna, which can be used both as an independent antenna and the feed antenna for dielectric spheres lens antenna, has the advantages of small size 24 mm × 7 mm × 7 mm, and low standing wave ratio, and perfect circular polarization.

## ANTENNA DESIGN

Figure 1 shows the geometry and configuration of the proposed dielectric rod antenna. The material of the circular waveguide is set to copper and of the septum is FR4. The relative permittivity of the dielectric rod is 2.53.

The detailed dimensions are depicted in Table 1. As expressed in Figure 1, a dielectric septum is inserted in the middle of the circular waveguide. An incident wave oriented at 45° relative to the dielectric septum is assumed, which can be decomposed into two equal orthogonal projections $E_x$ and $E_y$, respectively, parallel and perpendicular to the dielectric septum. The two components then propagate through the septum region with little reflection due to the small septum discontinuity. In the septum region, propagation constant $\beta_x$ of the $E_x$ component is strongly perturbed by the dielectric septum because the electric field line is parallel to the septum. On the other hand, propagation constant $\beta_y$ of the $E_y$ component is weakly perturbed because the electric field line is perpendicular to the septum. As a result, this polarizer can be implemented by choosing a suitable septum length L so as to realize a 90° phase difference at the output port. As the result, circular polarized modes will be transmitted through the output port of the waveguide. The length L is determined by

$$\Delta \varphi = \left( \beta_x - \beta_y \right) \times L = 90^0 \tag{1}$$

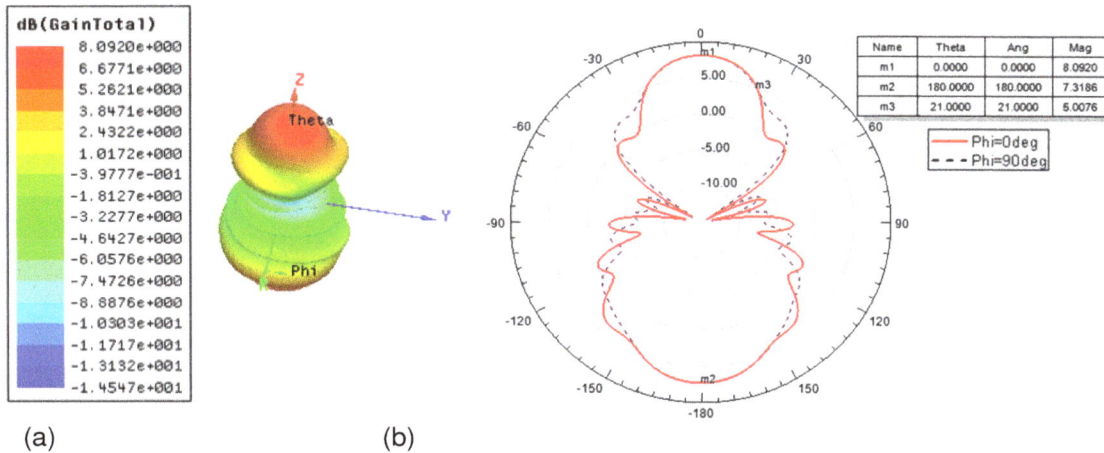

**Figure 2.** The simulated radiation pattern (a) three-dimensional (b) two-dimensional.

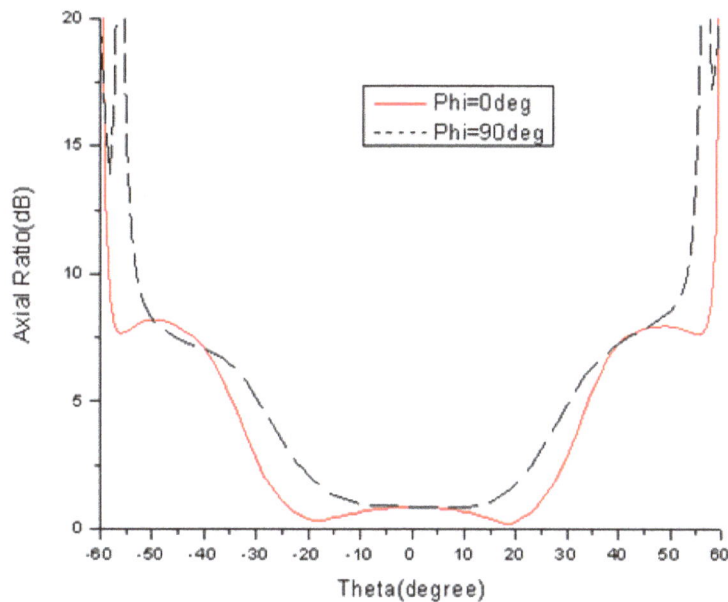

**Figure 3.** The enlarged axial ratio.

Where $\beta_x$ and $\beta_y$ can be approximated as (Stephen, 2006). The phase difference $\Delta\varphi$ is close to 90°, which means that the axial ratio is similar to 0 dB. In generally, when the axial ratio is less than 3 dB between the half power beam widths, it shows that the circular polarization property is good.

## SIMULATION RESULTS AND DISCUSSION

The Ansoft HFSS software is used to simulate the model. The antenna works on the 8 mm band, the size of the antenna is only 24 mm $\times$ 7 mm $\times$ 7 mm. The radiation pattern shown in Figure 2 presents that the gain of the antenna is approximately 8.1 dB, which is 2 dB higher than the traditional circular horn antenna. The enlarged axial ratio (defined by value less than 3 dB between the half power beam widths) is shown in Figure 3. The simulation VSWR shown in Figure 4 is small, which indicates that the antenna is an excellent independent antenna.

Tables 2, 3 and 4 exhibit the impacts of various Rout, Ru and Hu have on the gain, the first side lobe level (FSLL), the half power beam width (HPBW) and the axial ratio. As shown, Rout has an important influence on the axial ratio. And the larger Rout is, the smaller the Gain or HPBW is.

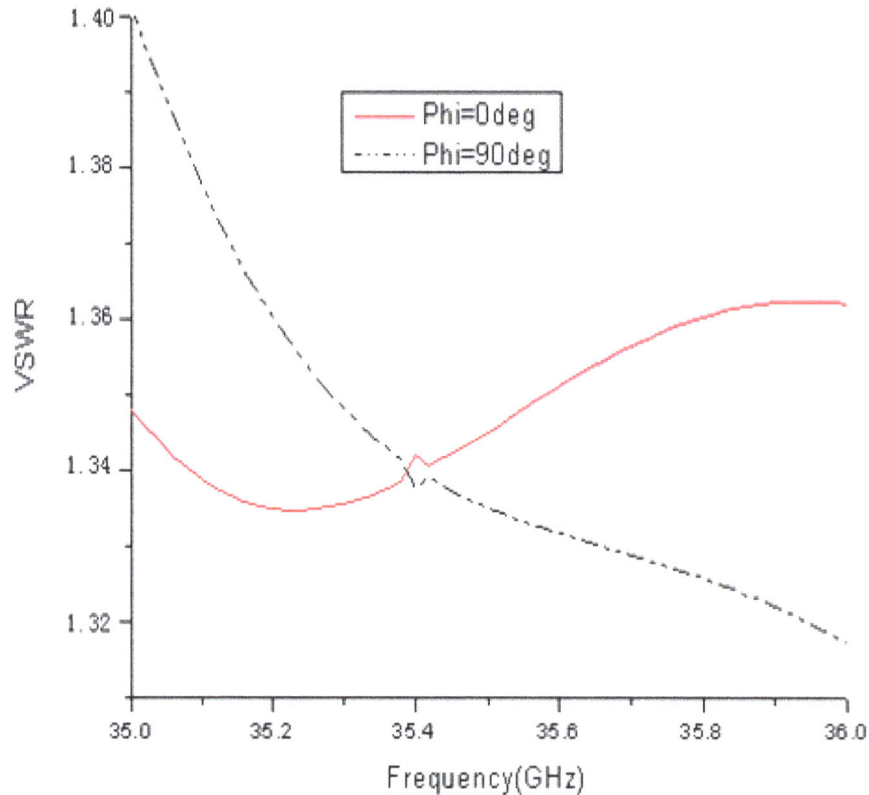

**Figure 4.** The VSWR.

**Table 2.** Rout --the radius of the circular waveguide.

| Rout (mm) | Gain(dB) | FSLL(dB) | Axial ratio(dB) | HPBW(°) |
|---|---|---|---|---|
| 2.9 | 9.2716 | -14.0763 | 12.9344 | 48 |
| 3.0 | 8.0920 | -14.1164 | 0.8834 | 42 |
| 3.1 | 7.9082 | -12.0040 | 9.0276 | 40 |

**Table 3.** Ru-- the top radius of the cone.

| Ru (mm) | Gain(dB) | FSLL(dB) | Axial ratio(dB) | HPBW(°) |
|---|---|---|---|---|
| 2.8 | 7.9793 | -13.9605 | 1.0530 | 42 |
| 2.9 | 8.0405 | -14.1643 | 0.9267 | 41 |
| 3 | 8.0873 | -14.2570 | 1.0309 | 41 |

**Table 4.** Hu--the height of the top cone.

| Hu (mm) | Gain (dB) | FSLL (dB) | Axial ratio (dB) | HPBW (°) |
|---|---|---|---|---|
| 6.1 | 8.0386 | -13.1895 | 3.1132 | 44 |
| 6.2 | 8.0920 | -14.1311 | 0.8834 | 42 |
| 6.3 | 8.1018 | -13.6293 | 1.1749 | 40 |

Not quite the same, the larger Hu is, the larger the Gain or the smaller the HPBW is.

## THE IMPROVED ANTENNA AND SIMULATION RESULTS

As shown in Figure 2, the disadvantage of this antenna is that its radiation pattern has a high back lobe level. This may has two reasons: the first is the unbalanced feed system leads to the leakage of the energy; the second is the septum resulting in energy loss. One of the known techniques to reduce the back lobe level of the radiation pattern is to use a quarter wavelength choke on the waveguide wall (Kishk and Lotfollah, 1987). The geometry and configuration of the improved antenna is shown in Figure 5, the material is set to copper and the detailed dimensions of the choke are indicated in the figure. As shown in Figure 6 - 8, the simulated results show that utilizing this approach the back lobe level reduces, on average, in excess of 2 dB. In addition, the directive gain of this dielectric rod with choke have somewhat increased. As shown in Figure 9, the peak gains of the proposed antenna illustrate stability in the given frequency range.

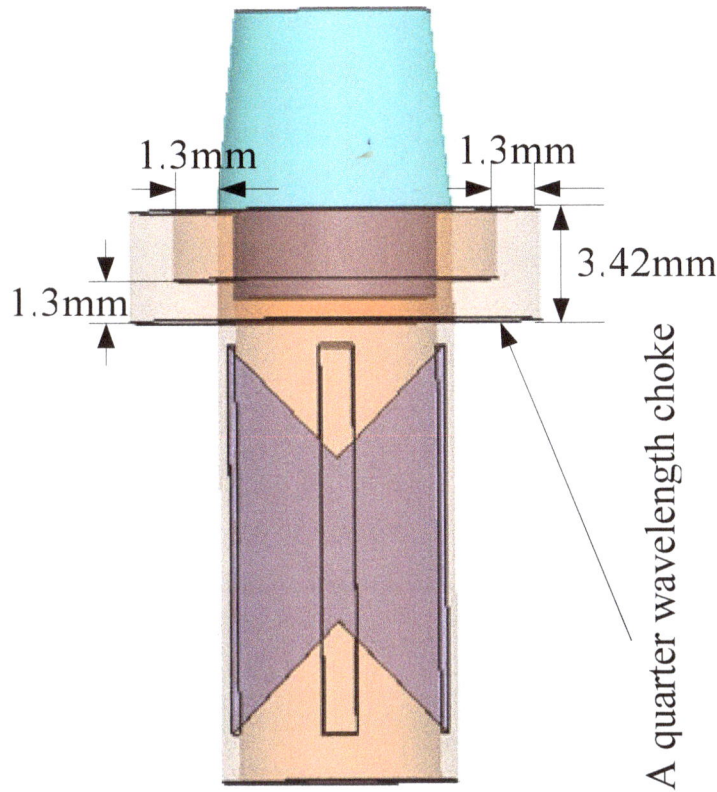

**Figure 5.** Short dielectric rod antenna with a quarter wavelength chokes.

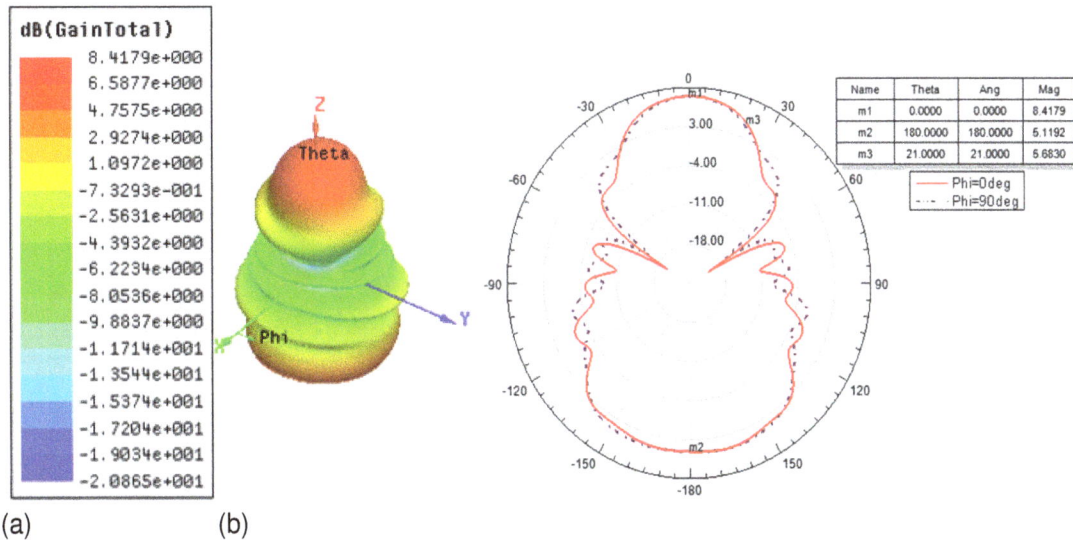

(a)                              (b)

**Figure 6.** The simulated radiation pattern with choke (a) three-dimensional (b) two-dimensional.

## AS THE FEED FOR THE DIELECTRIC SPHERE LENS ANTENNA

Taking account of the gain of the dielectric spherical lens and the simulation memory of the computer, the permittivity $\varepsilon_r$ = 2.53 and the radius 21 mm of the lens are selected. The simulation models are shown in Figure 10

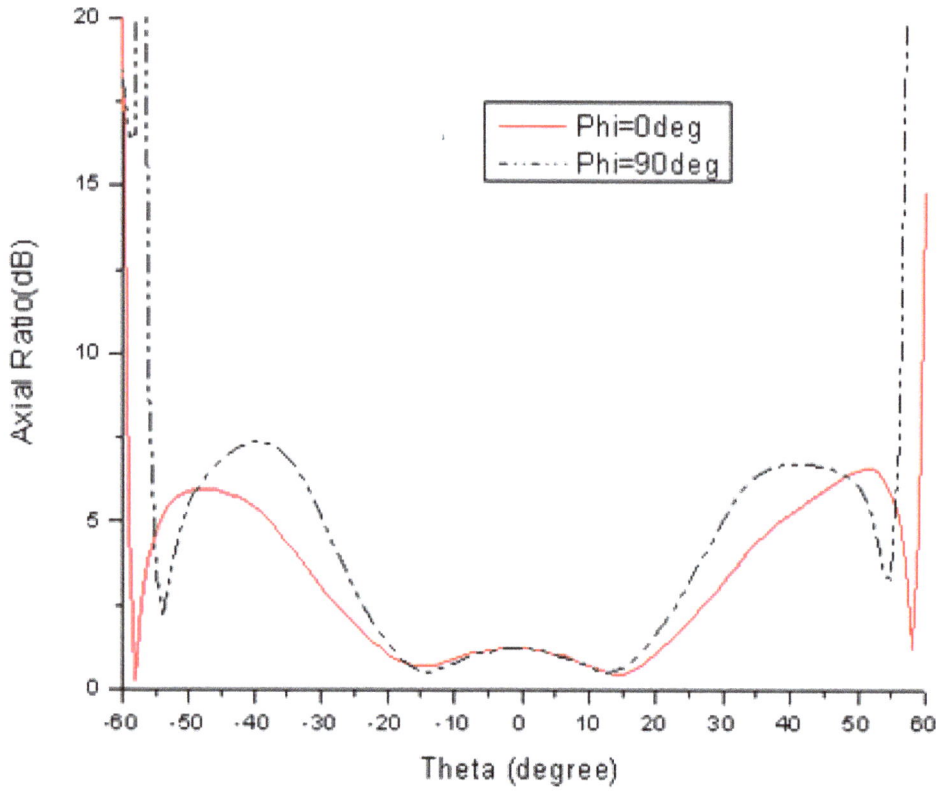

**Figure 7.** The enlarged axial ratio with choke.

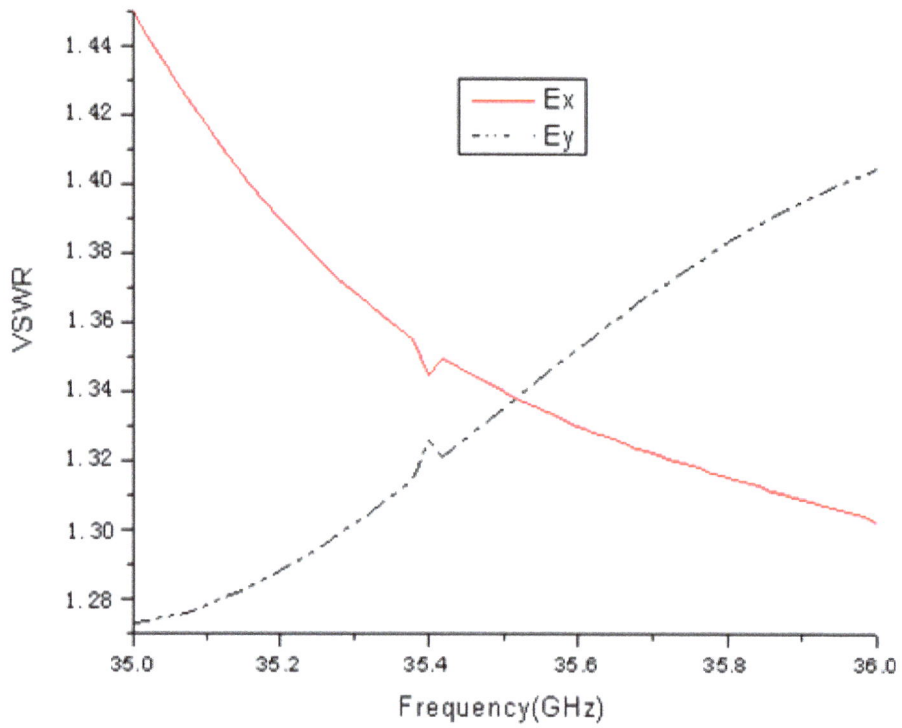

**Figure 8.** The VSWR with choke.

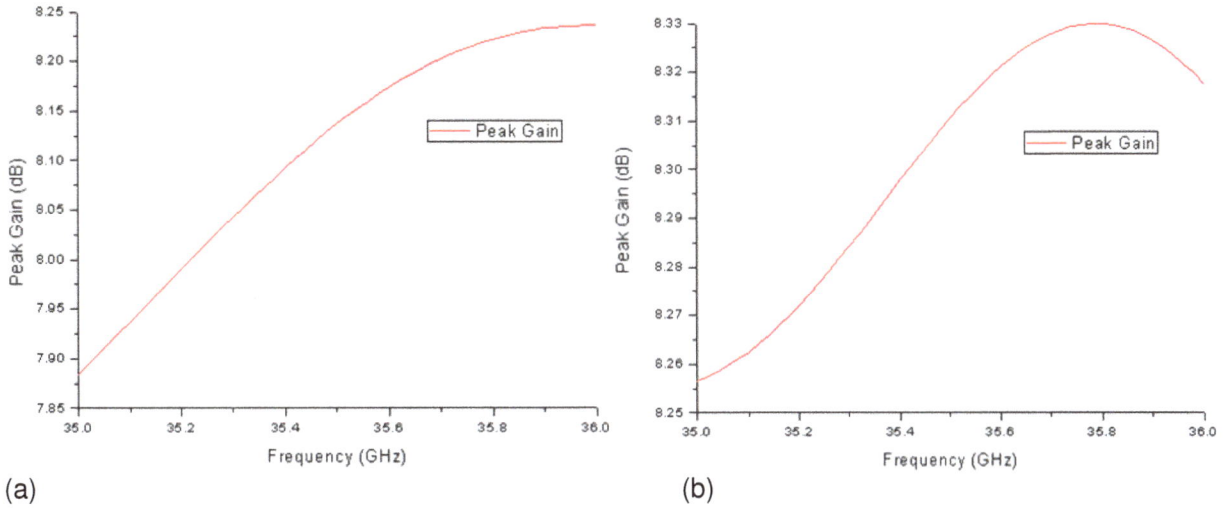

**Figure 9.** The peak gains of the proposed antenna (a) without choke (b) with choke.

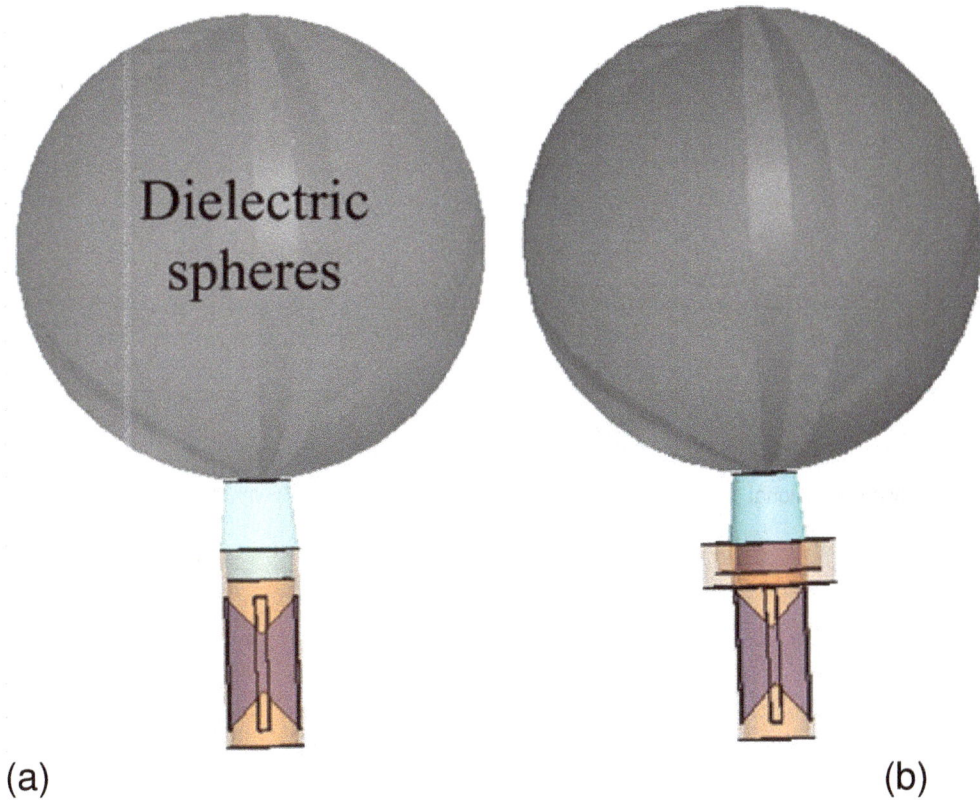

**Figure 10.** The circularly polarized short dielectric rod antenna feed for dielectric sphere lens antenna (a) without choke (b) with choke.

for both cases with and without choke. The simulation radiation pattern shown in Figures 11 and 12 show that the gain of the antenna without and with choke are 18.116 and 20.369 dB, respectively. And the beam widths are separately 14 and 16°. More importantly, the back lobe level of the antenna with choke becomes very small, which is only -0.15 dB, compared with the antenna without choke 6.46 dB. In addition, the enlarged axial ratio and VSWR are shown in Figure 13 and 14, respectively. Figure 15 demonstrates the stable peak

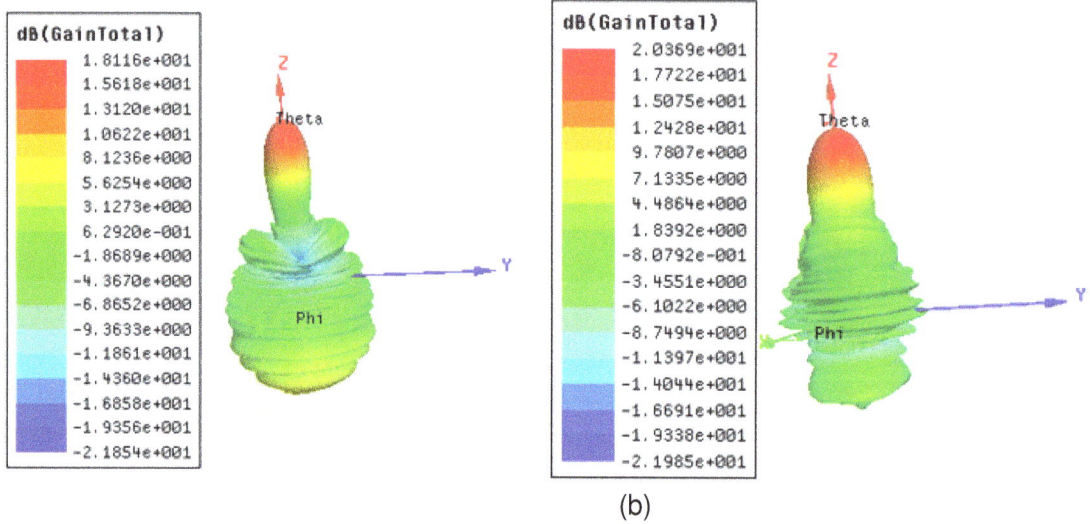

(a)

(b)

**Figure 11.** The three-dimensional simulated radiation pattern (a) without choke (b) with choke.

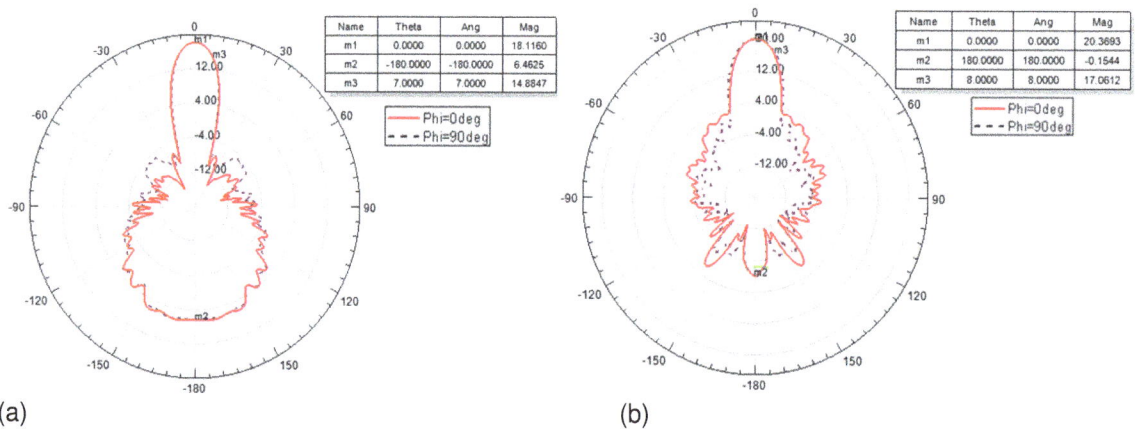

(a)

(b)

**Figure 12.** The two-dimensional simulated radiation pattern (a) without choke (b) with choke.

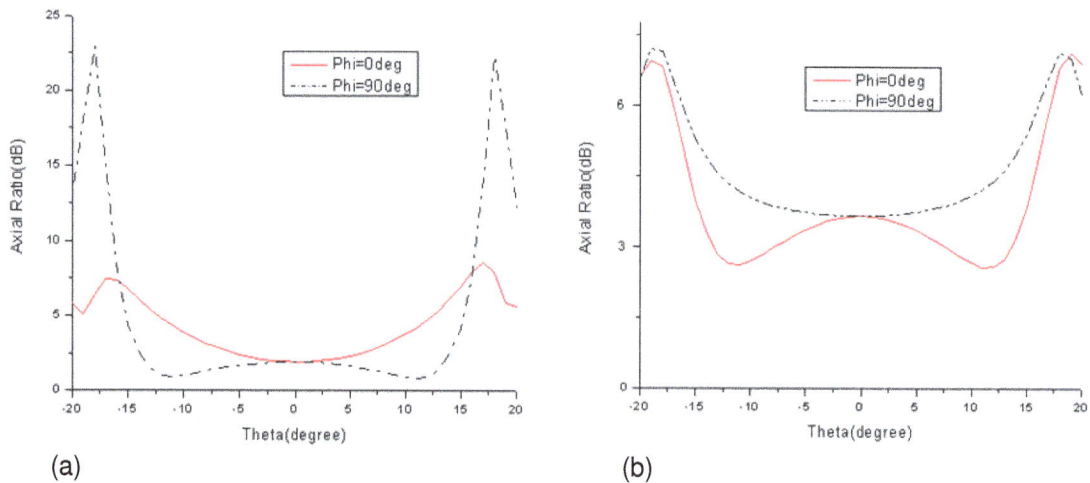

(a)

(b)

**Figure 13.** The enlarged axial ratio (a) without choke (b) with choke.

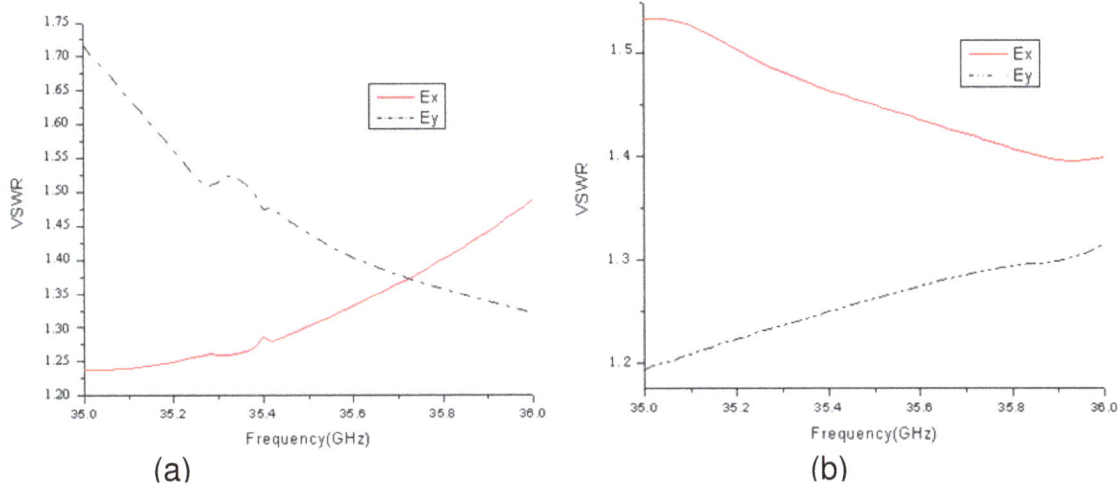

**Figure 14.** The VSWR (a) without choke (b) with choke.

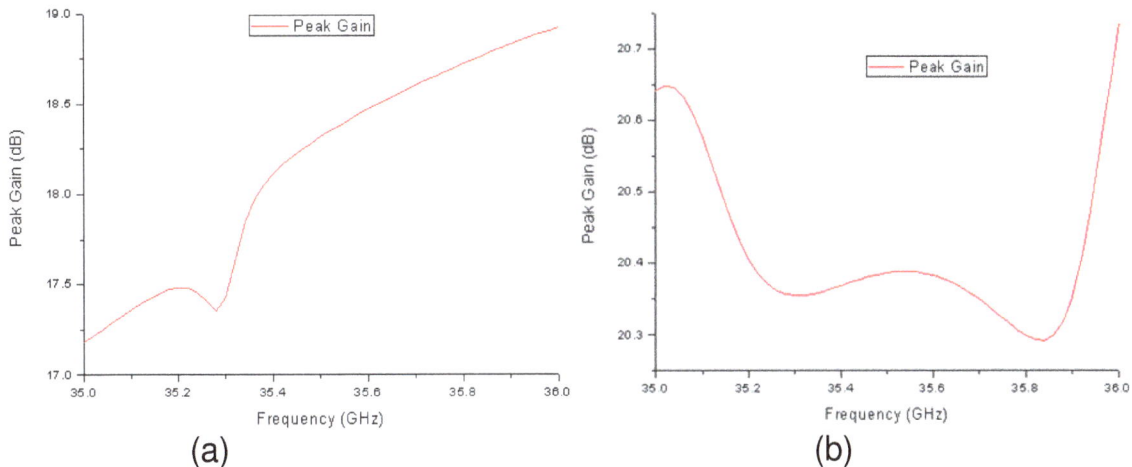

**Figure 15.** The peak gains of the spherical lens antenna (a) without choke (b) with choke.

gains of the spherical lens antenna without and with choke in the given frequency range. The simulation of the spherical lens takes a very large computer memory, while the computer configuration of working group can not meet the requirement, some points (Mosallaei and Rahmat-Samii, 2001; Boriskin and Nosich, 2002), in design of spherical lens antennas can not be in-depth discussion and research. It is only hoped that this part will provide an idea to the relevant researchers.

## CONCLUSION

The design of a circularly polarized short dielectric rod antenna, which can be used as an independent antenna and a feed antenna for dielectric spheres lens antenna, has been presented in the paper. The simulation results show that the design is reasonable and feasible. This

kind of antenna has a simple and compact structure which can be easily manufactured and large-scale produced. The results, such as the gain, the axial ratio, the bandwidth etc, show the antenna is a good feed antenna.

## REFERENCES

Bernhard S, Xidong W, Jim PE (2002). Wide-Scan Spherical-Lens Antennas for Automotive Radars", IEEE Trans. Microw. Theory Tech., 50(9): 2166 – 2175.

Bo X, Minmin H, Hongfu M, Sen C, Yunhua L, Wenbin D (2009). Multi-beam Antenna at Q Band, 34th International Conference on Infrared, Millimeter, and Terahertz Waves, IRMMW-THz.

Boriskin AV, Nosich AI (2002). Whispering-gallery and Luneburg-lens effects in a beam-fed circularly-layered dielectric cylinder", IEEE Trans. Antennas Propag., 50(9):1245–1249.

Kishk AA, and Lotfollah S (1987). Radiation Characteristics of the Short Dielectric Rod Antenna: A Numerical Solution", IEEE Trans.

Antennas Propag., 35(2):139-146.

Mosallaei H, Rahmat-Samii Y (2001). Nonuniform Luneburg and two-shell lens antennas: radiation characteristics and design optimization", IEEE Trans. Antennas Propag., 49(1): 60–69.

Naofumi Y, Hiroyuki M, Masao Y (2000). A design of novel grooved circular waveguide polarizers", IEEE Trans. Microw. Theory Tech., 48(12): 2446-2452.

Phillip RG, Bernhard S, Gabriel MR (2004). A 24-GHz High-Gain Yagi–Uda Antenna Array". IEEE Trans. Antennas Propag., 52(5): 1257-1261.

Pramendra KV, Raj K, Mahakar S (2008). Design and Simulation of Dielectric Tapered Rod as Feed for Dielectric Lens Antenna at 140 GHz", Proc. Int. Conf. Microw., pp. 233-235.

Qiu J, Wang N (2009). Optimized Dielectric Rod Antenna for Millimeter Wave FPA Imaging System, IST 2009. Int. Workshop Imaging Syst. Tech,, May 11-12, 10.

Shih-Wei W, Chih-Hung C, Chun-Long W, Ruey-Beei W (2004). A Circular Polarizer Designed With a Dielectric Septum Loading. IEEE Trans. Microw. Theory Tech., 52(7): 1719-1723.

Stephen D (2006). Targonski,A Multiband Antenna for Satellite Communications on the Move" , IEEE Trans. Antennas Propag., 54(10): 2862 - 2862.

Tian-ling Z, Ze-hong Y (2006). A Ka Dual-Band Circular Waveguide Polarizer.

Watson RB, Horton CW (1948). The radiation patterns of dielectric rods-experimental and theory. J. Appl. Phys., 19: 661-670.

Zhiru Y, Jin P (2009). A novel optimization strategy for the design of large tolerance circular waveguide septum polarizer". IEEE International Symposium on Antennas and Propagation and USNC/URSI National Radio Science Meeting, p. 4.

# 12

# High transmission bit rate of multi giga bit per second for short range optical wireless access communication networks

## Ahmed Nabih Zaki Rashed

Department Electronics and Electrical Communication Engineering, Faculty of Electronic Engineering, Menouf 32951, Menoufia University, Egypt. E-mail: ahmed_733@yahoo.com

In the present paper, the backbone and the broadband wireless access communication network technologies can increasingly provide unprecedented bandwidth capacities, the focus being gradually shifted toward broadband access technologies capable of connecting the customer premises to the local exchange. Optical wireless is increasingly becoming an attractive option for multi giga bit per second within short range (up to 5 km) links where laying optical fiber is too expensive or impractical. For such links, a tracking scheme is essential to maintain proper pointing of the transceivers at each other to establish error-free communication. The optical wireless technology is used mostly in wide bandwidth data transmission applications. Also, we have investigated the maximum transmission distance and data transmission bit rates that can be achieved within broadband wireless optical links for multi giga bit optical network applications. The wireless optical broadband access network architecture has been proposed as a flexible solution to meet the ever-demanding needs in access networks. At the wireless front end multi channel communication, with routers having multiple radio interfaces tuned to non overlapping channels, it can be used to improve network throughput in a cost effective way.

**Key words:** Optical wireless communications, short range, indoor links, data link, and outdoor links.

## INTRODUCTION

Optical wireless communication, also known as free-space optical (FSO), has emerged as a commercially viable alternative to RF and millimeter (Abd El-Naser et al., 2009) wave wireless for reliable and rapid deployment of data and voice networks. RF and millimeter wave technologies allow rapid deployment of wireless networks with data rates from tens of Mb/s (point-to-multipoint) up to several hundred Mb/s (point-to-point). However (Abd El-Naser et al., 2009; Shea and Mitchell, 2009), spectrum licensing issues and interference at unlicensed ISM bands will limit their market penetration. Though emerging license-free bands appear promising, they still have certain bandwidth and range limitations. Optical wireless can augment RF and millimeter wave links with very high (>1 Gbit/s) bandwidth. In fact, it is widely believed that optical wireless is best suited for multi-Gbit/s communication (Shea and Mitchell, 2009; Yong-Yuk Won et al., 2009). Optical and wireless networks were initially developed for different communication

scenarios (Ab-Rahman et al., 2009). Optical networks have been mainly used for high-bandwidth and long-distance communications while the wireless technology is used at wireless local networks with flexibility and low bandwidth needs. The present growing demand for bandwidth-intensive services, and the way people now communicate, are accelerating research on efficient and cost-effective access infrastructures whereas optical-wireless combinations are seen as promising approaches. The wireless optical broadband access network architecture has been recently proposed as a flexible solution to meet such ever demanding needs in access networks (Kedar and Arnon, 2006). Wireless optical broadband access network architecture provides a flexible and cost effective solution where fiber is provided as far as possible from the central office (CO) to the end users and then wireless access is provided at the front end. Because of such excellent compromise early versions are being deployed as municipal access

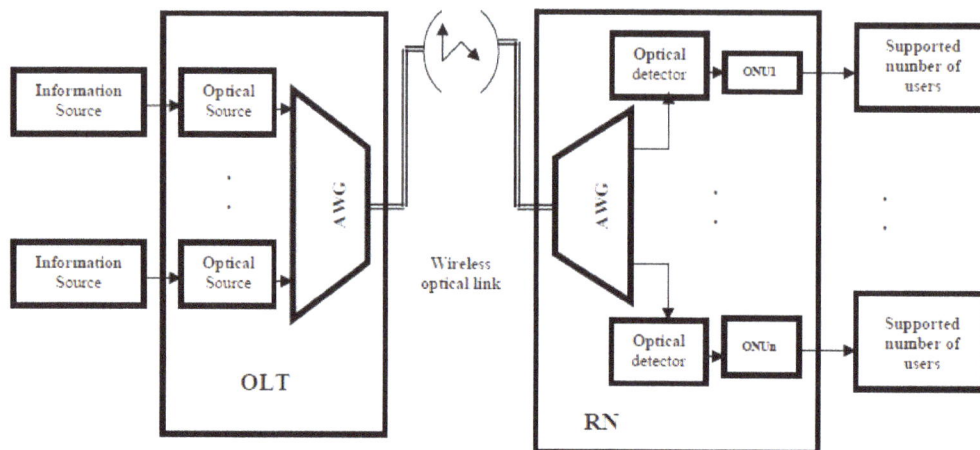

**Figure 1.** Simplified optical network architecture for wireless links.

solutions to eliminate the need for wired connection to every customer's wireless router thus saving on network deployment cost (Schuster et al., 2005).

Optical wireless communication also known as "lasercom," has become a fact of life over recent years in the sphere of urban wireless networks (Kiasaleh, 2005). There is a growing need for high data-rate transmissions in real time, and optical communications can provide the required bandwidth. Optic fiber backbone has been laid around the globe and reaches most major cities in the world, but the cramping bottleneck still restricts data flow over "the last mile," from the fiber backbone to the end user. Optical wireless communication can bridge the gap by enabling connection with and between offices, business facilities, and other targeted locations without relinquishing the performance parameters offered by optical communications and without the resource expenditure of laying optic fiber. In addition to the capital investment and running costs of fiber, it is not always possible to lay fiber at all, whereas, conversely, the rapid deployment of wireless systems facilitates their application when premises are temporarily or suddenly moved or communication is required in remote areas (Mietzer and Hoeher, 2007). In the present study, we have investigated multi giga bit per second over broadband wireless optical communication networks over wide range of the affecting parameters. Moreover, we have analyzed parametrically and numerically the maximum transmission distance and transmission bit rates and products that can be achieved within wireless optical links for optical networks.

## SIMPLIFIED OPTICAL NETWORK ARCHITECTURE FOR WIRELESS LINKS

The architecture model of passive optical network with different optical links is shown in Figure 1. PON consists

of many laser diodes as a source of optical signals which converts the electrical signal in the information source to optical signal, arrayed wave grating (AWG) multiplexer in the OLT, different optical fiber links, AWG demultiplexer, optical network unit (ONU) in the remote node (RN), optical detector which converts the optical signal to electrical signal for processing to ONU and connects to the supported number of users. In the transmission direction, the information source (electrical signal) is transmitted from the backbone network to the OLT and according to different users and location, optical source [laser diode or light emitting diode] convert it into optical signal and is transmitted into corresponding wavelength and multiplexed by Mux. When traffic arrives at RN, wavelengths are demultiplexed by Demux and sent to optical detector [Avalanch photodiode or PIN photodiode] convert the optical signal into electrical signal and then sent to ONUs which is distributed to different number of supported users (Abd El-Naser et al., 2009).

## MODELING AND EQUATIONS ANALYSIS

In the design of wireless optical link system, it is important to determine the link budget equation. The general link budget equation is given by Ackerman et al. (2008):

$$P_{received} = P_{transmit} \cdot \frac{57.295 A_{receiver}}{(\theta L)^2} e^{-\alpha L} \qquad (1)$$

Where $P_{received}$ is the power at receiver (watt), $P_{transmit}$ is the transmission power (watt), $A_{receiver}$ is the receiver effective area (m$^2$), $\theta$ is the beam divergence (degrees), L is the length of the optical link (m), and $\alpha$ is the atmosphere absorption (dB/Km). The total loss coefficient is determined by:

$$\sigma L = \sigma_{rain} L + \sigma_{fog} L + \sigma_{snow} L + \sigma_{sc\,int\,illation} \qquad (2)$$

where $\sigma_{rain}$ is the absorption due to rain (Km$^{-1}$), $\sigma_{fog}$ is the absorption due to fog (Km$^{-1}$), $\sigma_{snow}$ is the absorption due to snow (Km$^{-1}$), and $\sigma_{scin}$ is the absorption due to scintillation (Km$^{-1}$). A variety of models exist for the calculation of these absorption coefficients. In the case of fog, the Kruse model according to Kamalakis et al. (2008):

$$\sigma_{fog}(Km^{-1}) = \frac{3.912}{V}\left(\frac{\lambda}{\lambda_0}\right)^{-q} \qquad (3)$$

where V is the visibility at ($\lambda=\lambda_0$), Km, $\lambda$ is the actual wavelength of the beam, μm, $\lambda_0$ is the reference wavelength in μm for the calculation of V, and the exponent q is the size distribution of the scattering particles and is equal to 1.3 if 6 Km < V < 50 Km, and equal to 0.585 V$^{1/3}$ for low visibility V < 6 Km. Also to calculate the optical losses due to snow, the empirical formula can be used (Suman et al., 2007):

$$\sigma_{snow}(dB/Km) = A S^b \qquad (4)$$

Where S is the snow fall rate (in mm/hour), A=5.42x10$^{-5}$ $\lambda$+ 5.9458, and b= 1.38. In the same way, to calculate the optical losses due to rain, the empirical formula can be used:

$$\sigma_{rain}(dB/Km) = 1.076\,R^{2/3} \qquad (5)$$

Where R is the rain fall rate measure (in mm/h). Finally the optical loss due to scintillation is calculated using the following expression:

$$\sigma_{sc}^2 = 4.\left[23.17\left(\frac{2\pi}{\lambda}10^9\right)^{7/6}\right]C_n^2\,L^{11/6} \qquad (6)$$

Where $C_n^2$ is the scintillation strength (in m$^{-2/3}$). It should be noted that the case of wireless optical link system, fog induced absorption is the most impairment and can be significantly affect the performance of the system. A link budget for wireless optical link using one lens in the transmitter and one lens in the receiver is calculated. Different kinds of losses are calculated that may cause power losses during transmission (Jihui et al., 2005). Equation (7) shows that the ray losses of the system depend on the radius of the receiver lens and the beam radius at the receiver unit. A Gaussian beam intensity distribution is assumed (Suman et al., 2008):

$$F_s = 10\log\frac{P_{receiver}}{P_{total}} = 10\log\left(1 - e^{-\frac{2R^2}{w(L)}}\right) \qquad (7)$$

Where L is the wireless link length in km, F$_S$ is the ray losses, dB, P$_{total}$ is the total beam power at L, watt, R is the lens radius, mm, w (L) is the beam radius, mm. Geometrical losses occur due to the diverence of the optical beam. These losses can be calculated using the following formula [14]:

$$\frac{A_R}{A_T} = \left(\frac{57.295D_R}{D_T + 100.\,d.\theta}\right)^2 , \qquad (8)$$

where A$_R$ is the effective area of the receiver lens, A$_T$ is the effective area of the transmitter lens, D$_R$ is the diameter of the transmitting lens, D$_T$ is the diameter of the receiving lens, d is the distance between the wireless optical transmitter and receiver, θ is the divergence of the transmitted laser beam in degrees. Based on

curve fitting MATLAB program, the fitting equations between optical signal to noise ratio (OSNR), the operating signal wavelength for transmitter and receiver, and the wireless optical link length are (Ramachandran et al., 2006):

$$OSNR = 17.35 - 12.27L + 7.05L^2 - 5.87L^3 , \qquad (9)$$

$$OSNR = 3.85 - 10.73\lambda + 2.13\lambda^2 + 9.75\lambda^3 , \qquad (10)$$

The radio frequency transmission response provides the relative loss or gain in a wireless communication system links with respect to the signal frequency. Any signal attenuation due to the wireless communication links can be expressed:

$$Transmission(dB) = 10\log\left(\frac{P_{transmitter}}{P_{incident}}\right) , \qquad (11)$$

where P$_{transmitter}$ is the radio frequency power calculated at the output of the receiver, and P$_{incident}$ is the radio frequency power calculated at the input to the laser transmitter. Based on curve fitting MATLAB program, the fitting equations between transmission response, operating radio frequency, and amplification range are (Ramachandran et al., 2006):

$$Transmission(dB) = 10.82 - 2.05f + 7.42f^2 - 4.23f^3$$
(without amplification), $\qquad (12)$

$$Transmission(dB) = 3.09 + 13.65f - 2.56f^2 + 1.85f^3$$
(with amplification) $\qquad (13)$

The Shannon capacity theorem to calculate the maximum data transmission bit rate or the maximum channel capacity for the wireless optical links is as follows (Abd El-Naser et al., 2009):

$$C = B.W\log_2(1 + OSNR), \quad Gbits/\sec \qquad (14)$$

Then the Shannon bit rate-distance product can be determined by the following expression:

$$P_{Sh} = C.L , \quad Gbit.km/\sec \qquad (15)$$

Where L is the wireless link length in km.

## SIMULATION RESULTS AND DISCUSSION

The main objective of the wireless optical link design is to get as much light as possible from one end to the other, in order to receive a stronger signal that would result in higher link receive a stronger signal that would result in higher link margin and greater link availability. As shown in Table 1, the proposed wireless optical link parameters to achieve maximum both tranmission link distance, transmission data rate and transmission bit rate-distance product.

Based on the assumed set of the controlling parameters for wireless optical link design to achieve the best transmission bit rates and transmission distances and the set of Figures 2 to 9, the following facts are assured:

**Table 1.** Proposed wireless optical link design parameters.

| Operating parameter | Value |
|---|---|
| Power transmitted ($P_T$) | 100 mWatt |
| Operating wavelength range ($\lambda$) | 0.85 to1. 55 $\mu$m |
| Transmitter beam divergence ($\theta$) | 100 degree |
| Receiver diameter ($D_R$) | 0.1-0.5  m |
| Link distance range | 0.1 to 10 Km |
| Receiver sensitivity ($S_R$) | 5 $\mu$Watt |
| Transmitter and receiver losses ($\eta$) | 50 % |

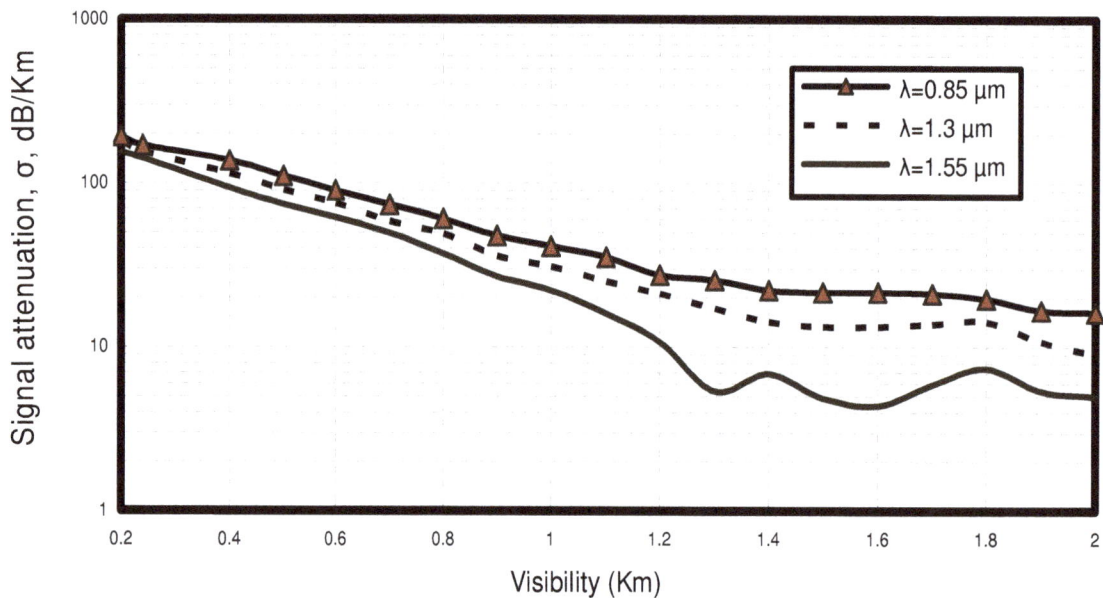

**Figure 2.** Variations of the signal attenuation with visibility for different laser diode wavelengths at the assumed set of parameters.

(i) Figure 2 has indicated that as the transmission distance (visibility) increases, the signal attenuation decreases at the same optical signal wavelength. While as the optical signal wavelength increases, signal attenuation decreases at the same transmission distance.
(ii) As shown in Figure 3, as the beam diameter at receiver increases, the ray losses also increases at the same lens diameter. While as the lens diameter increases, the ray losses decrease at the same beam diameter at receiver.
(iii) Figure 4 has demonstrated that as wireless optical link distance increases, the optical signal to noise ratio (OSNR) decreases at the same optical signal wavelength. Moreover, as the optical signal wavelength increases, the OSNR also increases at the same wireless optical link distance.
(iv) As shown in Figures 5 and 6, as optical signal to noise ratio increases, this leads to increase in Shannon bit-rate distance product at a constant of both wireless

link length and operating signal wavelength. Also, as both the wireless link length and operating signal wavelength increased, this result in increasing Shannon bit-rate distance product at a constant optical signal to noise ratio. Variations of Shannon bit rate-distance product against optical signal to noise ratio at the assumed set of parameters
(v) As shown in Figure 7, as the transmitted radio frequency increases, the signal transmission also increases for both amplification and non amplification techniques. But with amplification technique offered high signal transmission.
(vi) Figures 8 and 9 have indicated that as the transmitted radio frequency increases, the transmission data rate also increasesin both cases of amplification and non amplification techniques at the same wireless link distance. While, as the wireless link distance increases, the transmission data rate decreases at the same transmitted radio frequency. Moreover with amplification

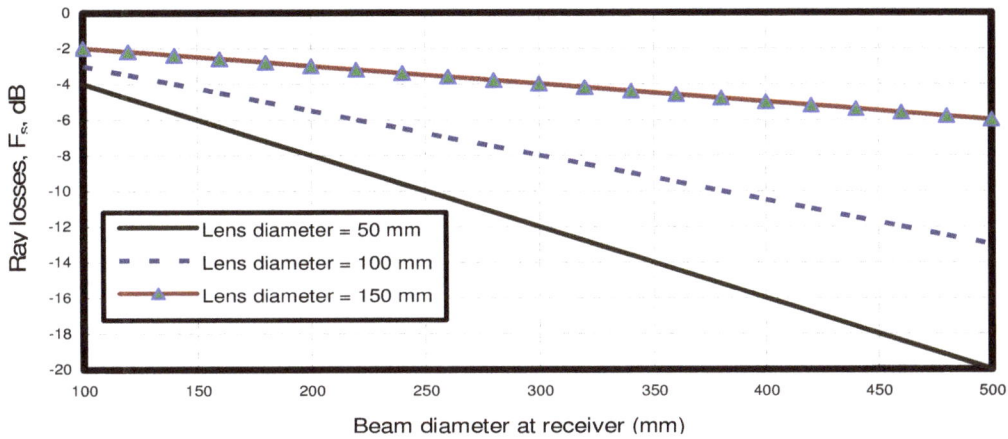

**Figure 3.** Variations of the ray losses with beam diameter at receiver for different lens diameter at the assumed set of parameters.

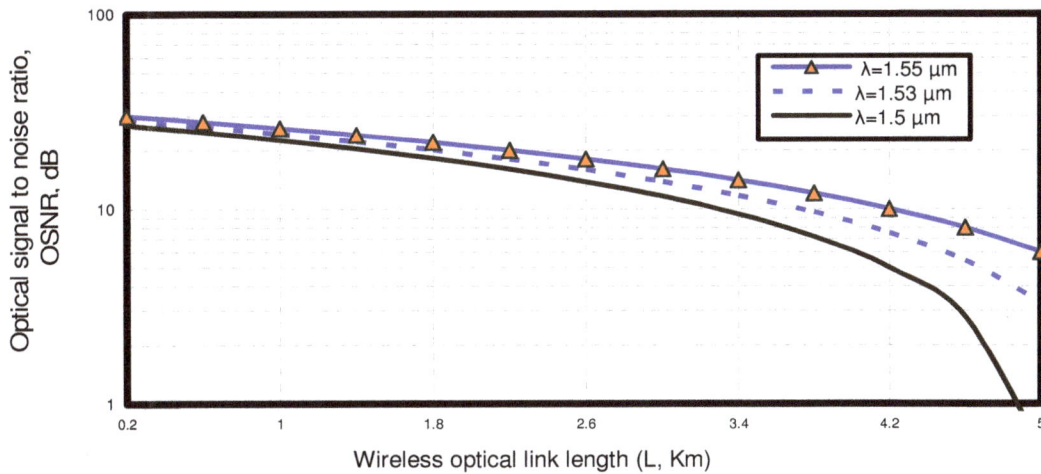

**Figure 4.** Variations of optical signal to noise ratio with wireless optical link distance at the assumed set of parameters.

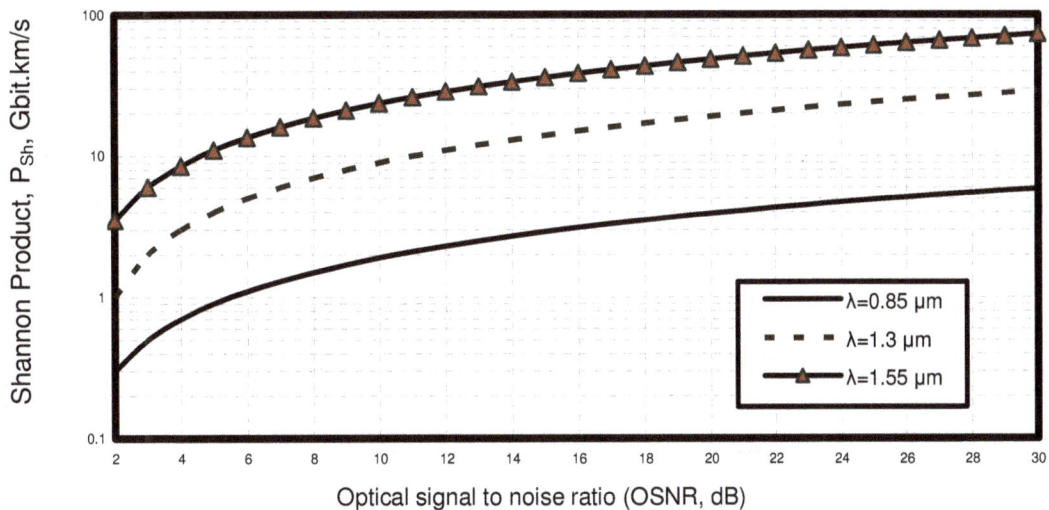

**Figure 5.** Variations of Shannon bit rate-distance product against optical signal to noise ratio at the assumed set of parameters.

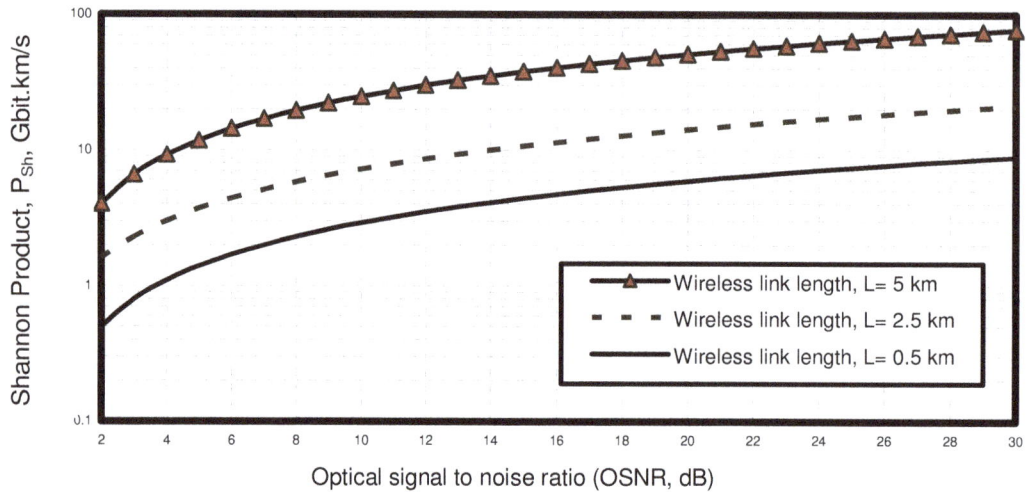

**Figure 6.** Variations of Shannon bit rate-distance product against optical signal to noise ratio at the assumed set of parameters.

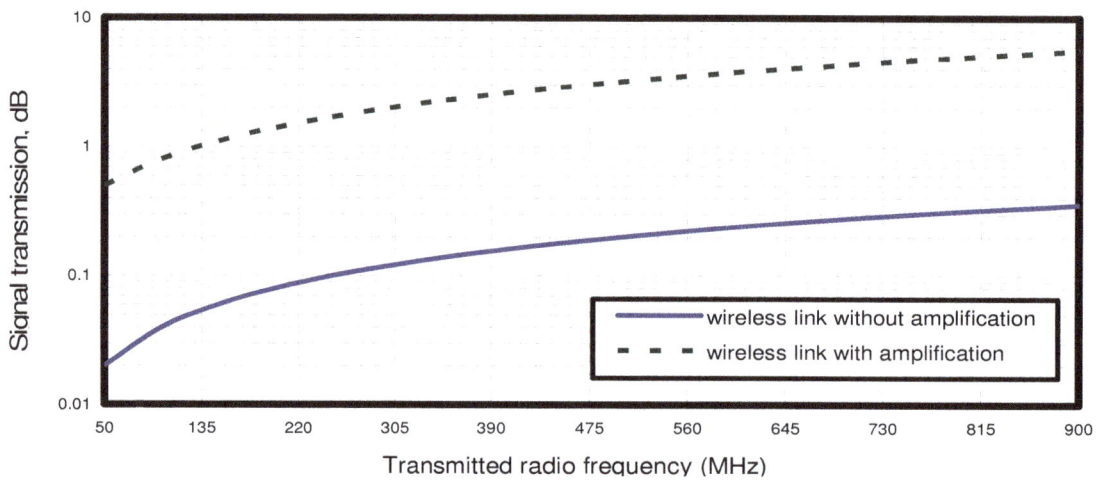

**Figure 7.** Variations of wireless transmission with transmitted radio frequency at the assumed set of parameters.

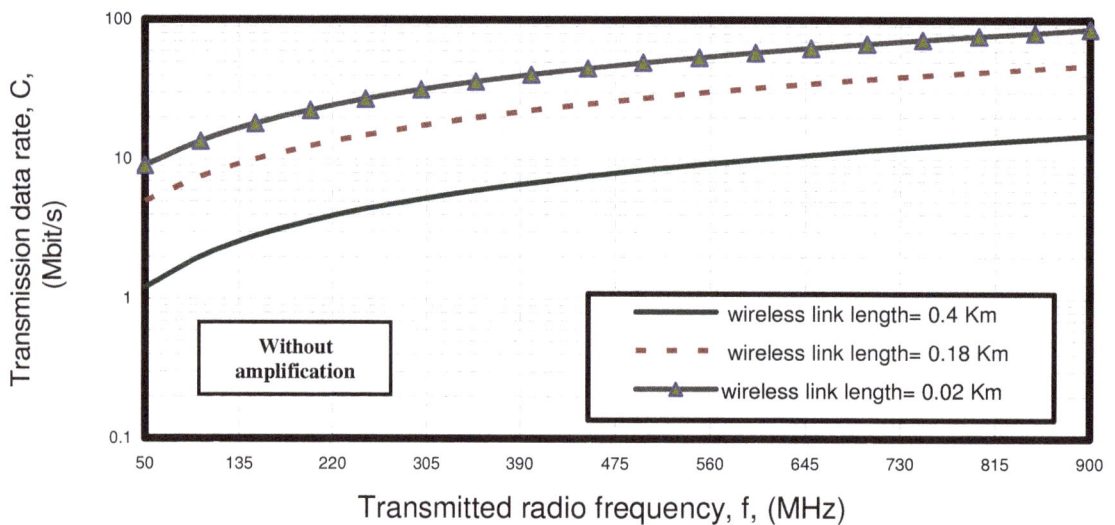

**Figure 8.** Variations of transmission data rate with transmitted radio frequency at the assumed set of parameters.

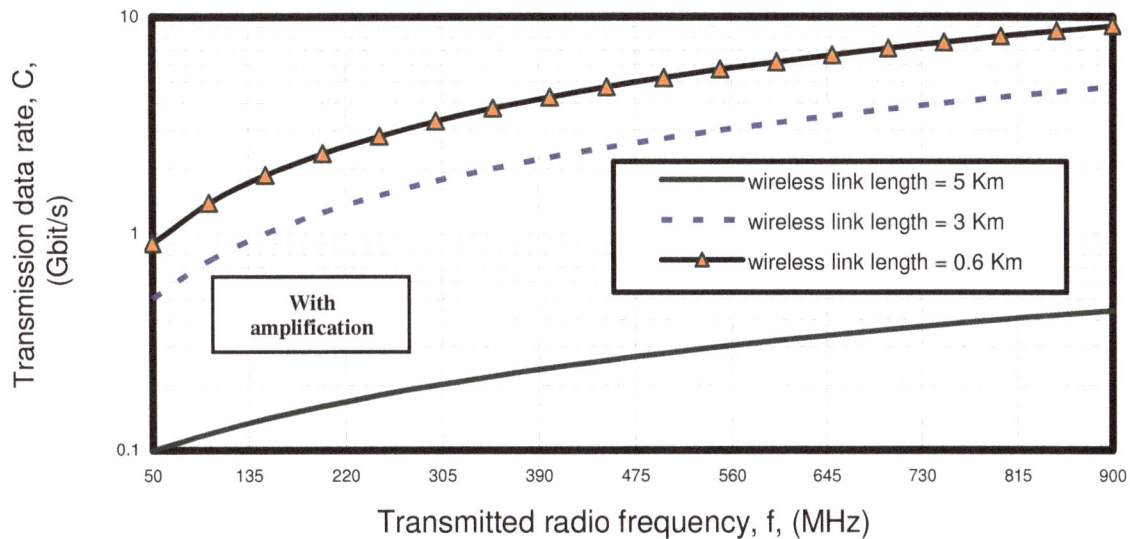

**Figure 9.** Variations of transmission data rate with transmitted radio frequency at the assumed set of parameters.

techniques offered both high transmission link diatance and transmission data rate.

## Conclusions

We have investigated the high transmission bit rate of multi giga bit per second for short range optical wireless access communication networks over wide range of the affecting operating parameters. We have demonstrated that the larger the optical signal wavelength, the higher the transmission distance for both wireless optical links. Moreover, we have demonstrated that with amplification techniques, which added additional costs to the wireless system, the wireless optical link offered both high transmission distances and transmission data rate. It is observed that the increased of optical signal to noise ratio, operating signal wavelength, and wireless link length for short range, the increased Shannon bit rate-distance product.

## REFERENCES

Abd El-Naser AM, El-Halawany MME, Rashed ANZ, Eid MM (2009). Recent Applications of Optical Parametric Amplifiers in Hybrid WDM/TDM Local Area Optical Networks." IJCSIS Int. J. Comput. Sci. Inf. Secur., 3(1): 14-24.

Abd El-Naser AM, Abd El-Fattah AS, Ahmed NZR, Mahomud ME (2009). Characteristics of Multi-Pumped Raman Amplifiers in Dense Wavelength Division Multiplexing (DWDM) Optical Access Networks." IJCSNS Int. J. Comput. Sci. Netw. Secur., 9(2): 277-284.

Abd El-Naser AM, El-Halawany MME, Ahmed NZR, El-Nabawy AM (2009). Transmission Performance Analysis of Digital Wire and Wireless Optical Links in Local and Wide Areas Optical Networks." IJCSIS Int. J. Comput. Sci. Inf. Secur., 3(1): 106-115.

Shea DP, Mitchell JE (2009). Architecture to Integrate Multiple Passive Optical Networks (PONs) with Long Reach DWDM Backhaul." IEEE J. Sel. Areas Commun., 27(2): 126-133.

Yong-Yuk W, Hyuk-Choon K, Moon-Ki H, Sang-Kook H (2009). "1.25-Gb/s Wire line and Wireless Data Transmission in Wavelength Reusing WDM Passive Optical Networks." Microw. Opt. Technol. Lett., 51(3): 627-629.

Ab-Rahman MS, Guna H, Harun MH, Jumari K (2009). Cost-Effective Fabrication of Self-Made 1x12 Polymer Optical Fiber-Based Optical Splitters for Automotive Application." Am. J. Eng. Appl. Sci., 2(2): 252-259.

Kedar D, Arnon S (2006). Urban Optical Wireless Communication Network: The Main Challenges and Possible Solutions." IEEE Comm. Mag., 42(3): 3-8.

Schuster J, Willebrand H, Bloom S, Korevaar E (2005). Understanding the Performance of Free Space Optics." J. Opt. Netw., 3(2): 34-45.

Kiasaleh K (2005). Performance of APD-Based, PPM Free-Space Optical Communication Systems in Atmospheric Turbulence." IEEE Trans. Commun., 53(2): 1455-1461.

Mietzer J, Hoeher P (2007). Boosting the Performance of Wireless Communication ystems: Theory and Practice of Multiple Antenna Technologies." IEEE Commun. Mag., 42(3): 40-46.

Ackerman EI, Betts GE, Burns WK, Campbell JC, Cox CH, Duan N, Prince JL, Regan MD, Roussell HV (2008). Signal-to-Noise Performance of Two Photonic links using Different Noise Reduction Techniques." IEEE Int. Microw. Sympos., pp. 51–56.

Kamalakis T, Tsipouras A, Pantazis S (2008). Hybrid Free Space Optical/Millimeter Wave Outdoor Links for Broadband Wireless Access Networks," the 18th Annual IEEE International Symposium on Personal, Indoor and Mobile Radio Communications (PIMRC), pp. 51-57.

Suman S, Prince JL (2007). Hybrid Wireless-Optical Broadband-Access Network (WOBAN): A Review of Relevant Challenges." IEEE J. Lightw. Technol., 25(11): 3329-3340.

Suman S, Betts GE (2008). Hybrid Wireless-Optical Broadband-Access Network (WOBAN): Network Planning and Setup." IEEE J. Lightw. Technol., 26(6): 12-21.

Jihui Z, Campbell JC (2005). "Joint routing and scheduling in multi-radio multi-channel multi-hop wireless networks." Broadonets, 1: 631-640.

Ramachandran KN, Duan N (2006). "Interference-Aware Channel Assignment in Multi-Radio Wireless Mesh Networks. IEEE INFOCOM, pp. 1-12.

# 13

# Design and development of fuzzy logic controller to control the speed of permanent magnet synchronous motor

## Lini Mathew[1] and Vivek Kumar Pandey[2]*

[1]Electrical Engineering Department, National Institute of Technical Teachers Training and Research Sector-26, Chandigarh, India.
[2]Electrical Engineering Department, Bharat Institute of Technology By-Pass Road, Partapur, Meerut, U.P., India.

The paper presents a fuzzy logic controller (FLC) for permanent magnet synchronous motor (PMSM). The fuzzy logic controller is used for speed control of this type of motor. The dynamic response of (PMSM) with the proposed controller is studied under different load disturbances. The effectiveness of the proposed fuzzy logic controller is compared with that of the conventional PI and PID controllers. The proposed controller is used in order to overcome the nonlinearity problem of PMSM and also to achieve faster settling response.

Key words: Speed control, plus integral controller, fuzzy logic controller, permanent magnet synchronous motor.

## INTRODUCTION

Traditionally commutator motors, also known as direct current (dc) motors were preferred for variable speed drives while induction motors were used for constant speed applications. Advances in solid state devices helped in development of suitable controllers possessing provision of vector control. Such controllers made it possible to incorporate in the induction motor almost all the characteristics of a dc motor. In vector control scheme, torque and flux are decoupled from each other like in dc motors (Benchouia et al., 2004). Industry automation is mainly developed around motion control systems in which controlled electric motors play as a heart of the system a crucial role. The high performance motor control systems thus, contribute to a great extent, to the desirable performance of automated manufacturing sector by enhancing the production rate and the quality of products. In fact the performance of modern automated systems, defined in terms of swiftness, accuracy, smoothness and efficiency, mainly depends on the motor control strategies (Sung et al., 2004). The advancement of control theories, power electronics and micro-electronics in connection with new motor designs and materials has contributed largely to the field of electric motor control for high performance systems. Newly developed permanent magnet synchronous motors (PMSM) with high energy permanent magnet materials particularly provide fast dynamics, efficient operation and very good compatibility with the applications but only if they are controlled properly (Nour et al., 2006). However, the ac motor control including control of PMS motors is a challenging task due to very fast motor dynamics and highly non-linear models of the machines. Therefore, a major part of motor control development consists of deriving mathematical models in suitable forms.

The dynamic models of the motors can be presented in different reference frames to lay down a basis for the motor control design. The mathematical formulations and the equivalent circuit models can be provided to help in better controller design for PMSM drives. There are two competing control strategies for ac motors viz vector control (VC) and direct torque control (DTC) for PMSM

*Corresponding author. E-mail: vikku1781@gmail.com.

(Rahideh et al., 2007). Vector control scheme with several benefits is the most applied control strategy. The decoupling of torque control and flux linkage control are the basis of vector control technique. The motor phasor diagrams can provide better understanding of different control schemes used for PMSM drives. It is the most straightforward approach for motor modelling and control. It also reduces the analytical burden (Cao and Fan, 2009). Even though equipped with modern solid state devices and suitable controller, the vector controlled induction motor drives are still not able to cope up with the more stringent requirements of loads used in some high performance applications. Pump, fan and compressor drive motors in industrial applications that operate large percentage of the time necessitate highly efficient and reliable service.

Also in aerospace applications as well as in robotics, superior power density ratio and reduction in size become prime requirements (Wang et al., 2007; Cao et al., 2008; Kamel et al., 2009). Moreover in some industries, the presence of dust particles are extremely damaging to the brushes and commutator of the dc motor. These requirements paved the way for the evolution of permanent magnet brushless motors. These motors use permanent magnet to produce the air gap magnetic field rather than field coils. These motors are showing increasing popularity in recent years for industrial drive applications (Wang et al., 2008; Cao and Fan, 2008).

The momentum of this popularity will considerably increase in the near future due to the recent availability of the high-energy low-cost neodymium-iron-boron (NdFeB) permanent magnet (Kung et al., 2009).

## MODEL OF THE PMSM

The model of PMSM without damper winding has been developed on rotor reference frame using the following assumptions:

1) Saturation is neglected.
2) The induced EMF is sinusoidal.
3) Eddy currents and hysteresis losses are negligible.
4) There are no field current dynamics.

Voltage equations are given by (Wang and Liu, 2009; Song and Peng, 2009; Sant and Rajagopal, 2009):

$$V_q = R_s i_q + \omega_r \lambda_d + \rho \lambda_q \quad\quad\quad (1)$$

$$V_d = R_s i_d - \omega_r \lambda_q + \rho \lambda_d \quad\quad\quad (2)$$

Flux linkages are given by:

$$\lambda_q = L_q i_q \quad\quad\quad\quad\quad (3)$$

$$\lambda_d = L_d i_d + \lambda_f \quad\quad\quad\quad (4)$$

Substituting Equations 3 and 4 into 1 and 2:

$$V_q = R_s i_q + \omega_r (L_d i_d + \lambda_f) + \rho L_q i_q \quad\quad (5)$$

$$V_d = R_s i_d - \omega_r L_q i_q + \rho (L_d i_d + \lambda_f) \quad\quad (6)$$

Arranging Equations 5 and 6 in matrix form:

$$\begin{bmatrix} V_q \\ V_d \end{bmatrix} = \begin{bmatrix} R_s + \rho L_q & \omega_r L_d \\ -\omega_r L_q & R_s + \rho L_d \end{bmatrix} + \begin{bmatrix} \omega_r \lambda_f \\ \rho \lambda_f \end{bmatrix} \quad (7)$$

The developed torque (Liu et al., 2009; Febin and Subbiah, 2009; Yang and Wang, 2009) motor is being given by:

$$T_e = 3/2 \ (P/2) \ (\lambda_d i_q - \lambda_q i_d) \quad\quad\quad (8)$$

The mechanical Torque equation is:

$$T_e = T_L + B \omega_m + J \frac{d\omega_m}{dt} \quad\quad\quad (9)$$

Solving for the rotor mechanical speed form Equation 9:

$$\omega_m = \left( \int \frac{T_e - T_L + B \omega_m}{J} \right) dt \quad\quad (10)$$

And

$$\omega_m = \omega_r \left( \frac{2}{P} \right) \quad\quad\quad\quad (11)$$

In the aforementioned equations $\omega_r$ is the rotor electrical speed where as $\omega_m$ is the rotor mechanical speed.

## SPEED CONTROL OF PMSM WITH THE PROPOSED CONTROLLER

For the speed control of PMSM, many controllers are used. In conventional P, PI and PID controllers, very fine tuning is required which cannot cope up with system's parameter variations. Also the performance of such controllers is affected due to variations in physical parameters like temperature, noise, saturation etc. Many control systems use adaptive controllers for PMSM, which can track only linear systems. Therefore, fuzzy logic based controller may be used to achieve more accurate and faster solutions and to handle complicated non-linear characteristics.

Figure 1 shows the block diagram of the proposed control system. A simple structure fuzzy logic controller (FLC) is used in the speed control loop to regulate the motor speed. The inputs to the FLC are the speed error (e) and the change of speed ($\Delta$e).

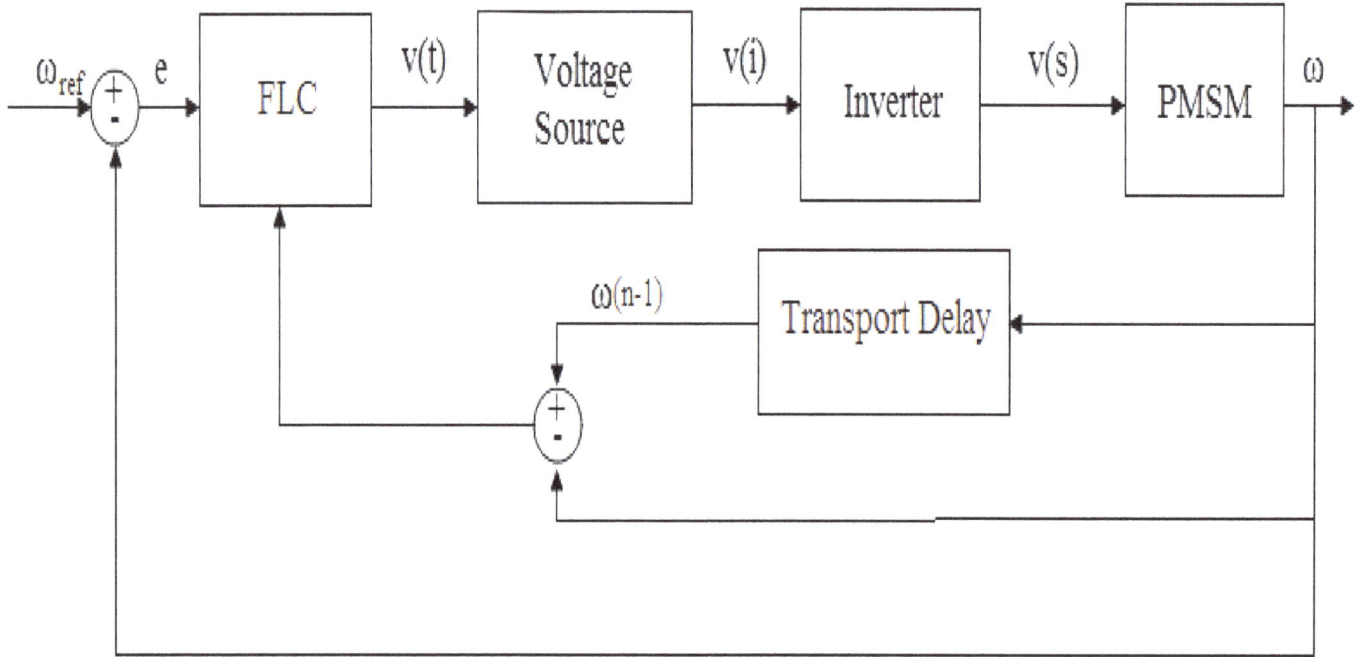

**Figure 1.** Speed control using FLC for PMSM.

The motor speed error (e) may be given by the following equation:

$$e(t) = \omega_{ref} - \omega(t) \qquad \text{...................} \qquad (12)$$

The change of error (Δe) may be described by the following equation:

$$\Delta e = e(t) - e(t\text{-}1) \qquad \text{...................} \qquad (13)$$

Where $e(t\text{-}1) = \omega_{ref} - \omega(t\text{-}1)$ and $\Delta e = \omega(t\text{-}1) - \omega(t)$.

### The conventional PI controller

The proportional plus integral (PI) controller is one of the famous controllers used in a wide range in the industrial applications. The output of the PI controller in time domain is defined by the following equation:

$$v_c(t) = k_p e(t) + k_c \int_0^t e(t)\, dt \qquad \text{.................} \qquad (14)$$

Where $v_c(t)$ is the output of the PI controller, $k_p$ is the proportional gain, $k_i$ is the integral gain, and e(t) is the instantaneous error signal. The main advantage of adding the integral part to the proportional controller is to eliminate the steady state error in the controller variable. However, the integral controller has the serious drawback of getting saturated after a while if the error does not

change its direction.

This phenomenon can be avoided by introducing a limiter to the integral part of the controller before adding its output to the output of the proportional controller. The input to the PI is the speed error (e), while the output of the PI is used as the input of controlled voltage source inverter. And finally the controlled voltage obtained from inverter is fed to the motor for controlling its speed. The dynamic response of the PMSM driven by the PI controller is shown in Figure 2.

### PID controller

In PID controller scheme only change is the replacement of PI controller block with PID controller block. The dynamic response of the PMSM driven by the PID controller is shown in Figure 3.

### The fuzzy logic controller (FLC)

In FLC scheme the output of the FLC is used as the input of the controlled voltage source which converts the input signal into an equivalent voltage in order to regulate the motor speed. All membership functions are iteratively adjusted and the result of the FLC corresponds to the minimum training error. The resultant MAMDANI-type FIS has only 21 rules which was found to provide sufficient accuracy after optimization. The membership functions of

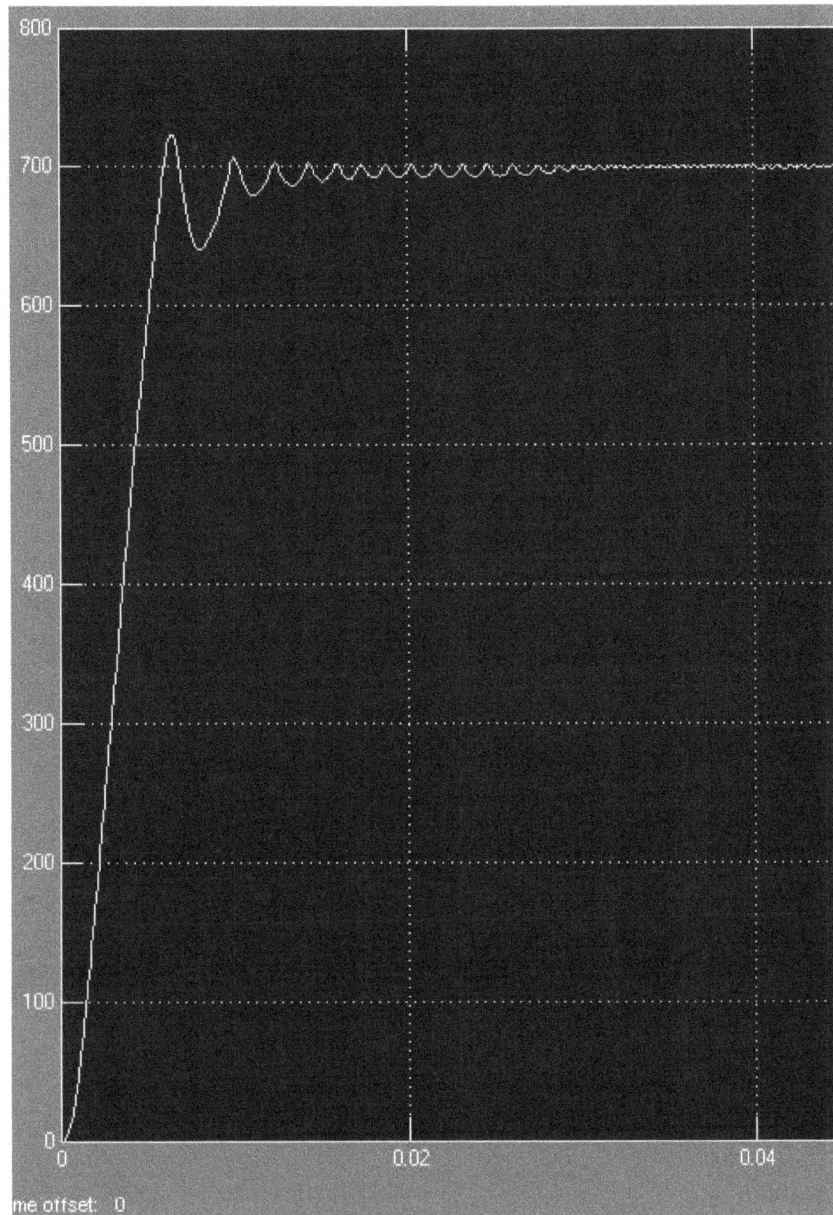

**Figure 2.** The dynamic response of the drive system using a PI controller.

the inputs are shown in Figures 4, 5 and 6. The dynamic response of the PMSM driven by the Fuzzy Logic Controller is shown in Figure 7.

## PERFORMANCE EVALUATION

### For PI controller

The PMSM starts in closed-loop because speed and current control are in cascade. The load torque applied to the machine's shaft is set to its nominal value. Two control loops are used: the inner loop regulates the motor's stator currents and the outer loop controls the motor's speed. Using a PWM inverter, a noise is observed in the electromagnetic torque waveform. However, the motor's inertia prevents this noise from appearing in the motor's speed waveform (Figure 2). The stator currents are quite "noisy," which is to be expected when using PWM inverters. The rotor speed increases fast to its synchronous value after few oscillations and preserves its value (Figure 2). The current takes initially a high value in order to develop the kinetic energy to accelerate the rotor. After a certain delay the current

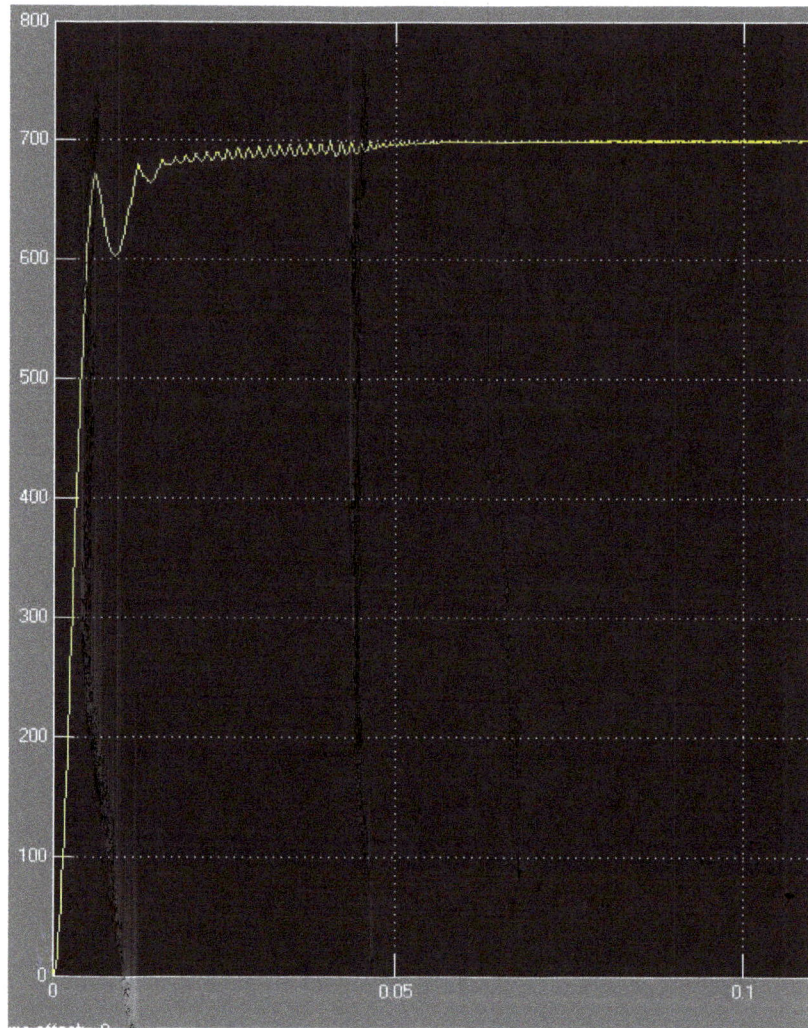

**Figure 3.** The dynamic response of the drive system using a PID controller.

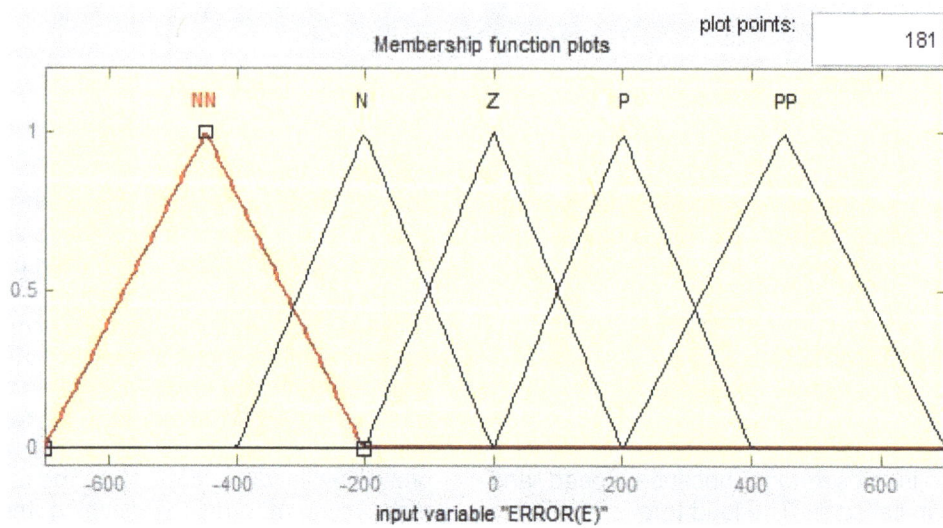

**Figure 4.** Error (E).

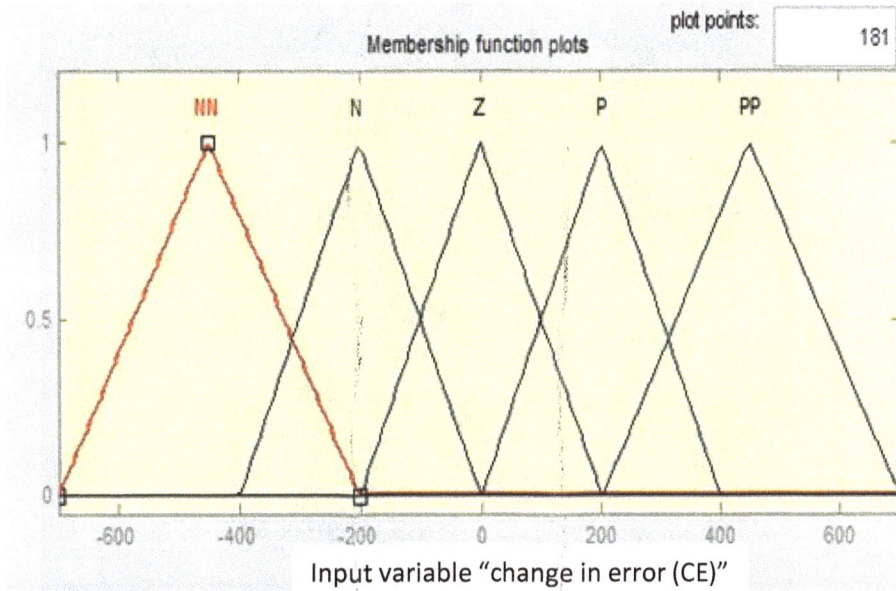

**Figure 5.** Change in error (CE).

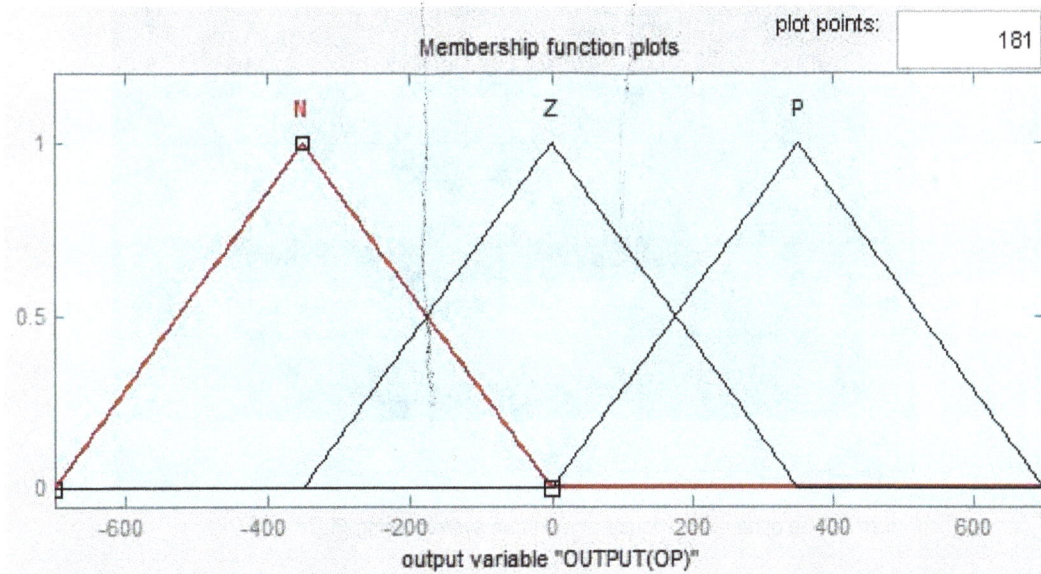

**Figure 6.** Output (O).

stabilizes to their nominal value. The reference speed of the motor is set at 700 rad/s. As soon as the speed reaches the reference speed the PI speed controller forces the speed to remain steady at reference speed. During starting period torque climbs to maximum capability of the motor after that it settles down to steady state value of 3 N-M. Also the current rises to its maximum value and after that comes back to nearly 2 A. A detailed Simulink model for a PMSM drive system with field oriented control has been developed and operation below and above rated speed has been studied using one current control scheme. Simulink has been chosen from several simulation tools because of its flexibility in working with analog and digital devices. Mathematical models can be easily incorporated in the simulation and the presence of numerous tool boxes and support guides simplify the simulation of large system as compared to other simulation tools. Simulink is capable of showing real time results with reduced simulation time and debugging. In the present simulation measurement of

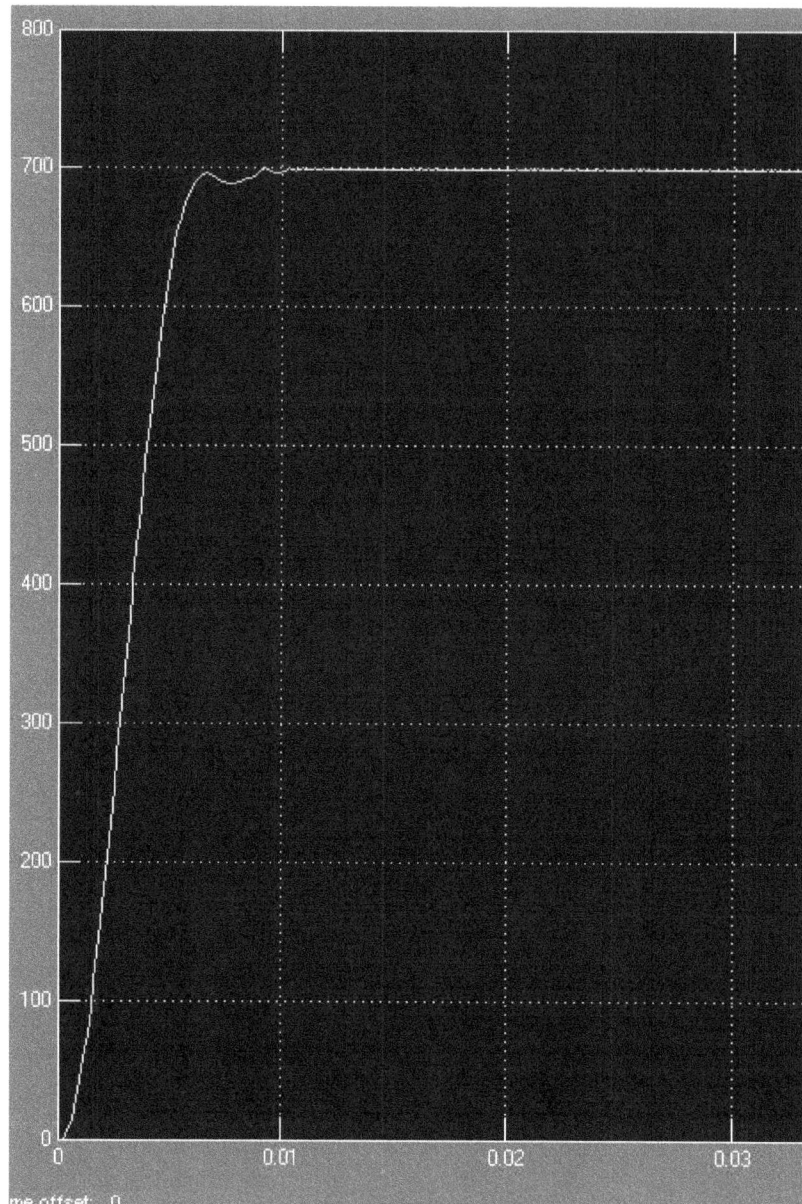

**Figure 7.** The dynamic response of the drive system using FLC.

currents and voltages in each part of the system is possible, thus permitting the calculation of instantaneous or average losses, efficiency of the drive system and total harmonic distortion. Usually in such a drive system the inverter is driven either by hysteresis or by PWM current controllers.

## For PID controller

In this control scheme, proportional control ($K_p$) is used to improve the rise time, derivative control ($K_d$) to improve the overshoot, an integral control ($K_i$) to eliminate the

steady-state error. The stator currents are here less "noisy", as compared to PI speed controller, which is to be expected when using PWM inverters. The rotor speed increases fast to its synchronous value after a few oscillations and preserves its desired value (Figure 3). The current takes initially a high value in order to develop the kinetic energy to accelerate the rotor. After a delay, the current stabilizes to their nominal value. The present model takes into consideration the movement equations in the PMSM model and the operating principle of the PWM voltage converter associated with the current controller. The reference speed of the motor is set at 700 rad/s. As soon as the speed reaches the reference speed the PID speed controller forces the speed to remain steady at reference speed. During starting period torque climbs to maximum capability of the motor after that it settles down to steady state value of 3 N-M. Also the current rises to its maximum value and after that comes back to nearly 2 A.

## For fuzzy logic controller

The FLC has exhibited high performance in tracking the speed reference. The stator current, flux and torque response of PMSM with fuzzy controller is faster than PI and PID controllers during start up and during a step change in torque. At the same time, dynamic behaviour of PMSM with fuzzy logic controller is more stable than PI and PID controllers. A performance comparison between the fuzzy logic controller and conventional PI and PID controllers has been carried out by several simulations confirming the superiority of the Fuzzy logic controller.

## Comparison of different types of speed controllers

In this work, three different types of speed controllers namely, PI, PID and FLC are used to investigate the dynamic behaviour of the PMSM drive. On the basis of simulated results, following important observations are made which are listed as:

1. Among all the speed controllers considered in this study, PI takes maximum starting time. Though PID controller takes nearly 39 ms to reach a speed of 698.79 rad/s but finally settles for 698.54 in nearly 55 ms. Fuzzy takes nearly 7.96 ms to reach 692.42 rad/s and settles for 699.75 in nearly 11 ms.
2. In case of PI controllers, there is some overshoot present in the speed response while FLC has very small steady state error.
3. The spikes during transient period in speed response are more with PI while less in PID and minimum in FLC. Also in FLC, minimum oscillations are observed.
4. The speed recovers very quickly when Fuzzy logic controller is used during the operation.

## CONCLUSIONS

The overall idea of this work is to compare the performance of different speed and current controllers for PMSM drive. For this purpose, the motor drive system simulation model was developed and verified through the SIMULINK toolbox of the MATLAB software package. For current controller, PWM is examined and for speed controller, performance of PI, PID and Fuzzy controllers are compared. The simulated results confirmed the viability of the model used in this work and it has been

shown that the model is suitable for transient as well as steady state condition. These results also confirmed that the transient torque and current never exceed the maximum permissible value. Among all the speed controllers discussed, Fuzzy logic controller makes the system robust as there is no speed overshoot and also minimum pulsation in torque and current are observed.

**REFERENCES**

Benchouia MT, Zouzou SE, Golea A, Ghamri A (2004). Modeling and Simulation of Variable Speed Drive System with Adaptive Fuzzy Controller Application to PMSM", IEEE International Conference on Industrial Technology (ICIT), pp. 683 – 687.

Cao X, Fan L (2008). Vector Controlled Permanent Magnet Synchronous Motor Drive Based on Neural Network and Multi Fuzzy Controllers, IEEE Fifth International Conference on Fuzzy Systems and Knowledge Discovery, pp. 254 – 258.

Cao X, Fan L (2009). A Novel Flux-weakening Control Scheme Based on the Fuzzy Logic of PMSM Drive", IEEE International Conference on Mechatronics and Automation, pp. 1228 – 1232.

Cao X, Fan L, Huang J (2008). An Implementation of Neural Network and Multi-Fuzzy Controller for Permanent Magnet Synchronous Motor Direct Torque Controlled Drive", IEEE International Conference on Automation and Logistics, pp. 498 – 503.

Ding C, Wang J, Wen Q, Xiao H (2008). Application of Fuzzy Sliding Mode Controller in Permanent Magnet Synchronous Motor", IEEE 7th World Congress on Intelligent Control and Automation, pp. 4727 – 4730.

Febin DJL, Subbiah V (2009). A Novel Fuzzy Logic Based Robust Speed Controller for Permanent Magnet Synchronous Motor Servo Drive", IEEE International Conference on Region 10, pp 1 – 4,

Kamel HM, Ibrahim HEA, Hasanien HM (2009). Speed Control of Permanent Magnet Synchronous Motor Using Fuzzy Logic Controller", IEEE Int. Conference Electric Machines Drives Syst., pp. 1587 – 1591.

Kung YS, Wang MS, Huang CC (2009). DSP-based Adaptive Fuzzy Control for a Sensorless PMSM Drive", IEEE International Conference on Control and Decision, pp. 2379 – 2384.

Liu YP, Wan JR, Yuan CH, Li GY (2009). Fuzzy Based Direct Torque Control of PMSM Drive Using An Extended Kalman Filter", IEEE Eighth Int. Conference Machine Learning Cybernetics, pp. 647 – 651.

Nour M, Aris I, Mariun N, Mahmoud S (2006). Hybrid Model Reference Adaptive Speed Control for Vector Controlled Permanent Magnet Synchronous Motor Drive" IEEE Power Electronics Drive Syst., pp. 618 – 623.

Rahideh A, Rahideh A, Karimi M, Shakeri A, Azadi M (2007). High Performance Direct Torque Control of a PMSM using Fuzzy Logic and Genetic Algorithm", IEEE Int. ConferenceElectric Machines Drives Syst., pp. 932 – 937.

Sant AV, Rajagopal KR (2009). PM Synchronous Motor Speed Control Using Hybrid Fuzzy-PI With Novel Switching Functions", IEEE Trans. On Magnetics, 45(10): 4673 – 4675.

Song L, Peng J (2009). The Study of Fuzzy – PI Controller of Permanent Magnet Synchronous Motor", IEEE International Conference on Power Electronics Motion Control, pp. 1863 – 1866,

Sung Yu J, Mo Hwang S, Yuen Won C (2004). Performances of Fuzzy – Logic – Based Vector Control for Permanent Magnet Synchronous Motor Used in Elevator Drive System", The 30th Annual Conference IEEE Ind. Electronics Society, pp. 2679 – 2683.

Wang J, Liu H (2009). Novel Intelligent Sensorless Control of Permanent Magnet Synchronous Motor Drive", IEEE Ninth Int. Conference Electronic Measure. Instruments, pp. 953 – 958.

Wang J, Wang HH, Yuan XL, Lu TH (2008). Novel Intelligent Direct Torque Control for Permanent Magnet Synchronous Motor Drive", IEEE Fifth Int. Conference on Fuzzy Syst. Knowledge Discovery, pp. 226 – 203.

Wang L, Tian M, Gao Y (2007). Fuzzy Self-adapting PID Control of

PMSM Servo System", IEEE Int. Conference on Electric Machines and Drives Syst., pp. 860 – 863.

Yang M, Wang X (2009). Fuzzy PID Controller Using Adaptive Weighted PSO for Permanent Magnet Synchronous Motor Drives" IEEE Second Int. Conference on Intelligent Comput. Technol. Automation, pp. 736 – 739.

**APPENDIX**

The specifications of the permanent magnet synchronous motor are listed in Table 1.

**Table 1.** The motor specifications.

| | |
|---|---|
| No of phases | Three |
| Rated Torque | 3 N-M |
| Rated Speed | 3000 RPM |
| The Stator Phase Resistance | 2.8750 ohm |
| The Stator Phase Inductance | 8.5 mH |
| The Motor Inertia | 0.00008 kg-m$^2$ |

# 14

# System reconfiguration and service restoration of primary distribution systems augmented by capacitors using a new level-wise approach

T. Ananthapadmanabha[1]*, R. Prakash[2], Manoj Kumar Pujar[3], Anjani Gangadhara[1] and M. Gangadhara[4]

[1]The National Institute of Engineering, Mysore, India.
[2]HMS IT, Tumkur, India.
[3]KPTCL, Bangalore, India.
[4]NIE, Mysore, India

Ensuring continous, economical and quality power supply to customers is the primary objective of suppliers. This implies quick restoration of supply in case of faults, to maximum possible load in minimum time with minimum loss. Power quality and network loss can be improved by providing reactive power compensation using capacitors. This paper proposes a new method for system reconfiguration and fault restoration of a distribution network. The approach is applicable to both single and multiple fault cases. The hierarchy of the methodology is downward from the feeder to the bus level. Additionally, an iterative method is proposed for evaluating capacitor size and location by calculating the loss sensitivity index (LSI) of the buses. Compensation is provided to the initial network. Faults are then simulated and the effect of the existing compensation after fault restoration using the proposed method is studied. Later, a new compensation scheme is derived by altering the original scheme. Suitable compensation solution can be decided based on either minimum loss or cost restraints. The proposed method is applied to a 44-bus distribution network of R. K. Nagar, KPTCL, Karnataka, India. Comparison of the network parameters and results after fault restoration without compensation, with initial compensation and after applying the new compensation scheme shows the method is very effective in responding to faults, in practical time periods.

**Key words:** Single fault, multiple faults, feeder fault, network line fault, reconfiguration, supply restoration, load shedding, capacitor placement.

## INTRODUCTION

The distribution network must be optimally configured to incur minimum annual cost (capital investment + overheads + running cost) while satisfying all the requirements and constraints like 1) Radial configuration, 2) All loads must served, 3) Lines, transformers other equipment should operate within current capacity limits, 4) Overcurrent protective devices must be coordinated, 5) Voltage magnitudes must be within limits. Reconfiguration means re-arranging the load, subject to availability of physical infrastructure on the ground in order to meet the above mentioned requirements. Consumer demands vary with time of day, day of the week, and season, therefore, feeder reconfiguration also enables load balancing transfers between regions. Furthermore, online configuration quickens management and distribution automation when remote-controlled switches are employed. Faults cause power outages to loads of the network. They affect the network in mainly two ways. One is feeder fault, where the feeder breaker trips causing all load supplied by the feeder to blackout. Another is network line fault causing power outage to only a portion of total feeder load. These faults can either occur at single or at multiple locations that is, single and multiple faults respectively. Multiple faults are more severe and more probable. Restoration

*Corresponding author. E-mail: drapn2008@yahoo.co.in.

means restoring the supply back to maximum possible loads in the affected area with acceptable voltage and balanced distribution of loads as best as possible, in minimum time with minimum loss.

Reactive (capacitive) power injection/compensation into the network improves the power factor and voltage of the load buses / nodes by nullifying the load point's inductive power demand, therefore decreasing the network loss and increasing feeder spare capacity. Hence, it helps reconfiguration and restoration procedure give better results. As placing capacitors at each bus is expensive, their effective locations and sizes have to be determined such that there is substantial loss recovery at minimum investment. Distribution networks are usually radial for simplifying overcurrent protection. Network reconfiguration/restoration depends on existing supply routes, source substations and load locations. Most feeders have several interconnecting tie switches to neighbouring feeders. To restore power to customers following a fault, configuration must be altered by changing the status of network switches (open/close), in such a way that radial nature, voltage and current limits are always maintained. Re-evaluating compensation and protective schemes of the new configuration is then necessary.

Since a typical distribution system has hundreds of switches, combinatorial analysis of all possi-ble options is not practical. Choosing the appropriate solution becomes a problem in itself. The radiality constraint and the discrete nature of switch values limit the use of classical optimization techniques to solve the reconfiguration/ restoration problem. Also approaching the exact required values of all result variables, that is network loss, voltage profile and line current is not possible. While one variable may be approaching the required value, another may be driven away from its value. Compromises will, therefore, have to be made using some criteria. In order to save time and effort, heuristic search techniques using knowledge-based engines are used instead of analytical techniques involving complex equations and boundary conditions.

Recent papers use expert system approach like fuzzy logic (Seong-II et al., 2006) and techniques like Genetic Algorithm, Ant Colony System Algorithm (Gómez et al., 2004) and Simulated Annealing technique (Young-Jae et al., 2002) for finding optimal network reconfiguration. Seong-II et al. (2006) proposed a service restoration methodology for multiple faults by classifying feeders as simple and compound interconnected feeders depending on the number of outage areas a feeder is associated to. Young-Hyun et al. (2000) proposed a left child / right sibling tree structured database using non-directional data for speeding up the tracing algorithm. Dash et al. (1991) used Artificial Neural Networks to determine an optimal compensation solution. Srinivasa and Narasimham (2008) used the Plant Growth Simulation Algorithm by defining and using a "loss sensitivity" factor of the buses in determining optimal capacitor locations and sizes. The above mentioned techniques and algorithms have also been applied in solving the compensation problem (Branko and Milos, 2004; Chung-Fu, 2008).

This paper proposes a new mechanism for system reconfiguration during normal operation and service restoration after single or multiple fault occurrences. A methodology for capacitor placement is also included and its effect on reconfiguration/restoration is studied. The compensation solution is again varied to optimize the system. These methods are iterative approaches and all the requirements and constraints mentioned above are complied with to obtain best possible practical results. The new approach reduces unnecessary load shedding, balances load on different feeders, reduces network loss to minimum level, and improves spare capacity availa-bility on feeders, if sufficient infrastructure in terms of switches, branches and parallel supply routes are available.

This methodology and programming is applicable to any network. Large data collection is necessary. That is, number of feeders, branches and nodes in each feeder, bus and line data, generator and feeder capacities, number, location and status of switches etc. Identification and rectification of faults is not included in this paper. Since the network load keeps changing with time, it is assumed here that the load is constant till the entire proposed method is run and the network changes are implemented. Faulty feeders and lines are assumed to be priorly known and are inputs to this approach. The mutual coupling between conductors is neglected and all the phases are assumed to be balanced. The possible values for compensation and its variation are in discrete steps. Stepped variable compensation is done through installation of on-off capacitor banks. The approach is practical and close to actual practice. It uses simple terminology. Extensive mathematics is avoided, other than what is needed in MATLAB, hence is easy to understand and implement. The results give clear and simple operator instructions for implementation with sufficient time to spare. Any contingency can be tackled with ease.

## PROPOSED METHODOLOGY

### The new level wise approach

Faults are classified, depending on the location as:

1. Feeder fault
2. Network line fault

Depending on the number of faults occurring simultaneously, as:

1. Single fault
2. Multiple faults

Single faults can be either feeder or network line fault. Multiple faults are a combination of both at multiple locations. For convenience, only two faults are considered in this paper but this method can accommodate any number of faults.

## Network plot for analysis

The existing geographically spread network of supply and load is converted into a plot to be used by MATLAB/load flow program. Such a plot will facilitate uni- and bi- directional load re-arrangement. It should be noted that the uni-directional / bi-directional terms have meaning only to the MATLAB program and bears no resemblance to the physical network. Since the network plot represents existing network, it has to be prepared meticulously.

## Data base

Network data describing the network has to be carefully entered in a data file, which will be frequently accessed by the program. The required data are as follows:

1. Busdata (node data) gives the real, reactive load and compensation at each bus.
2. Linedata gives each line's resistance, reactance and the two nodes they connect.
3. Dummy node, dummy feeder and dummy lines data.
4. Feeder data tables give details of loading capacity, number of branches.
5. Tie data tables list all the initial tie lines and the two feeders they connect.
6. Tables of switching sequence for shifting feeder branches by priority.
7. List of all possible standard capacitor values.

## Method 1

The approach is divided into five levels:

1. Feeder shifting
2. Branch shifting
3. Node shifting and shedding
4. Load picking
5. Result display and Operator Instructions

Service restoration of feeder faults starts at the feeder shifting level and system reconfiguration starts at the branch shifting level. Network lines fault restoration starts at the node shifting and shedding level. Load picking level is applicable only if there are any nodes shed in the third level. In the result display level, the feeders are checked for overload, the loss difference is measured and the operator instructions for implementing configuration changes are displayed.

## *Feeder shifting level*

This level is applicable only for feeder faults. When feeder fault occurs, the feeder main breaker trips and the entire load supplied by the feeder are blacked out. The entire load has to be transferred onto a healthy feeder. There should be multiple tie lines for every feeder to facilitate inter-feeder load transfer. The best host feeder is decided based on any one of the criteria listed below after evaluating all possible feeder combinations:

a) Spare capacity of the healthy feeder: The healthy feeder with largest spare capacity is chosen as the host for the blacked out feeder load. After the load transfer, the new total load of the host feeder must be less than its permissible thermal loading to avoid damage due to overheating. The effect of load transfer on the host feeder loading is reduced in the subsequent shifting operations.

b) Minimum loss after each tie line connection: All tie lines connecting the faulty feeder to all the healthy feeders are listed. For each tie line closed, the network loss is determined. Healthy feeder of that tie line connection giving the least network loss is chosen as the host. This method is helpful when there are more than one tie lines from the same healthy feeder. The tie lines closer to the feeder in the radial branches are preferred as the length and the amperage load of the branch through which current has to flow decreases, lessening the I2R losses.

After performing the feeder shift using any of the above criteria, host feeder overloading and minimum voltage of the new configuration is checked. If none of them are beyond specified limits, then the resulting network is considered as "restored after feeder fault", the result display level/operator instructions are implemented and restoration ends in a positive note, else it proceeds with the resulting configuration to the next level that is, branch shifting.

## *Branch shifting level*

Branch shifting transfers only part of the load (contained in the shifting branch) from one feeder to another. This kind of shifting redistributes the network load over all the feeders, making better use of their available spare capacity. Sending feeder is one which has the node with the least voltage in the network and receiving feeders are those which have tie line connections to the sending feeder. Host feeder is the receiving feeder chosen to host the shifted branches. This level gives better results when large number of nodes must be moved. There are two ways of branch shifting. First is uni-directional branch shifting (e.g. left). In a simple network plot, if each feeder has tie line connections with other feeders on either side, the branches of the first feeder can be shifted in both directions, but only left will be chosen (similarly for the right).

Tables listing the switching sequence to be executed for shifting each branch from the sending feeder to a receiving feeder are created for all its branches with priority indicated for all directions. The priority list is arrived at based on the network physical structure and the direction of shifting. It is the creation of these tables that make the method network specific as the switching sequence for each network has to be prepared manually. For the sample network described above, two tables, one for right shifting and another for left shifting has to be created. If an acceptable configuration is not obtained in uni-directional shifting, then the second type that is, bi-directional shifting is attempted. The resulting configuration will be the one yielding the highest minimum voltage. If this type of shifting also fails, the mechanism proceeds to the third level.

a) Uni-directional branch shifting: For the first iteration, the host feeder hosting the minimum unacceptable voltage-node in the resulting configuration obtained from the previous level, becomes the new sending feeder. The receiving feeder which has the largest spare capacity becomes the new host for shifted branches. The direction of shifting is thus determined. Branches from the sending feeder will be shifted in priority order by implementing the switching sequence entered in the shifting table prepared for the corresponding direction. After shifting each branch, load flow is run and the next minimum voltage-node is obtained. The feeder of this node will become the new sending feeder for the next iteration.

b) Bi-directional branch shifting: The initial network and the sending feeder is same as in uni-directional branch shifting for first iteration. In one iteration, a branch from the sending feeder is trial-shifted to all the receiving feeders, one at a time and load flow is run to get the minimum voltage-node of the network. The shift giving higher "minimum voltage of the network" is chosen. The minimum voltage-node of one iteration determines the sending feeder for the next iteration.

For both types of shifting:

1. In a network, the numbers of branches are usually limited to maximum three or four, out of which one branch gets the supply either from the feeder's generator or through a tie line. This branch cannot be shifted as it would cut off supply to the unshifted load of the sending feeder.

2. During service restoration of feeder faults, if the initial faulty feeder itself is the sending feeder, and its supporting healthy feeder itself is to host the shifted branches, then a branch of the supporting host feeder is shifted in the same direction instead of the faulty sending feeder. This is done to avoid disconnection between the faulty and the supporting feeder.

3. A reasonable number of iterations are pre-specified. If an acceptable configuration satisfying the voltage and current constraints is obtained in any iteration within that number, then the restoration process ends and operator instructions are displayed for the resulting configuration. Else with the best configuration of all iterations, method proceeds to the third level that is, node shifting and shedding.

### Node shifting and shedding level

#### Line fault affects only part of a branch:

For node shifting, all those nodes of the resulting configuration prevailing in the previous level which have voltage less than the specified minimum voltage limit are listed. All alternate connections possible to these nodes are tabulated. A node is then shifted over each possible alternate connection and its new voltage is checked. The alternate connection giving voltage higher than the specified minimum and with least network loss is selected for that node. If no connection is possible, then the node is shed. This procedure is repeated for all the listed nodes.

At the end of this level, the voltage constraint will be satisfied, but feeders may be overloaded (only theoretically, because in practice the generator voltage drops due to over- load). If none of the feeders are overloaded and no nodes are shed, the resulting configuration is considered as "restored after node shifting" and the result display level gives the operator instructions for implementation.

a) **Dummy lines and dummy feeder (to effectively use Load Flow program):** Load flow converges when all the nodes in the network are supplied. Shedding causes isolated nodes that prevent the load flow program from converging. To bypass this problem, a dummy generator node (feeder) and dummy lines (initially open) connecting shed/isolated nodes directly to the dummy feeder must be created. Nodes to be shed are disconnected from all network nodes and attached to this dummy node. To get more accurate results, the impedence of the dummy lines should be as minimum as possible and the load on the shed nodes should be made zero. This also keeps the voltage of the shed nodes at 1 P.U avoiding confusion from the network voltage values.

### Load picking level

Load picking is the last standard procedure in this new approach, both for reconfiguration and restoration. It is necessary that at least one node should be shed for load picking to take place. First, a list of all possible reconnectable lines for each shed node is tabulated. Let the node at the other end of these lines be called "pair nodes". For a shed node, if the pair-node is part of the network, and their connecting line is not faulty, then such lines get shortlisted into "possible reconnection table", else if the pair-nodes are also shed,

then those lines are closed but not shortlisted. Next, for every node, taken one at a time, each of these shortlisted lines is closed one at a time and the minimum voltage of the network is determined. The reconnection resulting in "shed-but-now-reconnected-node" voltage being above minimum specified limit and least network loss are selected. If no connection satisfies the voltage constraint, then the node is permanently shed. This picking procedure is repeated for all the shed nodes.

The resulting network at the end of this level will have all its node voltages' above minimum limit. But feeders may be overloaded, in which case the third and fourth level are repeated till the loading on all the feeders are within their thermal limits and a dead end is reached beyond which no improvement is possible. Restoration of the permanently shed nodes is possible only after rectification of the fault. The network is reconfigured or restored at this point and the resulting configuration is the final configuration. The network loss of the final configuration obtained after fault restoration could be higher than that of the original network. The aim is to minimize the increase in loss.

### Result display level and operator instructions

At the end of the whole methodology, this new approach prints out the operator/technician instructions that is, the list of switching changes that are to be effected in order to realize the restoration or reconfiguration in the field. This instruction is the link between the MATLAB program and the actual network in the field. This list is prepared by the program by comparing the status of all the switches in the final configuration with that of the initial/existing configuration and tabulating the changes.

The new proposed mechanism definitely decreases the network loss since load gets re-distributed more equitably. Service after fault is restored quickly with maximum load pick up as all possible search options are executed and with minimum loss difference as least number of switching changes are done in the network. In case of multiple faults, feeder fault is taken up first, then network line fault is restored. While the first fault is being restored, the nodes affected by the second fault are temporarily shed that is they are attached to the dummy feeder to obtain convergence of the load flow program. In the flowchart shown in Figure 1, "V" implies voltage limits, that is, if voltages of all nodes are above the specified minimum voltage limit, and "I" implies feeder loading that is, if any feeder is overloaded. For convenience, thermal limits of network lines are assumed to be high and neglected.

### Reactive power compensation

With assumptions stated in the introduction section, finding the accurate compensation solution is possible but time consuming. Hence a compromise has to be made between loss reduction and the time taken to obtain and implement the solution. Following the assumptions, a compensation solution specific to the network configuration is achieved. But after fault restoration, the network configuration changes and some of the buses and capacitors may be shed. Initial reactive power injected by capacitors at some buses may not be appropriate. Hence a new compensation solution will have to be arrived at after each of the different fault restoration scenarios. And to achieve this, variable capacitors must be installed at both generator/feeder and load buses in the network (Figure 2).

### Method 2

Variable capacitors are inducted at the generator buses. If the load flow program does not facilitate this addition, four fictitious nodes must be introduced right beneath the generator buses, before the

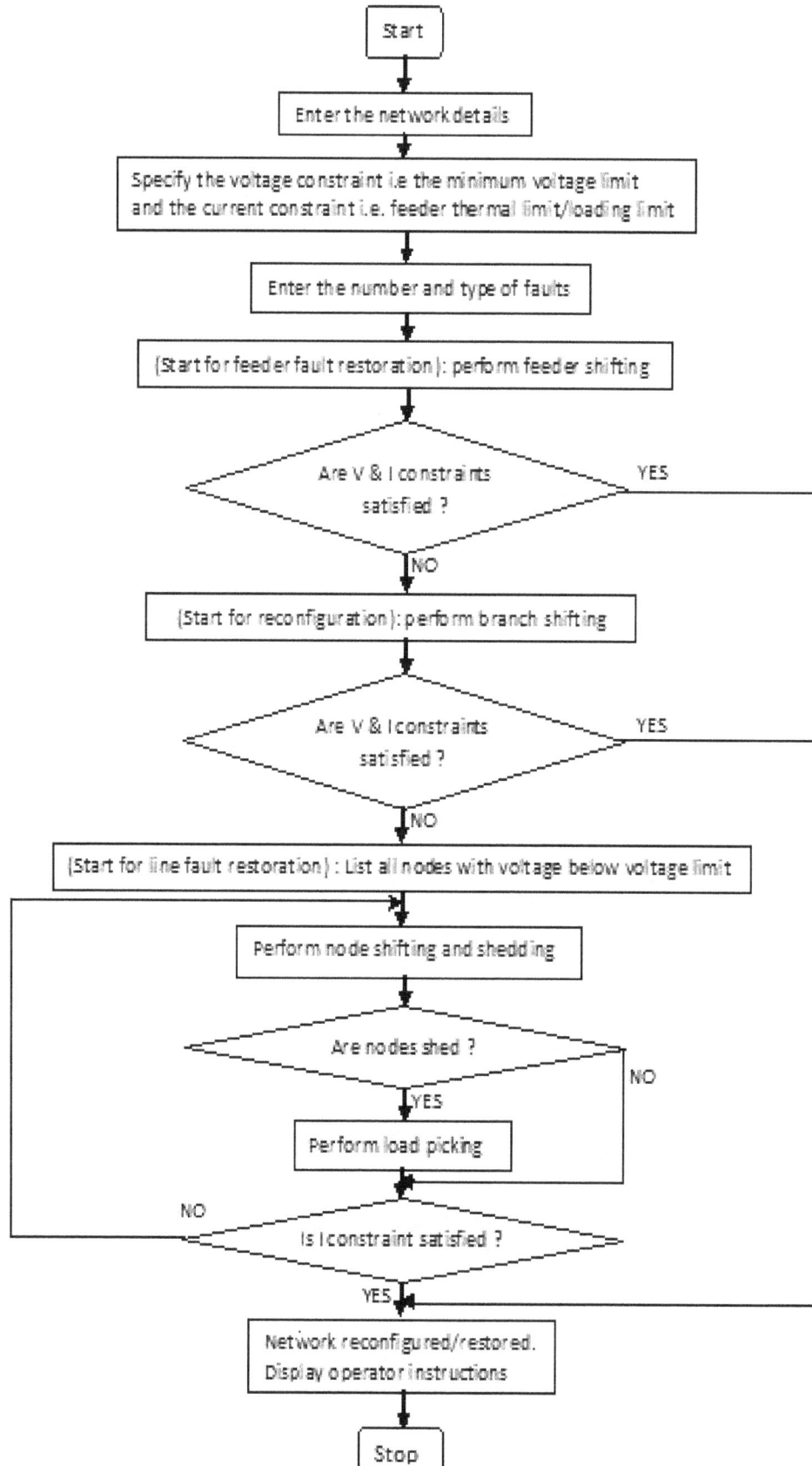

**Figure 1.** The new level-wise approach applicable to all types of faults

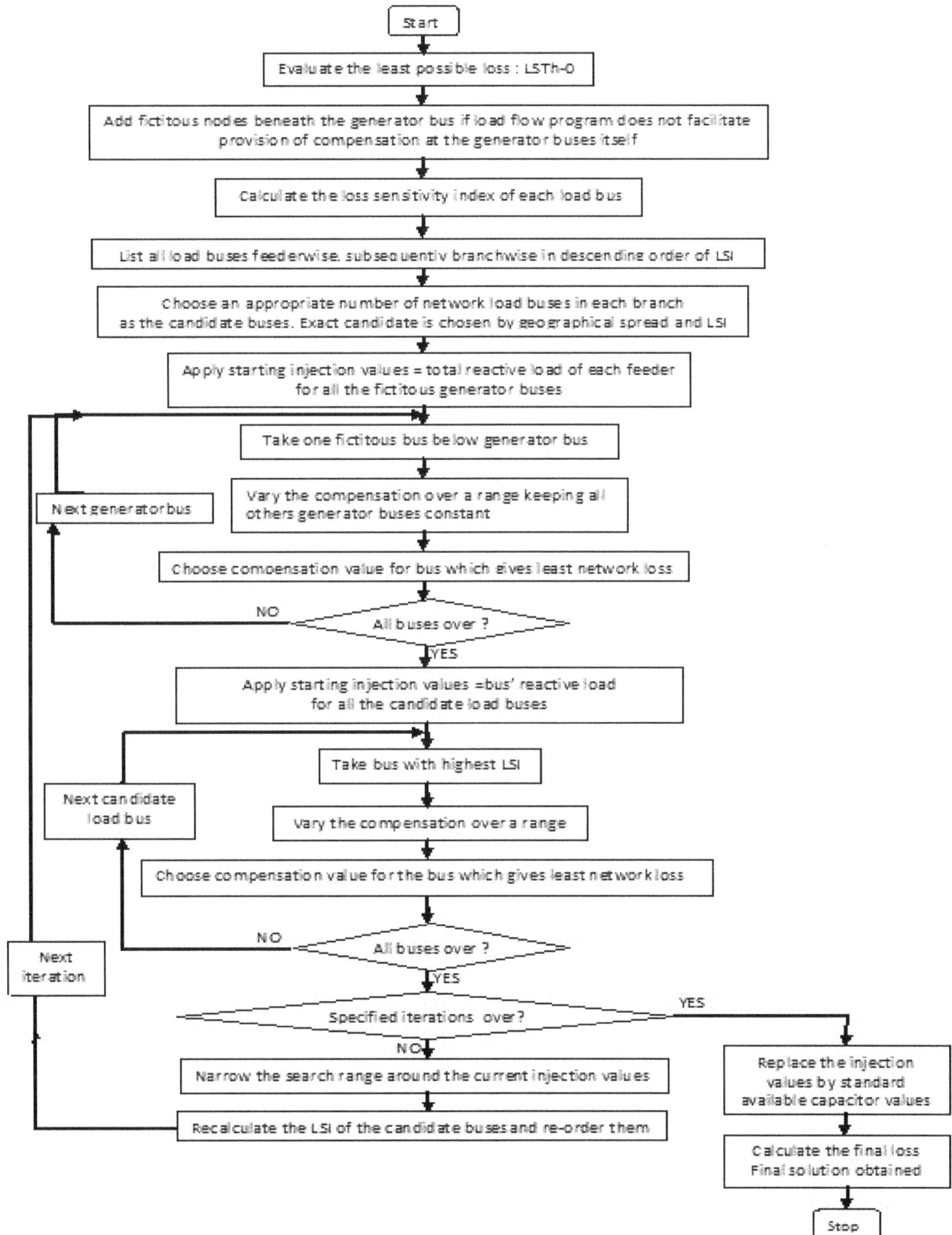

**Figure 2.** Proposed methodology for reactive power compensation of a distribution network

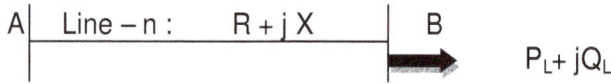

**Figure 3.** Line section of a distribution network.

load buses. The load at these fictitious buses must be zero and impedence of the lines connecting these buses to the generator buses must be as low as possible for better accuracy. The theoretical least possible loss (LSTh-0) in the network can be obtained by nullifying the reactive load at each bus and reactance of every line in the uncompensated network. This is done by injecting reactive power equal to the reactive load at each bus and capacitors at both ends of each line. Injecting reactive power higher than these values will result in increased losses. Again achieving LSTh-0 is not practical due to capacitor cost constraints. Hence, only a few numbers of buses must be identified for compensation based on some criteria like loss sensitivity index (LSI). LSI is the sensitivity of network loss (LSN) to incremental change in the reactive load of each of the network buses. The LSI of each bus is calculated and buses are tabulated in descending order of LSI. It is prudent to inject reactive power at few buses with higher LSI. But, the candidate buses for compensation must be geographically distributed in the network to cater for the loss of buses due to shedding during outage. Therefore, the buses are arranged feederwise and branchwise in descending order of their LSI, and two buses of each branch of every feeder are chosen as the candidate buses. The chosen number of buses that is, two is user specified for convenience and/or the buses can be chosen based on geographical spread.

Once the locations are determined, selected buses are compensated such that the network loss LSN approaches LSTh-0. The injection values are determined first for the fictitious buses near the generator buses, and then for the chosen network nodes to achieve LSN closer to LSTh-0. The proposed method is an iterative process for obtaining the optimal compensation solution. For the first iteration, starting values equal to the total reactive load of their respective feeder are applied to each fictitious bus. Thereafter, power injection values are varied over a range, for one bus at a time, keeping the starting values of other buses constant. The injection value resulting in least LSN is selected for that bus. Compensation for the next and the remaining buses are similarly determined. Alternatively, compensation values for each fictitious bus are obtained by permutation and combination. The range is initially from zero to the maximum allowed injection. The maximum is the sum of the network reactive load. The initial search steps should be large enough for this range in order to obtain results quickly, but the results will be not exact yet. Next, starting injection value equal to its reactive load is applied to each candidate load bus. The candidate buses are taken up one at a time in descending order of their LSI and their injection values are varied over the same range. The compensation values are obtained in the same manner as above.

At the end of the first iteration, an approximate compensation solution is obtained and the network loss will have either remained same or decreased. This process is repeated first with the generator buses and then the candidate load buses over multiple iterations. In each subsequent iteration, the search range and steps have to be narrowed down in order to approach accurate injection values. Also the sensitivity of the load buses change after each iteration due to change in its net reactive load. Hence, the LSI has to be re-evaluated for all the candidate load buses and again taken in its decreasing order. The injection values so obtained need not be the standard values of commercially available compensation. The number of iterations can be limited to an expert-user-specified value or till the difference between the losses values of two

consecutive iterations is less than specified tolerance limit. At the end of the iterations, the loss value will have approached the least network loss possible that is, LSTh-0, but the bus injection values obtained might not be standard available commercial values. If continous compensation variation is possible, these values can be used, else they have to be replaced by nearest standard compensation values. For the latter case, new loss sensitivity list of the candidate buses are obtained and for each bus taken in decreasing order of the LSI, the immediate higher and lower standard compensation values are tested and the value giving lesser network loss is chosen for the particular bus. In this case, the network loss may increase due to the approximation made in the compensation values, but it will be acceptable in view of the time and cost constraints.

After fault restoration, the network configuration changes and some buses may be shed. In such a case, the compensation solution may not reduce the network loss to its lowest, or instead it might add to the losses. Hence, the optimal solution will have to be re-evaluated. Again, the same iterative procedure is applicable. The candidate buses remain the same, only their injection values have to be altered. The existing injection values serve as the starting values for both the fictitious generator buses as well as the candidate load buses. The initial search range of injection values will be wide i.e. from zero to the maximum allowable compensation value. The maximum allowable compensation value has to be re-calculated for the new configuration

### Calculation of loss sensitivity index (LSI) of a bus

Consider a distribution line (line - n) with impedance R+jX and load of PL+ jQL connected between buses 'A' and 'B' as shown here (Figure 3):

Active power loss in the nth line is given by

$$P_{LINELOSS}[n] = \frac{(P_L^2[B] + Q_L^2[B]) * R[n]}{(V[B])^2}$$

....... (1)

Similarly, reactive power loss in the nth line is given by

$$Q_{LINELOSS}[n] = \frac{(P_L^2[B] + Q_L^2[B]) * X[n]}{(V[B])^2}$$

....... (2)

Where,

PL [B] = Total effective active power supplied beyond the bus 'B'.
QL [B] = Total effective reactive power supplied beyond the bus 'B'.
V [B]   = Voltage at bus 'B'.

Network loss includes both real and reactive loss. Reactive loss of the network is directly proportional to its net reactive load. But, sensitivity of the network real loss needs to be studied. Hence for purpose of studying the effect of reactive power compensation, network loss can safely refer to only the network real loss. Therefore:

Total network loss (LSN) = $\sum_n P_{LINELOSS}[n]$

...... (3)

LSI of a bus B is now the sensitivity of network real loss to incremental change in reactive load at bus B. It is obtained by differentiating LSN w.r.t QL [B].

$$LSI[B] = \frac{\partial (LSN)}{\partial (Q_L[B])} = \frac{2 * Q_L^2[B] * R[n]}{(V[B])^2}$$

...... (4)

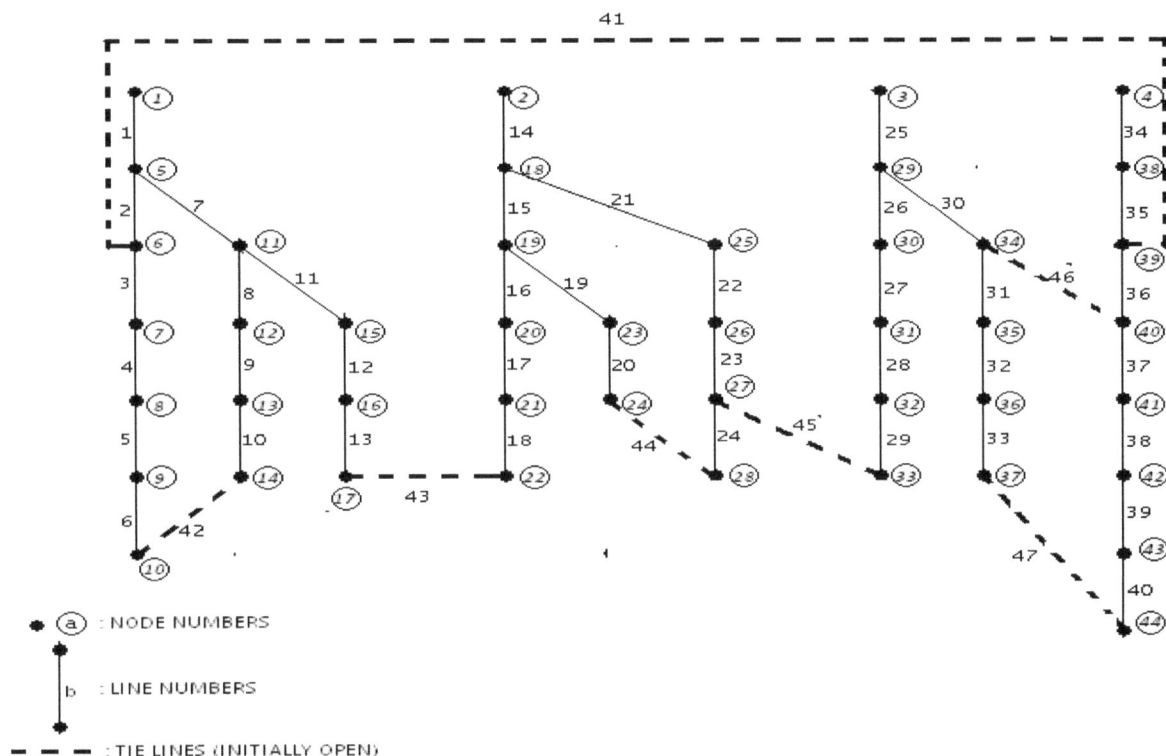

**Figure 4.** Sample distribution network Nodes 1-4 are generator buses. The remaining are load buses.

**Table 1.** Standard available capacitor values.

|  | 1 | 2 | 3 | 4 | 5 | 6 |
|---|---|---|---|---|---|---|
| Capacitor rating (MVAR) | 0.15 | 0.30 | 0.45 | 0.6 | 0.9 | 1.2 |

Once the LSI of all buses are calculated, they are arranged in descending order of LSI and used.

## Sample system

A practical primary distribution system of R. K. Nagar, Karnataka Power Transmission Corporation Limited (KPTCL), of 11 KV having 44 buses, 40 sectionalizing switches and 7 tie lines have been taken as the sample system. The line resistance, reactance and voltage are measured in P.U. Real, reactive and apparent power are measured in MW, MVAR and MVA respectively. The network plot shown in Figure 4 is derived from the practical network, drawn suitable to processing by MATLAB. No changes are made in the circuit.

Programming was done in MATLAB 7.6 in Windows XP on a 1.5 GHz processor. The load flow program used is based on GAUSS-SIEDEL method also written in MATLAB. Newton-Raphson method can also be used, but the variables must be carefully tapped out. Presenting existing network to MATLAB is very important in order to obtain correct results and so also is interpreting the results to the operator. All the relevant data and tables are created and stored in a data file in MATLAB. The maximum feeder capacity is taken as 15 MW and the minimum voltage limit is entered as 0.98 P.U (user defined). For convenience, only the feeder's maximum current (load) capacity is considered. The capacity of the network lines is

assumed to be sufficiently high for all scenarios. The permissible feeder overloading is taken as 40%. The voltage profile, total network loss, switch positions (that is, switch status on/off), generation and spare capacity of all the feeders are recorded at all the stages of the program from the initial to the final network.

The standard available capacitor values are given in Table 1. All possible capacitor combinations are made from these standard values and are applied to obtain the compensation solution. For convenience, the capacitor cost constraint is neglected for this case study and solution for minimizing the network loss is the primary objective.

## RESULTS

A single feeder fault is simulated at feeder 2. Service is restored to the outage load using the new approach without compensation. Later, a compensation scheme to reduce network loss is determined for the initial sample network. The same fault is simulated and restoration is attempted again. After restoration, the existing compensation is then altered to minimize the network loss. The voltage profile, network loss, number of switching changes, generation and spare capacity of the network

**Table 2.** Operator instructions without and with compensation.

| | Before fault | | After fault | | |
|---|---|---|---|---|---|
| | Without comp | With comp (A) | Without comp | With comp (A) | Altered comp (B) |
| Number of switching changes | - | - | 9 | 9 | - |
| Switches to close | - | - | 41,43,44,45 | 41,43,44,45 | - |
| Switches to open | - | - | 2,6,16,20,21 | 2,6,16,19,21 | - |
| Generation ( MW ) | 30.9757 | 30.9267 | 27.1894 | 28.4211 | 27.1398 |
| Spare cap ( MW ) | 29.0243 | 29.0733 | 32.8106 | 31.5789 | 32.86 |
| Loss (In MW ) | 0.3669 | 0.3155 | 0.3391 | 0.3436 | 0.2902 |
| Reduction in loss as % of initial network loss + increase / - decrease | | -14.01 % | -7.565 % | -6.346 % | -20.90 % |
| Load to be compulorily shed | - | - | - | - | - |
| Nodes shed permanently | - | - | 10,18,19,23 | 10, 18, 19 | 10, 18, 19 |
| Actual load shed | - | - | 4.90 | 3.752 | 3.752 |
| Minimum voltage (Node) | 0.9730 (22) | 0.9773 (10) | 0.9811 (25,28) | 0.9829 (22) | 0.9834 (23) |
| Feeders overloaded | - | - | - | - | - |

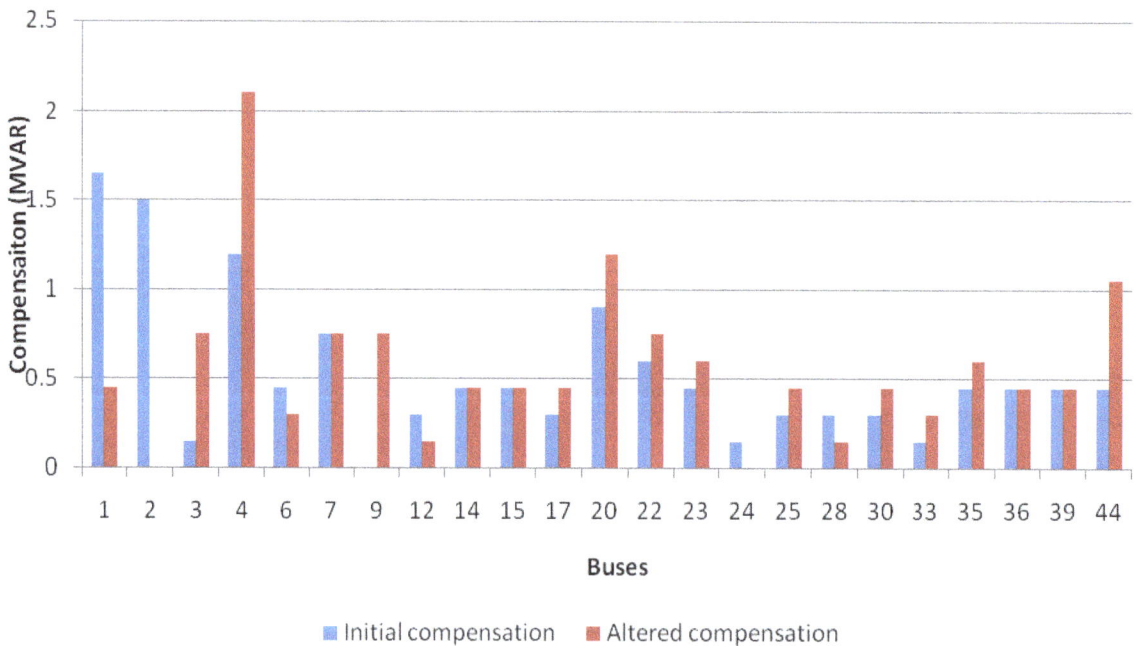

**Figure 5.** Initial and altered compensation values applied at compensated buses of the sample distribution network.

are tabulated in Table 2 and compared before and after restoration, without and with compensation.

Fault type: Feeder 2 fault
Outaged load: 9.73 MW

Minimum possible loss for:

Initial network = 0.310 MW

Network after fault restoration = 0.2883 MW

As seen in Table 2, fault restoration using capacitors results in lesser load shedding and better voltage profile. Voltage profile improves further and loss reduction is enormous after the compensation is altered at the same locations. Figure 5 shows the original and altered injection values at all the compensated buses. The improvement in voltage profile is evident in Figure 6.

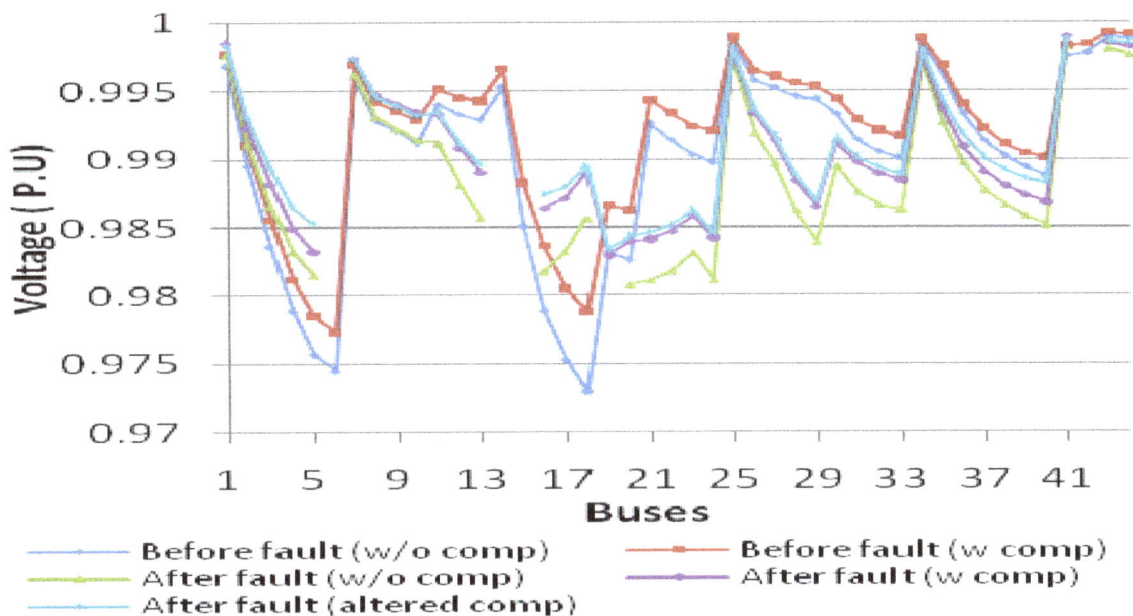

**Figure 6.** Voltage profiles of the sample distribution network before and after fault restoration for both without and with compensation cases.

## SCOPE FOR IMPROVEMENT

This program can be extended to include real time fault identification and rectification. Feedback control to detect fault location and to periodically check the network condition can also be incorporated. Including more details like maximum network line capacities and varying load condition makes the mechanism more practical. This method can be used to plan new networks. Capacitor costs can be included while determining the optimal solution and payback/break-even time period.

## Conclusion

This paper gives a new, fast and effective method for system reconfiguration, service restoration and providing reactive power compensation of a primary distribution network. An iterative method is used to determine an optimal compensation solution which also enables variation of the solution when the network configuration changes. Application of the new approach decreases the network loss enormously with lesser switching changes while maintaining the voltage profile and line loading within limits. Applying compensation using the proposed method helps in saving additional power, improves the voltage profile further and causes lesser load shedding during restoration. Since the methodologies are iterative, the results obtained are optimal for given conditions.

## REFERENCES

Branko DS, Milos SN (2004). Solving the problem of general capacitor placement in radial distribution systems with laterals using Simulated Annealing", Sci. Tech. Rev. Vol. 54 (3-4).

Chung-Fu C (2008). Reconfiguration and capacitor placement for loss reduction of distribution systems by Ant Colony Search Algorithm", IEEE transactions on Power Systems, 23(4) :1747 - 1755

Dash PK, Saha S, Nanda PK (1991). Artificial Neural Net Approach for capacitor placement in power system", Energy research centre, Dept. of Electrical Eng., R. E. College, Rourkela 769008, India. .

Gómez JF, Khodr HM, De Oliveira PM, Ocque L, Yusta JM (2004). R. Villasana and A. J. Urdaneta,"Ant Colony System Algorithm for the planning of primary distribution circuits", IEEE Trans. on Power Syst. 19(2): 996 – 1004.

Seong-Il L, Myeon-Song C, Dong-Jin Lim (2006). Service restoration methodology for multiple fault case in distribution systems", IEEE Trans. on Power Syst. 21 (4): 1638 – 1644.

Srinivasa RR, Narasimham SVL (2008). Optimal capacitor placement in a radial distribution system using Plant Growth Simulation Algorithm", Intern. J. Electrical Power Energy Syst. Eng. Spring 2008.

Young-Hyun M, Byoung-Hoon C, Ho-Min P, Heon-Su R, (2002). Fault Restoration Algorithm using fast tracing technique based on tree structured database for the distribution automation system", IEEE Trans. on Power Syst. pp. 411-415

Young-Jae J, Jae-Chul K, Jin-O. Kim, Joong-Rin S, Kwang YL (2002). An efficient Simulated Annealing Algorithm for network reconfiguration in large scale distribution systems", IEEE Trans. on Power Delivery, 17(4): 1070-1078

# 15

# A Tabu search algorithm for multi-objective purpose of feeder reconfiguration

**Tilak Thakur[1] and Jaswanti Dhiman[2*]**

[1]Department of Electrical Engineering, Punjab Engineering College, Deemed University, Chandigarh, India.
[2]Department of Electrical Engineering, Chandigarh College of Engineering and Technology (CCET), Chandigarh, India.

This paper presents a new method which applies the application of Tabu Search (TS) as a meta-heuristic method for network reconfiguration problems in radial distribution system. The objective of the paper presented in this work is to make a Tabu Search based algorithm for multi-objective programming to solve the network reconfiguration problem in a radial distribution system. With the appearance of the Tabu search, by Fred Glover in 1986, diverse applications have arisen from the procedure to solve diverse problems as for the classic problem of the route of the vehicle (also known as the travelling agent problem) and the allocation of a plant. Here six objectives are considered in conjunction with network constraints. The main objective of the research is allocation of optimal switches to reduce the power losses of the system. It is tested for 33 bus systems. Simulation results of the case studies demonstrate the effectiveness of the solution algorithm and proved that the TS is suitable to solve this kind of problems.

**Key words:** Combinatorial optimization, distribution system, energy loss minimization, genetic algorithm, simulating annealing, Tabu search.

## INTRODUCTION

The electric power distribution usually operates in a radial configuration, with tie switches between circuits to provide alternate feeds. The losses would be minimized if all switches were closed, but this is not done because it complicates the system's protection against over-currents (Tilak and Jaswanti, 2006). Whenever components fail, some of the switches must be operated to restore power to as many customers as possible. As loads vary with time, switch operations may reduce losses in the system. All of these are applications for reconfiguration (Venkatesh, 2004).

The reconfiguration problem is combinatorial problem, which precludes algorithms that guarantee a global optimum. Most existing reconfiguration algorithms fall into two categories (Hayashi and Matsuki, 2004). In the first, branch exchange, the system operates in a feasible radial configuration and the algorithm opens and closes candidate switches in pairs. In the second, loop cutting, the system is completely meshed and the algorithm

opens candidate switches to reach a feasible radial configuration. Reconfiguration algorithms based on neural network, heuristics, genetic algorithms, Tabu Search and simulated annealing have also been reported, but not widely used (Fukuyama, 2000). These existing reconfiguration algorithms work with a simplified model of the power system, and they handle voltage and current constraints approximately, if at all (Teng, 2003). TS explore the whole solution space definitely based on the local search in which controlled up-hill move is admitted.

The roots of the TS go back to the 1970's; it was first presented in its presentable form by Glover (1986) later it was formalized by him; the basic idea was sketched by Hansen. Up to now, TS is a strategy with more functions for solving combinatorial optimization problem and is applied to various fields to obtain high quality solutions within reasonable computing time. The TS method is built upon a descent mechanism of a search process, which biases the search toward, points with lower objective function values (Ramon, 2000). However special features can also be added to avoid being trapped in the local minima. Basic component requirement to implement the TS are: Moves and Selection, Tabu List Aspiration Criterion

---

*Corresponding author. E-mail: jaswanti98@yahoo.co.in

Intensification and Diversification. A large number of papers has been published so for on Tabu search algorithm for various combinatorial optimization solution (Ramon, 2001).

This paper presents a new method which applies the TS as meta-heuristic method for network reconfiguration problems in radial distribution system. This work has been tested on 33-bus RDS. System has five tie lines. The main advantage of TS with respect to conventional Genetic algorithm and Simulation annealing lies in the intelligent use of the past history of the search to influence its future search procedures.

## LITERATURE REVIEW

One of the most relevant phases in the study concerns the analytic formulation of the objectives of the problem. For the six objective and the two constraints considered in the proposed formulation, in what follows, the analytical expressions and the relevant calculation hypotheses are reported.

### Minimize power losses

The power losses vary with the network configuration and with the compensation level. They are associated with the resistive elements of lines and of HV/MV transformers. Assuming for the loads a constant current model, losses at MV/LV transformers can be neglected since they are not varying with current. Other losses terms like those due to insulation of lines and capacitors can be neglected too.

The minimization of the real power losses arising from feeders can be calculated as follows:

$$\text{Min } f_1(\overline{X}) = \sum_{i=1}^{N_b} r_i \frac{p_i^2 + q_i^2}{v_i^2} \qquad (1)$$

### Ensuring voltage quality

The regular supply of the loads is guaranteed if the voltage value at the terminal nodes is as close as possible to the rated value. Bus voltage is one of the most significance security and service quality indices, which can be described as follows:

$$\text{Min } f_2(\overline{X}) = \max |V_i - V_{Rate}|, \qquad (2)$$
$$i = 1,2,3 \dots, N_b$$

Where $N_b$ is the total number of buses; $V_i$ and $V_{Rate}$ are the real and rated voltage on bus i, and $f_2(\overline{X})$ represents the maximal deviation of the bus voltage in the system of interest. Lower $f_2$ values indicate a higher quality voltage profile and better security of the considered system.

In order to quantify the extent of violation of limits imposed on voltages at buses in a RDS, the following Voltage Deviation Index (VDI) has been defined.

$$VDI = \sqrt{\frac{\sum_{i=1}^{NVB}(V_{Li} - V_{LiLIM})^2}{N}}$$

(3)

Subject to

$$V_{jMIN} \leq V_j \leq V_{jMAX} \qquad j \in 1 \text{ to } N$$

Where NVB is the number of buses that violates the prescribed voltage limits and $V_{LiLIM}$ is the upper limit of the $I^{th}$ load bus voltage if there is upper limit violation or lower limit if there is a lower limit violation.

During reconfiguration, if the state of the system has voltage limit violations; the given solution must try to minimize the index VDI and thereby improve the power quality.

### Service reliability assurance

The objective of network reconfiguration is to reduce power losses and improve reliability of power supply by changing the status of existing sectionalizing switches and ties. From the operator's prospective, service reliability in operating distribution systems refers to the ability to support unexpected increasing loads and to relieve other feeds following faults:

$$\text{Min } f_3(\overline{X}) = 1 - \min_i \left\{ \frac{I_{iRate} - I_{iLoad}}{I_{iRate}} \right\},$$
$$i = 1,2,3 \dots, N_l \quad (4)$$

$$\text{Min } f_4(\overline{X}) = 1 - \min_i \left\{ \frac{S_{iRate} - S_{iLoad}}{S_{iRate}} \right\},$$
$$i = 1,2,3 \dots, N_l \quad (5)$$

Selecting a specific index for ensuring reliability of service is utility-dependent and would not alter the basic formulation. It is necessary of substation to select the most suitable configuration satisfying the reliability requirement of customers.

## Minimizing switches operation

Distribution systems can be reconfigured via a series of switches operations in order to reduce multi-objective problem. If any of the switch status is incorrect, then the application functions that use this data base will also produce incorrect results. One of the major obstacles in identifying errors in the switch statuses is the lack of sufficient real time measurement taken from the system. So it is better to use the minimize switches operation objective.

Its solution scheme starts with a meshed network by initially closing all switches in the network. The switches are then open one at a time until a new radial configuration is reached. In order to accomplish the transition from the initial configuration to the optimal configuration with minimum switch operations, an effective switch plan needs to be developed such that unnecessary switch operations in the switch sequence can be avoided. Minimizing the number of switch operations can be denoted as follows:

$$\text{Min } f_5(\overline{X}) = \sum_{i=1}^{N_s} |S_i - S_{oi}| \qquad (6)$$

Where Ns represents the total number of switch; $S_i$ and $S_{0i}$ are the new and original states of switch i, respectively; and $f_5(\overline{X})$ represents the number of switch operations under state $(\overline{X})$. A lower $f_5(\overline{X})$ value implies that less time is needed during the network reconfiguration process.

## Lesser solution time

Especially with the introduction of remote control capability to the switches, lesser computational time configuration management becomes an important part of distribution automation. A salient feature of the solution methodology is that it allows the designers to find a desirable, global non-inferior solution for the problem. An effective scheme to speed up the simulation technique have been presented and analyzed.

## Maximum loading

The whole load of the network should be divided in a balanced way among the transformers, on the basis of their rated power. In this way, the optimal working condition for the transformer is ensured and any over-loading situation due to fault occurrence can be promptly faced. In the literature on the topic, different formulations have been proposed for the Load Balancing Index.

For load balancing, we will use the ratio of complex power at the sending end of a branch, Si over its KVA capacity, $S_{imax}$ as a measure of how much that branch is loaded. The branch can be a transformer, a tie line with a sectionalizing switch or simply a line section. Then we define the load balance index for the whole system as the sum of these measures, i.e.

$$C_b = \sum \left(\frac{S_i}{S_i^{max}}\right)^2 = \sum \frac{P_i^2 - Q_i^2}{S_i^{max2}} \qquad (7)$$

This will be the objective function, Cb of load balancing.

When the general search algorithm introduced in section is used for load balancing, the calculations will be similar to that of the loss reduction case. The only difference will be in the calculation of the objective; for load balancing, we need to estimate the value of the new objective, load balance index, $C_b$ for every branch exchange considered during the search. Once the new power flow in the branches, Pi', Qi' are estimated then the new load balance index can be computed by employing Equation (7).

Having a network model, now we can express the power loss and measure the load balance in the system in terms of system variables.

## Constraints

Two constraints are considered in the formulation of the problem, although other constraints could also be taken into account within the proposed solution procedure:

1. The radial structure of the network must be maintained in each new structure.
2. All loads must be served.

The second constraint is considered in a situation without faults (in a normal case, in which reconfiguration yields multiple benefits).

### METHODOLOGY

The flow chart of the proposed Tabu Search based algorithm for the distribution network reconfiguration is shown in Figure 1 and described as follows:

1. Choose current operating network as an initial solution $S_{initial}$ and calculate the evaluation function of $S_{initial}$. Let the current solution vector $S_{currrent} = S_{initial}$ and the best solution vector $S_{best} = S_{initial}$. Initialize the Tabu List T and the aspiration function A. Set the iteration counter I=0.
2. If I is equal to pre-specified maximum permitted iteration number $I_{max}$, then output $S_{best}$ as the final result and stop. Otherwise, set I=I+1, and go to Step 3. Select a trial radial solution from the neighborhood of $S_{current}$ by the operator move, which will be defined later, and calculate the evaluation function f(S) of the corresponding solution S. Repeat the process until the specified neighborhood sampling number $N_{max}$ has been reached.

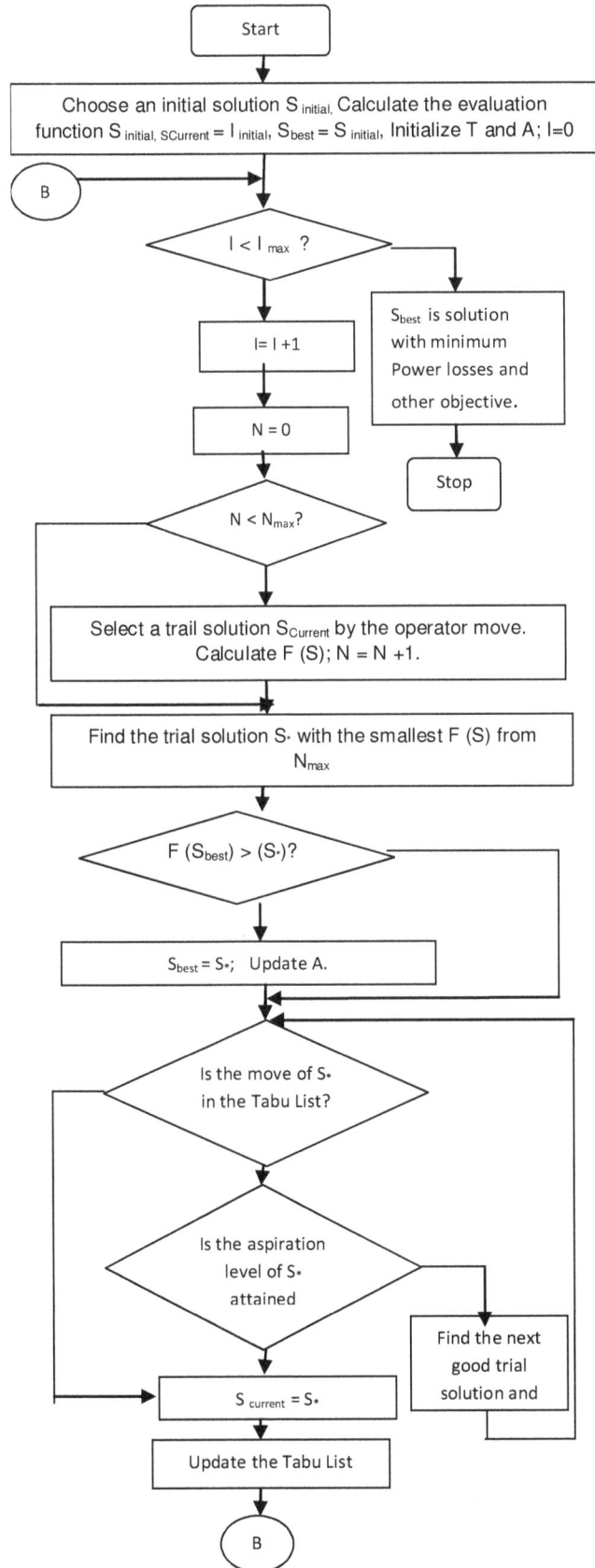

**Figure 1.** Tabu Search for network reconfiguration problem.

3. If $S_{best}$ is not better than the best trial solution that has the minimum evaluation function value, and then assign this best trial solution to $S_{best}$ and update the aspiration function A. Otherwise, go to Step 5.

4. $S_{current}$ is updated to the best trial solution that has the minimum evaluation function value as evaluated in the Step 3 if the corresponding move is not in the Tabu List or its aspiration level is attained. Then, include the move in the Tabu List and update the Tabu List T. Go to Step 2. If the best trial solution corresponds to a Tabu move and its aspiration level is not attained, then check the next trial solution, and repeat this step.

## RESULTS AND ANALYSES

Test sample system is considered for simulation, one 33 bus radial distribution system, 12.66 KV. Computer software has been developed for the test system in MATLAB to examine the efficiency of the proposed algorithm.

### Original configuration case

In test system, the power loss minimization technique is applied on a 12.66 KV, 1 MVA base hypothetical system as shown in Figure 2. It consists of 33 buses and 32 branches. The test system have five tie lines i.e. the system have five loops. Five tie switches exist between nodes (20,33), (14,34), (21,35), (32,36), and (28,37), which are normally open. Sectionalizing switches are also assumed to be associated with all other branches. Appendix A gives the data for the hypothetical system under consideration, with the values of load, branch impedances and voltage at nodes. Loads are converted into nodal current injection. Branch current are calculated by summing the nodal current from last node and moving towards the root node using the backward sweep. In the present condition the resistive line losses are coming out to be 203.059KW.

### TS reconfiguration case

Problem is to reconfigure the system to another form, so that the total resistive line losses are minimized without violating the current and voltage limit, no nodal load is isolated. Convert the system into a meshed one by connecting the tie lines in the system i.e. branch number 33, 34, 35, 36 and 37 between their respective nodes.

The next step is to carry out a load flow analysis of the system to determine the optimal flow pattern and find the line carrying the minimum current. Line 6 was carrying a minimum current of 9.3706 Amp in loop 1, so it was removed. Line 14 was carrying a minimum current of 0.4813 Amp in loop 2 and line 11 was carrying a minimum current 2.59 Amp in loop 3. But in loop 4 the line 31 was carrying a minimum current of 6.624 Amp and in loop 5 lines 28 was carrying a minimum current of

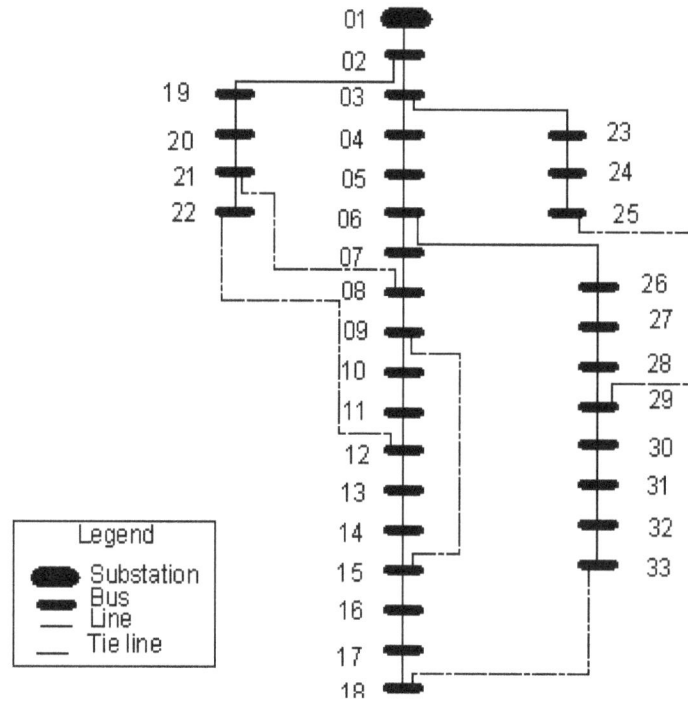

**Figure 2.** A 33-bus radial distribution system.

**Table 1.** Loop number after reconfiguration.

| S/n | Loop No. | Branch In | Branch Out | Current in branch out |
|-----|----------|-----------|------------|-----------------------|
| 1 | 1 | 33 | 6 | 9.3706 |
| 2 | 2 | 34 | 11 | 0.4813 |
| 3 | 3 | 35 | 31 | 2.5900 |
| 4 | 4 | 36 | 28 | 6.6240 |
| 5 | 5 | 37 | 14 | 0.8600 |

No. of loops = 5.

0.86 Amp as given in Table 1. So, branch 6, 14, 11, 31 and 28 were deleted and the tie lines of network now become the branch of the system. Now, the network has to be reconfigured for the change that place in the network. So, the branches current are again calculated after calculating the node voltage using the backward and forward sweep. Final losses that come out of the configuration are 136.4791 KW. Thus the losses are reduced from 203.0590KW to 136.4791 KW resulting in percentage loss reduction of about 14.82%.

### Results of multiobjectives problem

Figures 3 and 4 show the comparisons between voltage and losses respectively flowing through the lines in original and TS configuration cases. In TS configuration, lesser power is required because the losses have been decreased. This was the main objective to be achieved through reconfiguration.

### Comparison with other reference results

Figure 5 compares the results obtained for TS configuration and that given in the (Moussa, 2000) are considered. It is concluded from the results that voltages at the buses in case of TS configuration is better than that in the (Moussa, 2000) for the majority of buses. Few buses have lower voltages than that in (Moussa, 2000). It was also recorded that the voltage deviation value (VDI) improves from 0.0498 to 0.0253, and thereby

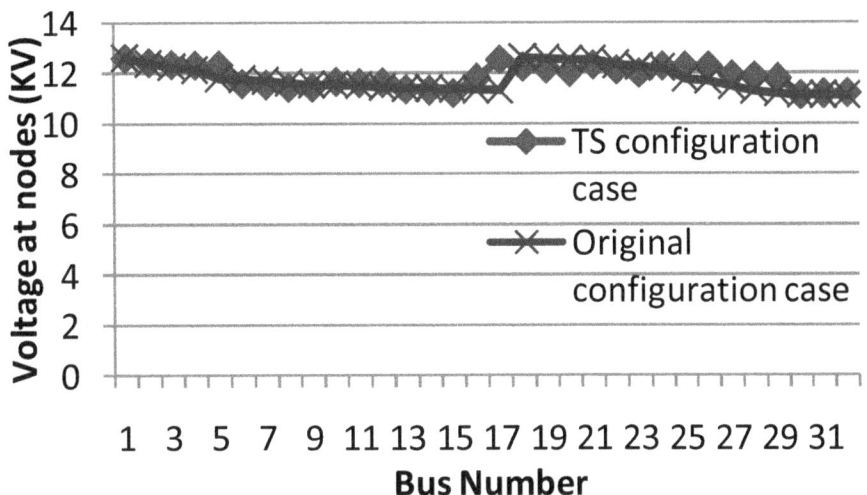

**Figure 3.** Voltage comparison of original and TS proposed configuration.

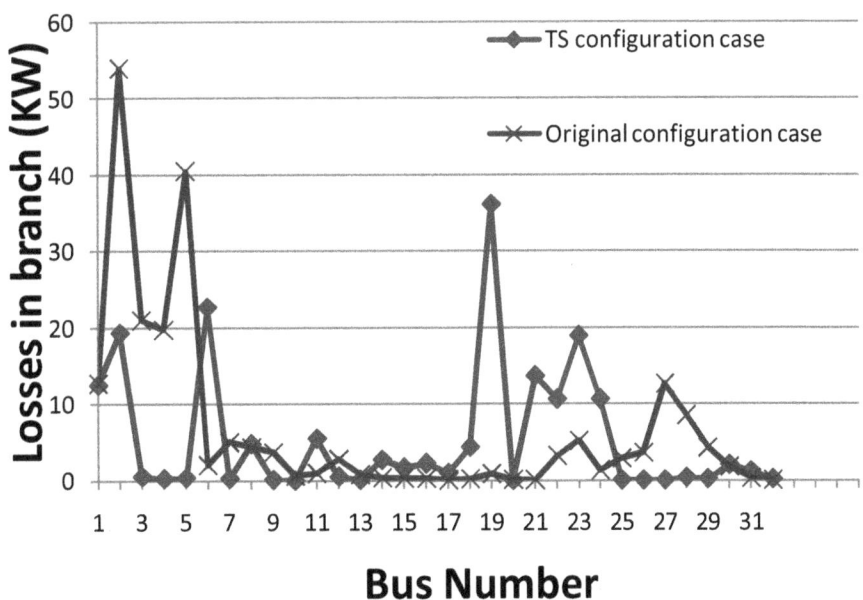

**Figure 4.** Power Losses comparison of original and TS proposed configuration.

improve the voltage and power quality. The capacity margin of feeders and transformer is taken as service reliability to specify the security of the distribution system as shown in Table 2.

The proposed algorithm also tries to minimize the number of tie-switches operations, but here, the branch having the lowest current has to be opened, eliminating one of the network loops. So, 6, 11, 14, 28, 31 switches were deleted and the tie lines of network now become the branch of the system.

In order to quantify the maximum loadability of the RDS, the total additional load that may be drawn from the RDS before it suffers a collapse is determined. This additional load is referred to as maximum loadability and is increased while retaining the existing power factor of the loads and load distribution in the RDS. In the original configuration case, maximum loadability value is 10368.10 KVA. After reconfiguration using TS approach, the maximum loadability value is increased to 16583.57 KVA.

The results obtained as shown in Table 3 by the TS re-configuration are very satisfying which encourage further

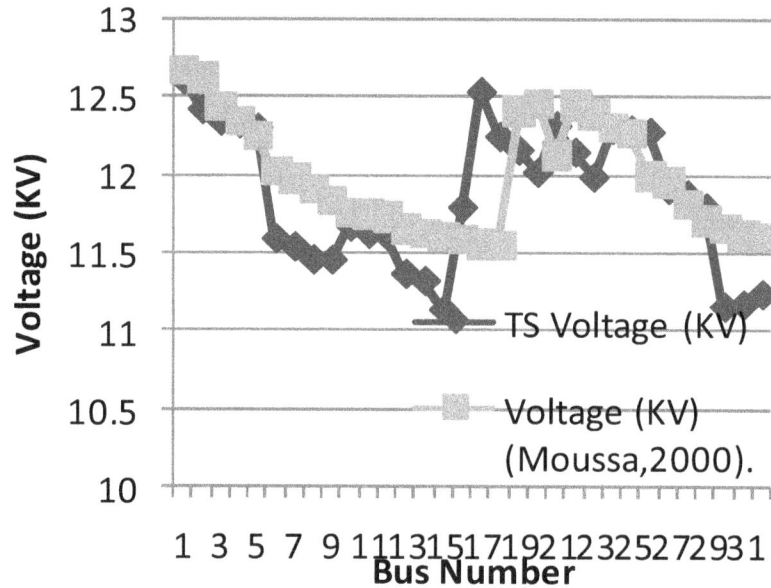

**Figure 5.** Voltage comparison of TS configuration and other (Moussa, 2000).

**Table 2.** Comparison of multi-objective results in 33 Bus System.

| Loading Level | 33 Bus system | |
|---|---|---|
| | Original configuration | TS reconfiguration |
| Power losses (KW) | 203.0590 | 136.4791 |
| Ensuring VOLTAGE QUALITY (KV) | 11.0757 | 11.1252 |
| Service reliability assurance : | | |
| Minimum Capacity | | |
| Margin among feeder (%) | 44.60 | 65.13 |
| Minimum capacity margin | | |
| Among transformers (%) | 54.12 | 56.39 |
| No. of switches operations | 33, 34, 35, 36, 37 | 6, 11, 14, 28, 31 |
| Maximum deviation of bus voltage (pu) | 0.0498 | 0.0253 |
| Maximum loadability | 10368.10 | 16583.57 |

**Table 3.** Simulation results comparison with other reference of 33-Bus system.

| Item | Tie switches | Power loss(KW) | Power loss reduction (%) | CPU Time(s) |
|---|---|---|---|---|
| Original configuration | 33, 34, 35, 36, 37 | 203.05 | -- | -- |
| Method (Ramon, 2001) | 7, 10, 14, 32, 37 | 137.37 | 30.21 | 25.3 |
| RGA (Moussa, 2000) | 7, 9, 14, 32, 37 | 139.5 | 31.2 | 13.8 |
| GA (Nara,1992) | 33, 9, 34, 28, 36 | 140.6 | 30.6 | 15.2 |
| TS Proposed configuration | 6, 11, 14, 28, 31 | 136. 4791 | 31.5799 | 0.25 |

research work with more multi objective functions. The proposed work is to be carried out with the evolution of Tabu Search method, which is best for finding out such multi objective solution as compared to other heuristic methods in respect to research space. Thus it may be concluded that the TS approach can be used to get the optimum configuration of any test system for the multi-objective purpose for electric power distribution system.

## Conclusion

This paper proposes the application of TS as meta-heuristic method for network reconfiguration multi-objective problems in radial distribution system. This work investigates six objectives that are: minimizing power losses, ensuring voltage quality, service reliability assurance, minimizing switches operation, lesser solution time, and maximum loadability for distribution system. The proposed methods have compared modern heuristic algorithms genetic algorithm, simulation annealing and Tabu search, for network reconfiguration. This work has been tested on 33-bus RDS with five tie lines.

Particularly nice feature of TS is that, like all approaches based on Local Search, it can quite easily handle the "dirty" complicating constraints that are typically found in real-life applications. It is thus, a really practical approach. It is not, however, a panacea: every reviewer or editor of a scientific journal has seen more than his/her share of failed TS heuristics. These failures stem from two major causes: an insufficient understanding of fundamental concepts of the method but also, more often than not, a crippling lack of understanding of the problem at hand. One cannot develop a good TS heuristic for a problem that he/she does not know well! This is because significant problem knowledge is absolutely required to perform the most basic steps of the development of any TS procedure, namely the choice of a search space and of an effective neighborhood structure. If the search space and/or the neighborhood structure are inadequate, no amount of TS expertise will be sufficient to save the day. All meta-heuristics need to achieve both depth and breadth in their searching process; depth is usually not a problem for TS, which is quite aggressive in this respect (TS heuristics generally find pretty good solutions very early in the search), but breadth can be a critical issue. To handle this, it is extremely important to develop an effective diversification scheme. So, a properly designed distribution system alone can render efficient and fault-free service to the consumers and at the same time reduce distribution losses to the minimum economically optimum level.

## REFERENCES

Glover F (1986). Future paths for integer programming and links to artificial intelligence." Comput. Oper. Res., 13(5): 533-549.

Glover F, Manuel L. Tabu Search". *www.Tabu search.online*.

Hayashi Y, Matsuki J (2004). Loss Minimum Configuration of Distribution System Considering N-1 Security of Dispersed Generators. IEEE Trans. Power Syst., 19(14): 636-642.

Moussa A, El-Gammal M, Abdallah EN, Attia AI (2000). A Genetic Based Algorithm For Loss Reduction in Distribution Systems, IEEE Trans. Power Delivery, 4(2): 447-453.

Ramon A, Gallego A, Aleir JM (2000). Tabu Search Algorithm for Network Synthesis." IEEE Trans,15(2): 490-495.

Ramon A. Gallego, Alcir Jose Monticelli and Ruben Romero, "Optimal Capacitor Placement in Radial Distribution Networks," IEEE Trans. on PWRS-16, No.4, Nov. 2001, pp. 630-637.

Teng J H (2003). A Direct Approach for Distribution System Load Flow Solutions. IEEE Trans. Power Delivery, 8(3): 882-887.

Tilak T, Jaswanti D (2006). Study and Characterization of Power Distribution System Network Reconfiguration," IEEE PES Transmission and Distribution conference and Exposition, Latin America.

Venkatesh B, Rakesh R, Gooi HB (2004). Optimal Reconfiguration of Radial Distribution Systems to Maximize Loadability. IEEE Trans. (19)1: 260-266.

## Appendix A

Table 1. Test data for 33 bus system [16].

| S/n | Branch Number | Receiving Node | Sending Node | Resistance of Branch (ohm) | Reactance of Branch (ohm) |
|-----|---------------|----------------|--------------|----------------------------|---------------------------|
| 1 | 1 | 0 | 1 | 0.0922 | 0.0470 |
| 2 | 2 | 1 | 2 | 0.4930 | 0.2511 |
| 3 | 3 | 2 | 3 | 0.3660 | 0.1864 |
| 4 | 4 | 3 | 4 | 0.3811 | 0.1941 |
| 5 | 5 | 4 | 5 | 0.8190 | 0.7070 |
| 6 | 6 | 5 | 6 | 0.1872 | 0.6188 |
| 7 | 7 | 6 | 7 | 0.7114 | 0.2351 |
| 8 | 8 | 7 | 8 | 1.0300 | 0.7400 |
| 9 | 9 | 8 | 9 | 1.0440 | 0.7400 |
| 10 | 10 | 9 | 10 | 0.1966 | 0.0650 |
| 11 | 11 | 10 | 11 | 0.3744 | 0.1238 |
| 12 | 12 | 11 | 12 | 1.4680 | 1.1550 |
| 13 | 13 | 12 | 13 | 0.5416 | 0.7129 |
| 14 | 14 | 13 | 14 | 0.5910 | 0.5260 |
| 15 | 15 | 14 | 15 | 0.7463 | 0.5450 |
| 16 | 16 | 15 | 16 | 1.2890 | 1.7210 |
| 17 | 17 | 16 | 17 | 0.7320 | 0.5740 |
| 18 | 18 | 17 | 18 | 0.1640 | 0.1565 |
| 19 | 19 | 18 | 19 | 1.5042 | 1.3554 |
| 20 | 20 | 19 | 20 | 0.4095 | 0.4794 |
| 21 | 21 | 20 | 21 | 0.7089 | 0.9373 |
| 22 | 22 | 21 | 22 | 0.4512 | 0.3083 |
| 23 | 23 | 22 | 23 | 0.8980 | 0.7091 |
| 24 | 24 | 23 | 24 | 0.8980 | 0.7011 |
| 25 | 25 | 24 | 25 | 0.2030 | 0.1034 |
| 26 | 26 | 25 | 26 | 0.2842 | 0.1447 |
| 27 | 27 | 26 | 27 | 1.0590 | 0.9337 |
| 28 | 28 | 27 | 28 | 0.8042 | 0.7006 |
| 29 | 29 | 28 | 29 | 0.5075 | 0.2585 |
| 30 | 30 | 29 | 30 | 0.9744 | 0.9630 |
| 31 | 31 | 30 | 31 | 0.3105 | 0.3619 |
| 32 | 32 | 31 | 32 | 0.3410 | 0.5302 |
| 33 | 33 | 7 | 20 | 2.0000 | 2.0000 |
| 34 | 34 | 8 | 14 | 2.0000 | 2.0000 |
| 35 | 35 | 11 | 21 | 2.0000 | 2.0000 |
| 36 | 36 | 17 | 32 | 0.5000 | 0.5000 |
| 37 | 37 | 24 | 28 | 0.5000 | 0.5000 |

# 16

# Design and implementation of a model (ADS-3G) of a traffic light using automated solar power supply

## D. A. Shalangwa

Department of Physics Adamawa State University, Mubi. Nigeria. E-mail: deshalangs3g@yahoo.com.

In this work, a model of an automated traffic light controller (ADS - 3G) had been designed, simulated, tested and implemented, using experimental techniques in electronic engineering, to manage the traffics at the busy four way junctions along Sahuda road in Mubi North Adamawa State, Nigeria. The designed was achieved with the help of 12V automated solar energy power supply, time base (555 timer), decade counter, D-flip-flop, timing sequence selector for red, green, amber and yellow light and relay circuit for switching the appropriate light. The average volume of vehicular traffics observed for the period of one week (4/4/07 - 11/4/07) for Masalachi, Stadium, Sahuda and Sarki roads are 2368, 1996, 1982 and 138, respectively which prompted the development of the model (ADS - 3G) to allowed 17.50, 14.00, 10.50 and 7.00 s accordingly. This model is capable of eliminating the inefficiency and likely error associated with human traffic controller by minimizing accident and unnecessary traffic jams at the junction.

**Key words:** Solar energy, traffic jam, traffic flow, traffic light controller and human traffic.

## INTRODUCTION

Mubi is the second largest town in Adamawa state of Nigeria, it lies between latitude 9° 30' and 11° North of the equator and longitude 13° and 13° 34' East of Greenwich meridians, Mubi has a land area of 4728.77 km$^2$ and a population of 759,045 (Adebayo, 2004). Mubi has many road junctions; but cases of accidents are more often recorded at Sahuda road junction especially from 2003 to date (Road Safety, 2005). Mubi has been experiencing increasing volume of vehicles/motorcycles traffic which leads to increased in the risk of accident occasioned by motorists contending over right of lay in the roads. There are also problems of traffic jam on the road. The situation becomes worse on daily basis, at the road junction and much more critical at the Sahuda road junction. Accident generally leads to loss of life, destruction of vehicle and bring unnecessary delay to vehicles and also the solar energy power supply is been introduced in this design because of the inconsistency of electrical power supply by the Power Holding Company of Nigeria (PHCN), sometimes the power supply is less than eight hours in a day which consequently renders the traffic light controller useless. It also has some advantages over electrical power such as it is free in nature, easy to maintain, relatively cheap, less hazardous and the problems of pollution in electrical power has been eliminated in the solar energy (Website, 2007).

To address these problems up front the demand for traffic light controller becomes necessary and this prompted the emergence of this work (Figure 1).

The traffic light controller is a device that manages free flow of traffics along three, or more road junctions. The device has a sequence selector for red, green, amber and a yellow light that indicates present state of the traffic flow. Here green colored light mean "Go", which permits entry into the intersection. Red coloured light means "Stop"; which prohibits entry into the junction. Amber light allows entry of traffic but requires clearance of intersection; and, yellow light means "Fault" indicating that there is a fault. Here, the yellow light will remain on until the fault is cleared. The device also consists of time base (555 timer), decade counter D-flip flop and relay circuit for switching the appropriate light. The automated traffic light controller was so designed on the basis of electronic instrumentation and experimental techniques in electronic engineering so as to allow more time of traffic flow for more busy roads across the junctions while less time for less busy roads, as others remain stand still to avoid collision. The traffic light controller ensures that the waiting vehicles/motorcycles are not unnecessarily delayed. This traffic light controller is capable of successfully managing the flow of traffics at the Sahuda road junctions in Mubi North in Adamawa state, Nigeria.

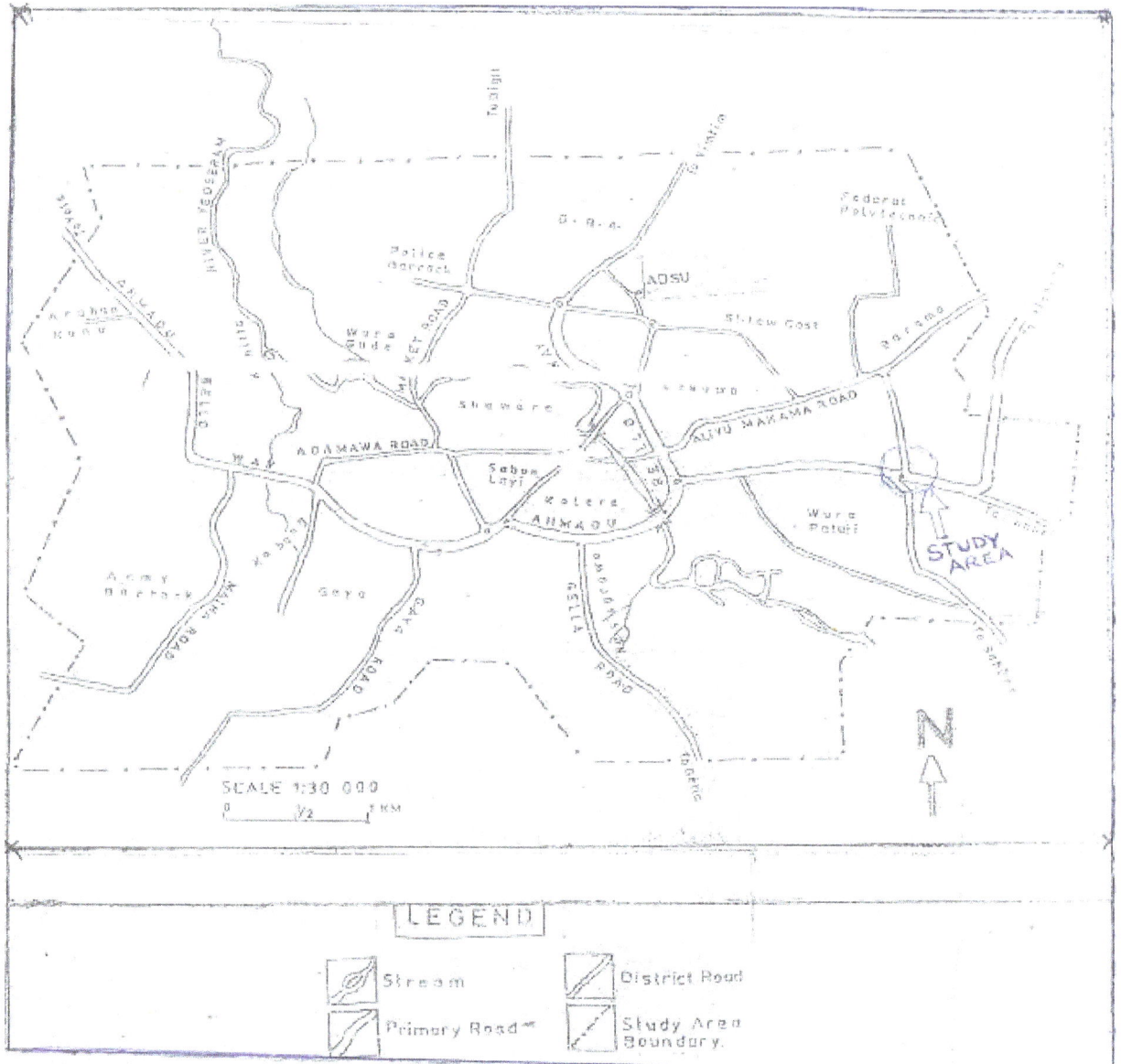

**Figure 1.** Map Mubi showing road network and the study area.

## THEORY AND DESCRIPTION OF ELECTRONIC COMPONENTS USED

### Power supply

The design used 12V rechargeable battery and solar panel from sunlight. The circuit consist of oscillator and a regulator transistor, the solar energy charge the battery when sunlight is bright enough. A diode is required between the panel and the battery as it leak 1 mA from the battery when it is not illuminated (Website, 2007).

The regulator transistor is designed to limit the output to 12V; this voltage will be maintain over the capability of the circuit, the transistor oscillator is a high current type as it is turned ON for a very short time of period to saturate the core of the transformer. The energy is then released as a high voltage pulse. These pulses are then passed to the electrolytic capacitor and appear as a 12V supply as a supply to the traffic light controller circuit. The supply can be made automatic by adding a 1 KΩ resistor and diode (INA 148) as in Figure 2(a). When the power supply is connected to the main circuit it starts operation satisfactorily (Frank, 2004) and circuit is further simulated as shown in Figure 2(b) to ascertain the technical function of the circuit.

### The oscillator

The 555 timer was used in the design as a stable mode

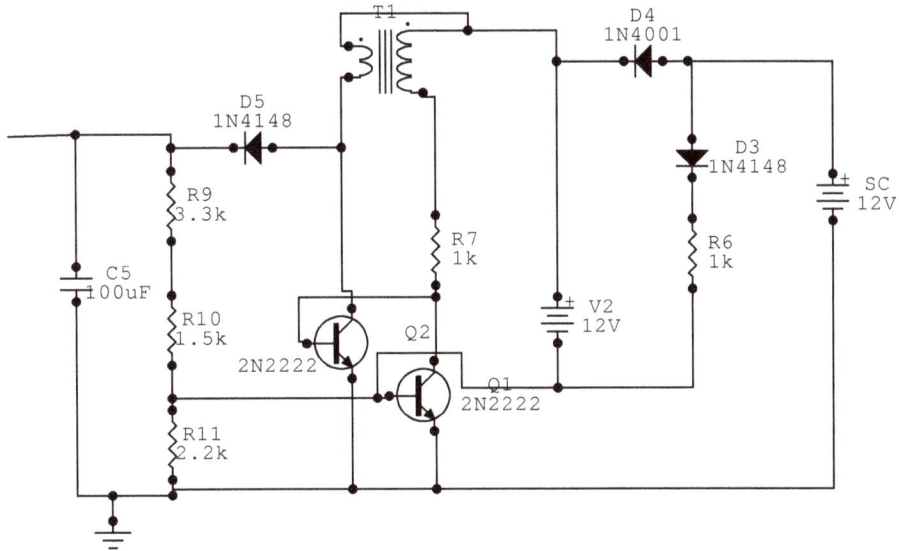

**Figure 2(a).** Solar energy power supply unit.

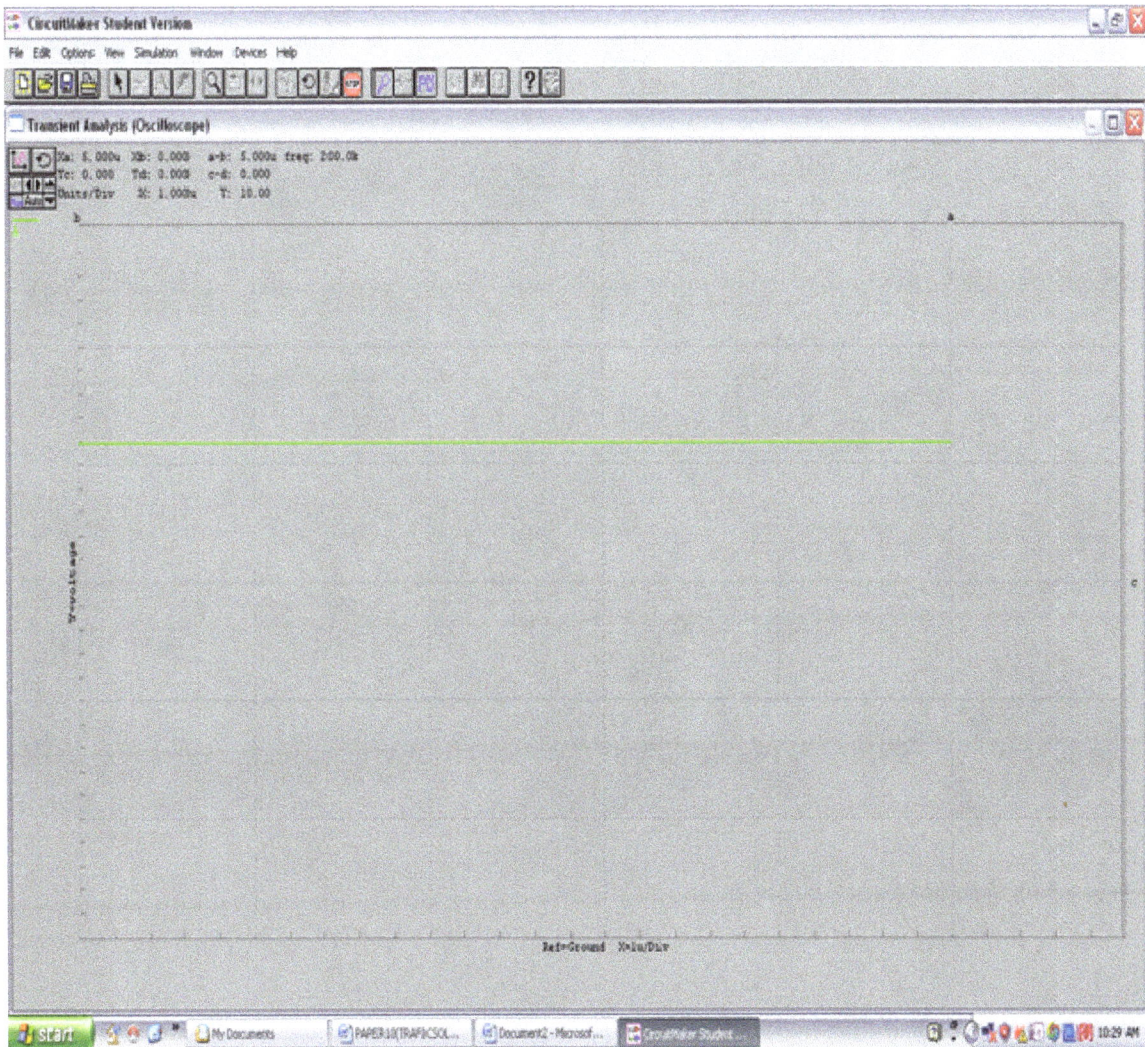

**Figure 2(b).** Simulated result of solar energy power supply (DC output Voltage).

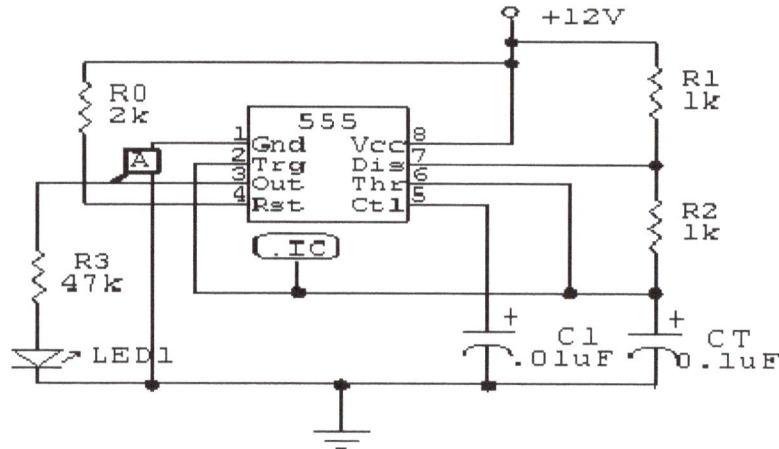

**Figure 3(a).** An oscillator circuit.

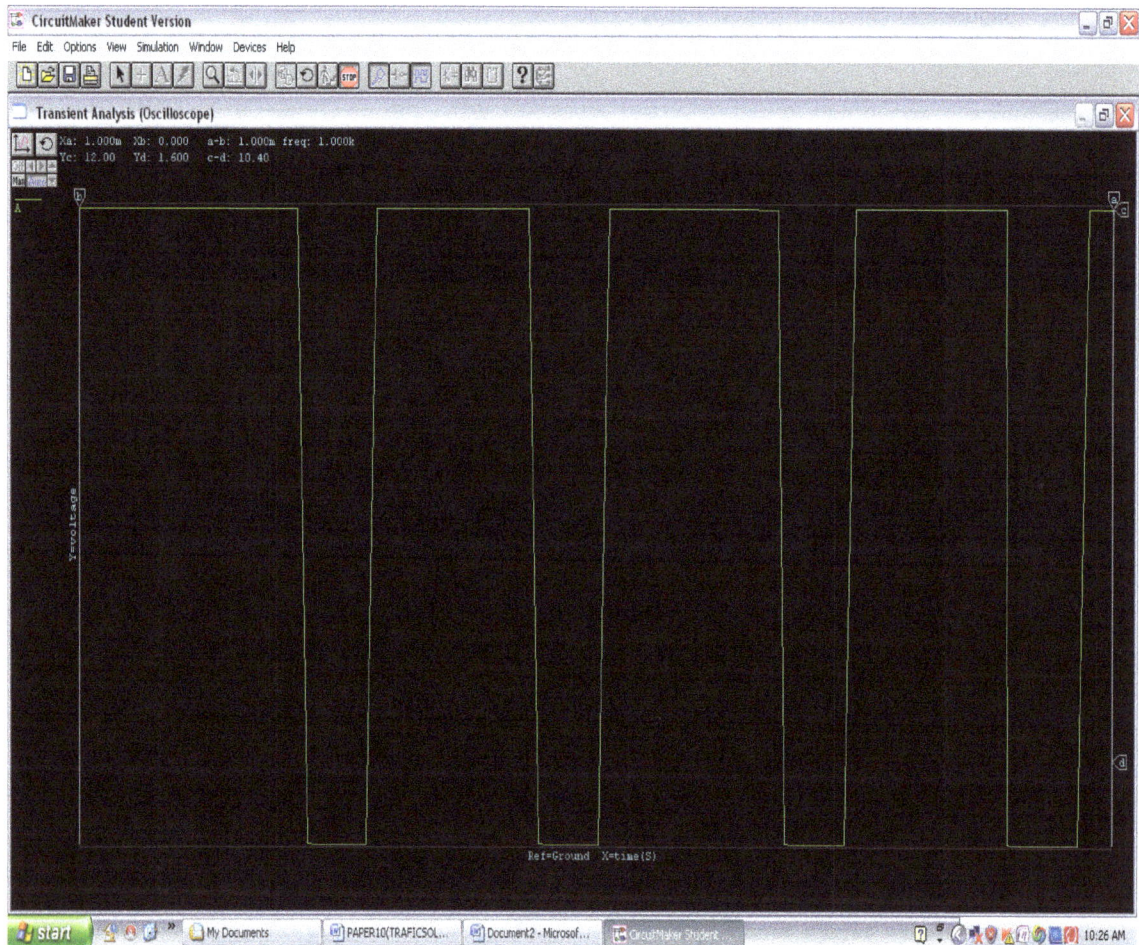

**Figure 3(b).** Simulated result of an oscilator circuit.

configured to operate as a multivibrator as shown in Figure 3a. The oscillator generates pulse by charging and discharging the capacitor C such that the charging time is

given by

$$T_1 = \ln 2 \, (R_1 + R_2) \, C_1 \qquad\qquad\qquad (1)$$

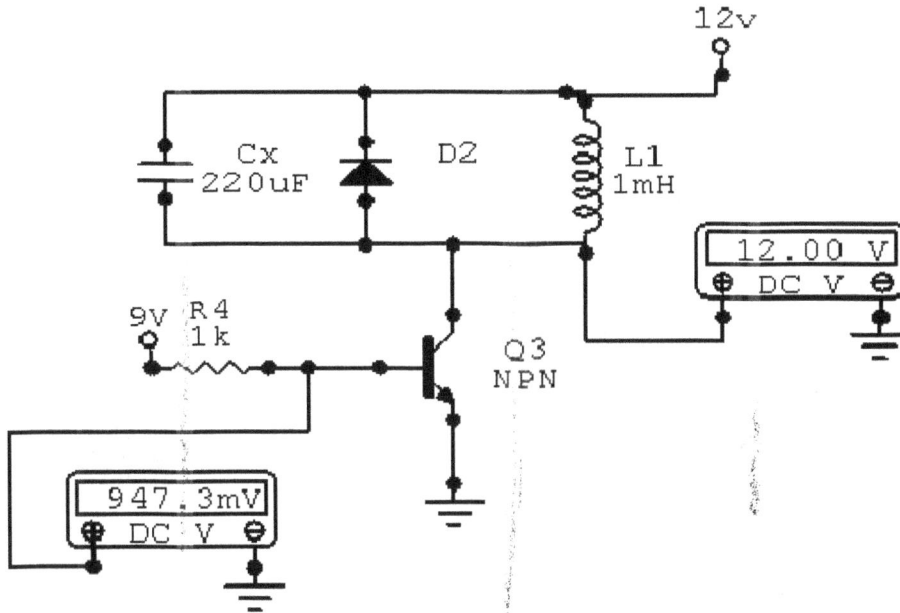

**Figure 4(a).** Switching and Interfacing circuit.

Similarly the standard discharging time is given by (Paul, 1995)

$$T_2 = ln2R_2C_1 \qquad (2)$$

Where the $T_1$ and $T_2$ stands for the charging and discharging time respectively (Tony, 2001; Ronald, 2001).

The total period taking by the capacitor to charged up completely and discharged is given by

$$T = T_1 + T_2 \qquad (3)$$

So that the frequency of the oscillation can be computed using the following expression as (Ali, 2007)

$$F = \frac{1}{T} \qquad (4)$$

### Indicator stage

The indicator stage consists of resistor $R_3$ and diode $D_1$ as shown in Figure 3a. Here the diode $D_1$ becomes "on" only when the clock pulse is generated. The value $R_3$ was obtained using simple ohm's law, given by

$$R_3 = \frac{V_{cc} - V_d}{I_d} \qquad (5)$$

Where $V_d$ is the diode drop; Vcc is the supply voltage $I_d$ is

the diode current (Charles, 1979; Loveday, 1984) and circuit is simulated as shown in Figure 3(b).

### Switching/interfacing circuit

The switching circuit as shown in Figure 4a is built on an NPN transistor with a $\beta = 40$; where $\beta$ represents the gain of NPN transistor, applying Kirchoff's voltage Law to the circuit it yields.

$$V_{CC} - I_B R_B - V_{BE} = 0$$

$$I_B = \frac{V_{CC} - V_{BE}}{R_B} \qquad (6)$$

$$I_C = \beta I_B \qquad (7)$$

Where $I_B$ is the base current; $\beta$ is the current amplifier and $I_c$ is the collector current is controlled by base current (Jones, 1993; Hughes, 2004).

### The interfacing circuit

The interfacing circuit in Figure 4(a) simply involves interfacing the switching circuit through the relay to the signal that indicates traffic flow controller condition at any particular point in time. The relay has two double contacts which are connected to the Red and Green light while the other double contact is connected to Amber and yellow

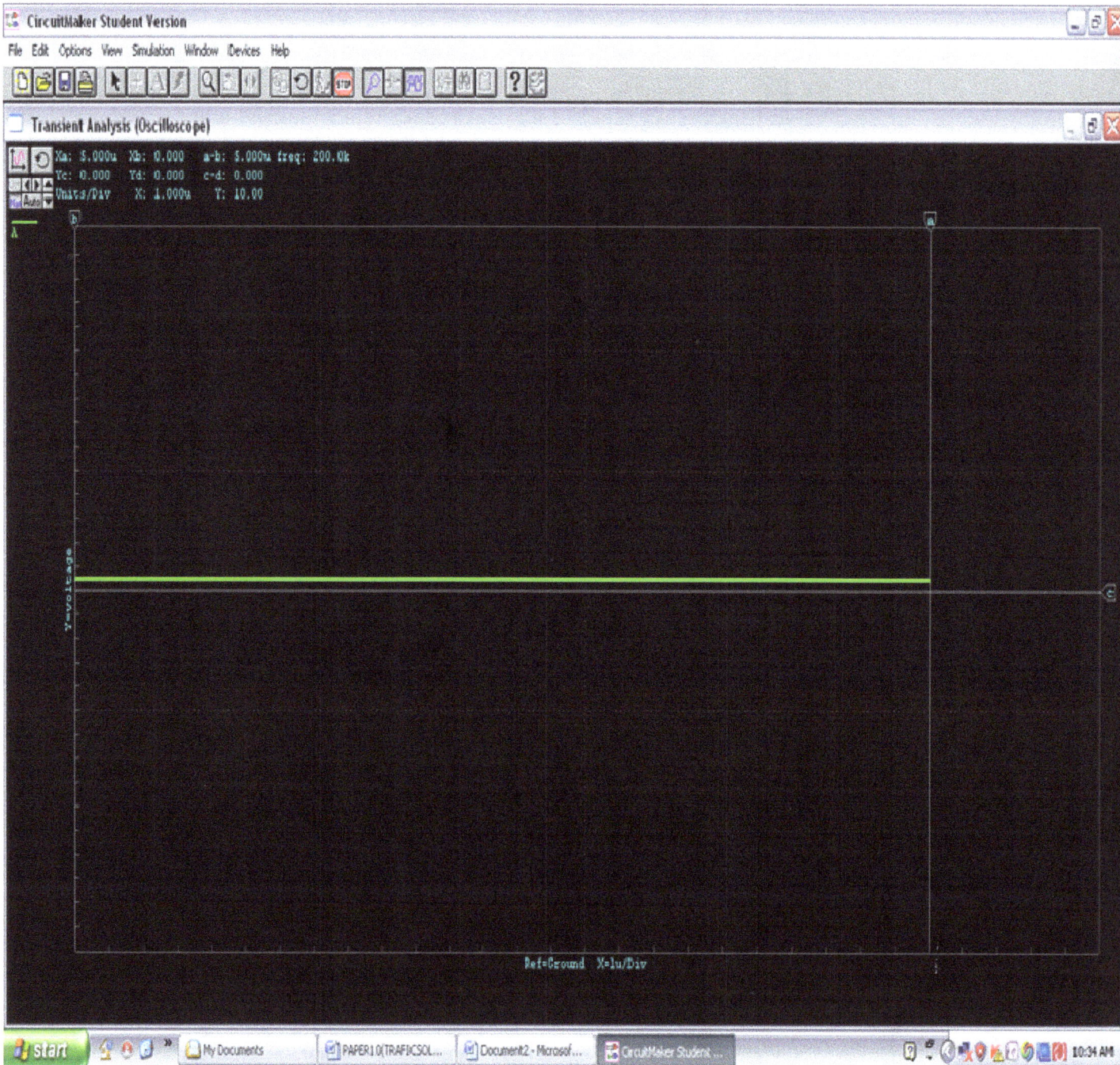

**Figure 4(b).** Simulated result of Switching and interfacing circuit (DC voltage).

light. On the double contact relay, the Red is connected normally to the closed path and Green to the opened path. When energized as a result of the transistor conducting at that instant, the Red path and the other path will open for the green light to come on. Similarly the other double contact will be energized for the Amber to come up when the transistor is conducting, (Morley, 1994; Ralph and Richard, 1992) and circuit is simulated as shown in Figure 4(b).

**The counter circuit**

The counter circuit as shown in Figure 5 is built on an IC (4017). Such that when the reset (pin 15) of the counter is taken HIGH, the counter will make the output "0" to go HIGH (1). When 'CLOCK INHIBIT' pin 13 is taken to HIGH, the counter will FREEZE on the output that is currently HIGH. The reset (pin 15) is connected to the supply via R5 and C3 to the earth, the reset is achieved

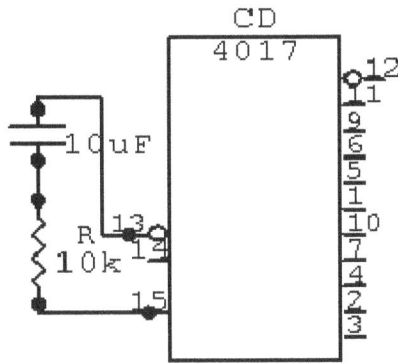

**Figure 5.** Counter circuit.

when C3 is supplied; since voltage across the capacitor cannot change instantaneously this makes the voltage across it to be zero. C3 then start to change and when fully charged creates an open circuit making R5 to take (pin 15) to ground; this process continues for the twenty (20) counting sequence (Onohaebi, 2006; Theraja and Theraja, 1997).

## DESIGN PROCEDURES

The design of the automated traffic light controller circuit took the following stages; as shown in Figure 6.

### Power supply

The only component that has to make in the transformer, is the core of a 20 mH choke is used and re-wound with two winding and remove the five winding, the first winding is 60 turns and the ends are connected to the pins at the end of the core. The other winding is 35 turns and has flying leads on the board. The 35 turns winding must be connected in a special way to provide a positive voltage to the base of the oscillator transistor, the operation of the circuit itself depend on the direction of the winding relative to the other.

### The oscillator

The value of the capacitor C that generated a pulse by charging and discharging was computed using Equation (1) as 225 µF with $T_1 = 3.5$s, $R_1 = 1$ kΩ and $R_2 = 10$ kΩ. But for practical purposes, the value of the capacitor $C_1$ chosen from data book was $C_1 = 220$ µF as the nearest available value (ECG, 2000). The time ($T_2$) taken for the capacitor $C_1$ to be discharged was computed using Equation (2) as 1.5 s and the frequency of the oscillation of the signal from Equation (4), was also computed to be 0.2 Hz.

### Indicator stage

The value of $R_3$ was computed from Equation (5) as $R_3 = 10$ kΩ with $V_d = 2$V and $I_d = 10$ mA.

### Switching circuit

The transistor base current was obtained from Equation (6) as $I_B = 10$mA with $V_{cc} = +12$V, $V_{BE} = 0.7$V and $R_B = 10$ KΩ while the collector current was computed from Equation (7) as $I_c = 0.04$A since $\beta = 40$.

## MATERIALS AND METHODS

The materials used for the design and its component rating is as follows 12V automated solar energy power supply; Diodes IN4001; Capacitors as filters; IC regulator (KA 7812); Oscillator (555 timer) as astable, +12V; Counter ($V_{cc}=12$V), +12V; Switching circuit consist of indicator with 12V, 20 mA and R=320Ω, NPN transistor ($h_{fe}=40$, $V_{be}=0.7$V); capacitor 220µf and diode at $V_{cc} = 12$V; Interfacing circuit with relay type Jzc20 (4088), 10A, 12V DC with coil resistance of 320 Ω and Stop watch

### Method of data collection

The traffic volume at the busy four - ways junction along sahuda Road in Mubi North, Adamawa state Nigeria was observed and documented at two-hourly intervals from (6:00 am - 10:00 pm) daily for the period of one week. The resulting volume of traffic across the junction obtained from the field survey was tabulated as shown in Table 1.

### Methods of design

The method employed in the design was an adaptation of the standard traffic light controller, although in this design the power supply utilizes an automated solar energy instead of the know electrical power supply that is more generally used in Nigeria.

In this work, a modeled area controlled by the traffic light was constructed on wooden board (100 cm in length and 70 cm in width), which shows the landmark of the four-ways junction. The direction of the traffic flow and the respective traffic light poles were erected on the side of each road. The height of the standard poles is 30 cm with holes were drilled to fix the bulb at the top end of the poles made of timber wood.

The test carried out involved the operation of the controlled traffic light model and observation of each bulb. In the test, the "on or off" times for each bulb are the corresponding "1 or 0" shown in Table 2.

### Operational principle of the traffic light controller

When the system is powered on, the oscillator starts producing pulses, which are used to clock the counter. The outputs of the counter are fed into a logic AND Gate selectively in a D flip-flop circuit. The output of the gate switches the bipolar transistor, which controls the light through the relays. The light is then addressed through the relay as it switches ON while the others remain OFF.

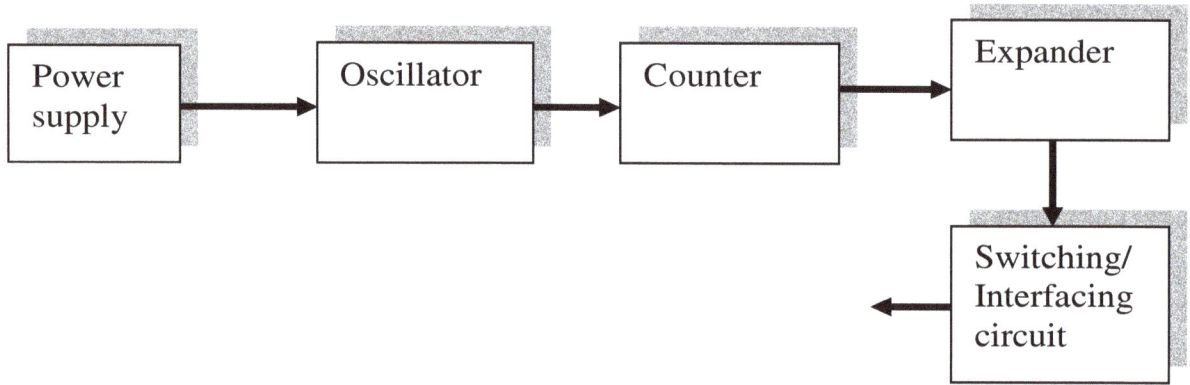

**Figure 6.** Block diagram of traffic light controller.

**Table 1.** The average volume of traffic across the four-way junction.

| Time | Sahuda | Masalchi | Stadium | Sarki |
|------|--------|----------|---------|-------|
| 6-8 am | 100 | 180 | 90 | 10 |
| 8-10 am | 420 | 586 | 480 | 30 |
| 10-12 noon | 318 | 310 | 324 | 20 |
| 12-2 pm | 207 | 182 | 200 | 17 |
| 2-4 pm | 188 | 206 | 192 | 32 |
| 4-6 pm | 412 | 571 | 418 | 12 |
| 6-8 pm | 218 | 233 | 222 | 11 |
| 8-10 pm | 80 | 100 | 70 | 06 |
| Total | 1982 | 2368 | 1996 | 138 |

Source. Field survey, 2007.

**Table 2.** Traffic light controller transmission of the designed model.

| Sarki Rd. | Stadium Rd | Sahuda Rd | Masalachi Rd |
|-----------|------------|-----------|--------------|
| RAGY | RAGY | RAGY | RAGY |
| 1 0 1 0 | 1 0 0 0 | 1 0 0 0 | 1 0 0 0 |
| 0 1 1 0 | 1 1 0 0 | 1 0 0 0 | 1 0 0 0 |
| 1 0 0 0 | 0 0 1 0 | 1 0 0 0 | 1 0 0 0 |
| 1 0 0 0 | 0 0 1 0 | 1 0 0 0 | 1 0 0 0 |
| 1 0 0 0 | 0 0 1 0 | 1 0 0 0 | 1 0 0 0 |
| 1 0 0 0 | 0 1 1 0 | 1 0 0 0 | 1 0 0 0 |
| 1 0 0 0 | 1 0 0 0 | 1 0 0 0 | 1 0 0 0 |
| 1 0 0 0 | 1 0 0 0 | 0 0 1 0 | 1 0 0 0 |
| 1 0 0 0 | 1 0 0 0 | 0 0 1 0 | 1 0 0 0 |
| 1 0 0 0 | 1 0 0 0 | 0 1 1 0 | 1 1 0 0 |
| 1 0 0 0 | 1 0 0 0 | 1 0 0 0 | 0 0 1 0 |
| 1 0 0 0 | 1 0 0 0 | 1 0 0 0 | 0 0 1 0 |
| 1 0 0 0 | 1 0 0 0 | 1 0 0 0 | 0 0 1 0 |
| 1 0 0 0 | 1 0 0 0 | 1 0 0 0 | 0 0 1 0 |
| 1 1 0 0 | 1 0 0 0 | 1 0 0 0 | 0 1 1 0 |

R = Red light, A = Amber light, G = Green light, Y = Yellow light, 1 = ON, 0 = OFF.

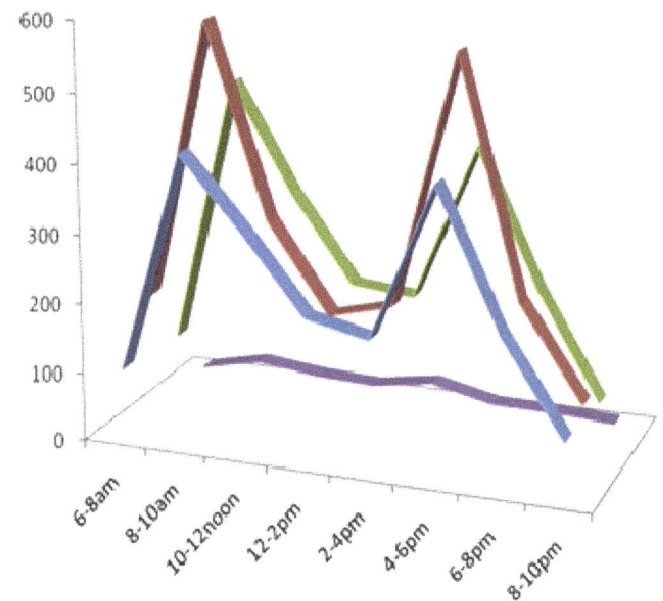

**Figure 7.** An analysis of average volume of traffic across the four – way junctions.

## RESULTS AND DISCUSSION

Figure 7 presents the average volume of traffic across the four-way junction in Sahuda road Mubi town. The observations of vehicular traffic at the junction made for the period of one week revealed that Masalachi road has the heaviest vehicular flow (2368) , followed by Stadium road (1996), Sahuda road (1982) and lastly with Sarki road (138) having the lowest flow ( with just $6 \pm 0.2\%$ of the highest flow). The traffic light controller was designed in such a way that more time is given to busy roads while less time for less busy road accordingly. The maximum time given to roads based on vehicular traffic flow follows: Masalachi road (17.50 s), stadium road (14.00 s),

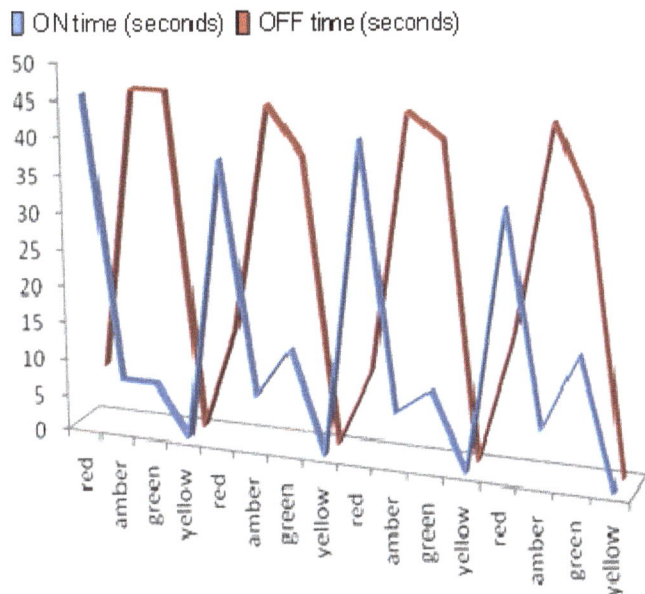

Figure 8. An analysis of timing for the traffic light controller.

Table 3. Simplification of timing traffic light controller.

| Signal (Head) | Colored Bulbs | On Time (sec) | Off time (sec) |
|---|---|---|---|
| Sarki road | Red | 45.50 | 7.00 |
| | Amber | 7.00 | 45.50 |
| | Green | 7.00 | 45.00 |
| | Yellow | 0.00 | 0.00 |
| Stadium road | Red | 38.50 | 14.00 |
| | Amber | 7.00 | 45.00 |
| | Green | 14.00 | 38.50 |
| | Yellow | 0.00 | 0.00 |
| Sahuda road | Red | 42.50 | 10.50 |
| | Amber | 7.00 | 45.50 |
| | Green | 10.50 | 42.00 |
| | Yellow | 0.00 | 0.00 |
| Masalachi road | Red | 35.50 | 17.50 |
| | Amber | 7.00 | 45.50 |
| | Green | 17.50 | 35.00 |
| | Yellow | 0.00 | 0.00 |

Sahuda (10.50 s) and Sarki (7.00 s).

Table 2 shows that each count of the controller had duration of 3.5 s after which a transmission was made to the next count. The transition of the counter from one count to another is continuous until the last count sequence was reached after which the counter returns to the start count again. There are fifteen (15) counts in all, implying that each count sequence has duration of 52.5s, based on this, the timing for the traffic light controller was developed for the four roads intersection at the four way - junctions as presented in Figure 8. The 3.5 s was chosen to ensure that sufficient number of waiting vehicles /motorcycles are passed.

This model has achieved simplification of timing for traffic light controller as shown in Table 3. Here the ON time and OFF time were computed based on the numbers of 1's or 0's with each count representing 3.5 s while 0 represents no time. In Masalchi road for instance there are five 1's. Therefore the ON time is (17.5 s) while the OFF time is 35.00 s making a total of 52.5 s. This was also applied to the other roads junction; for Stadium road ON time is 14.00 s, OFF time is 38.50 s; for Sahuda road the ON time is 10.50 s, while OFF time is 42.00 s; and Sarki road ON time is 7.00 s, OFF time 45.50 s.

Figure 9 compares the design value and practical value recorded when the test of the model was carried out as shown in Table 4. The test carried out involved the operation of the controller and observation of each bulb. In the test, the "ON and OFF" times for each bulb are the corresponding "1 or 0" with an insignificant error in reading of ± 0.05 counted in some values as shown in Table 5. This value of error has no effect on the efficiency of the system since it is very negligible.

The break down of the traffic light controller transmission design values in abnormal situation presented in Table 6 revealed that Red light of the traffic controller was completely ON for the whole 52.5 s while Yellow light flashes in every 3.5 s indicating the presence of faults in the traffic light controller.

Figure 10 gives the simplification of the traffic light controller in abnormal condition meaning that when any fault is developed in the traffic light controller, the Red light is completely ON for the period of 52.50 s while OFF will become 0.00, consequently, the Yellow light will keep flashing continuously at an interval of every 3.5 s, with an ON time of 24.50 s, while OFF time becomes 28.00 s. Here the Amber and Green lights will have their ON time as 0.00 seconds and OFF is 52.50 s.

## Conclusion

The design of automated traffic light controller was achieved successfully with the help of automated solar energy power supply, 555 timer connected in astable mode, decade counter, relay circuit and timing sequence selector for red, green, amber and yellow light. The Yellow light was also used in the circuit to indicate presence of faults in the automated traffic light controller. The automated 12V solar energy power supply.

The traffic light controller was developed to allow more time for busy roads while less busy roads attracted less time; this model was designed, implemented and tested with a satisfactory operation and performance efficiency.

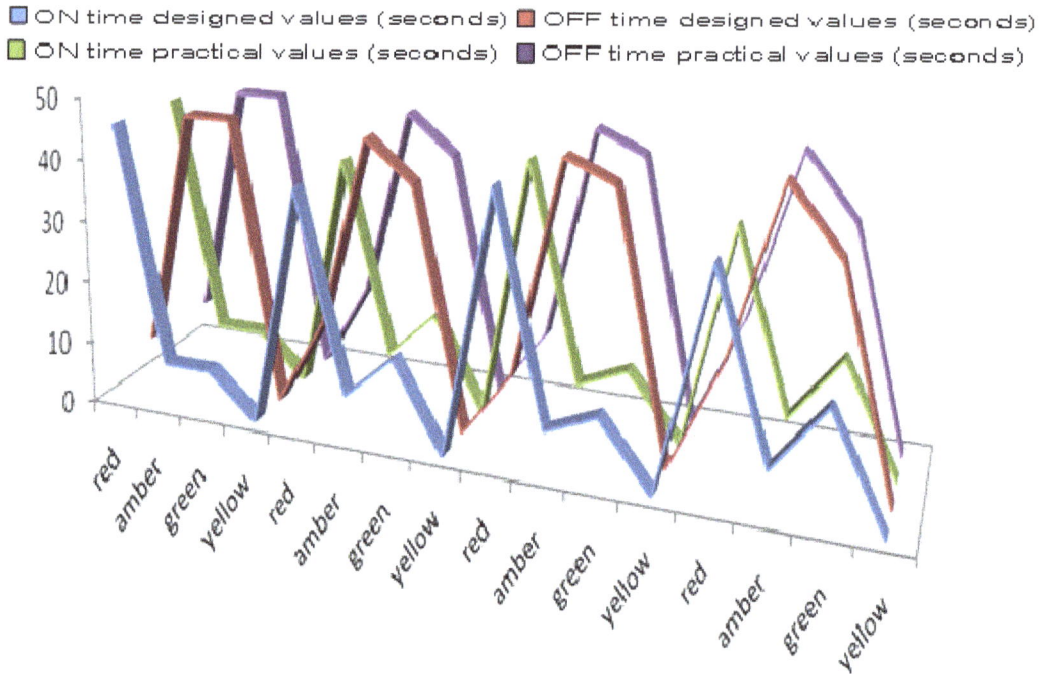

**Figure 9.** An analysis of the timing for each bulb.

**Table 4.** Break down of the timing for each bulb.

| Signal (head) | Colored bulbs | Designed values | | Practical values | |
|---|---|---|---|---|---|
| | | On time | Off time | On time | Off time |
| Sarki road | Red | 45.50 | 7.00 | 45.50 | 7.00 |
| | Amber | 7.00 | 45.00 | 7.00 | 45.50 |
| | Green | 7.00 | 45.50 | 7.00 | 45.50 |
| | Yellow | 0.00 | 0.00 | 0.00 | 0.00 |
| Stadium road | Red | 38.50 | 14.00 | 38.50 | 14.00 |
| | Amber | 7.00 | 45.50 | 7.00 | 45.50 |
| | Green | 14.00 | 38.50 | 14.50 | 38.00 |
| | Yellow | 0.00 | 0.00 | 0.00 | 0.00 |
| Sahuda road | Red | 42.50 | 10.50 | 42.50 | 10.00 |
| | Amber | 7.00 | 45.50 | 7.00 | 45.50 |
| | Green | 10.50 | 42.00 | 10.50 | 42.50 |
| | Yellow | 0.00 | 0.00 | 0.00 | 0.00 |
| Masalachi road | Red | 35.50 | 17.50 | 35.00 | 17.50 |
| | Amber | 7.00 | 45.00 | 7.00 | 45.50 |
| | Green | 17.50 | 35.00 | 17.50 | 35.00 |
| | Yellow | 0.00 | 0.00 | 0.00 | 0.00 |

It is recommended that further improvement on the system may be required to incorporate a device that can rectify the fault immediately, in case such situation occurs. This model will certainly eliminate the inefficiency associated with human traffic controller, also minimized incessant accident and unnecessary traffic jams at

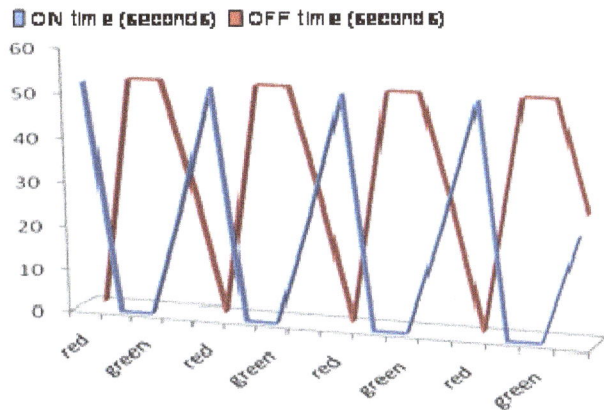

**Figure 10.** An analysis of traffic light in abnormal condition.

**Table 5.** Traffic light controller transmission design values in abnormal condition.

| Sarki Rd. | Stadium Rd | Sahuda Rd | Masalachi Rd |
|-----------|-----------|-----------|--------------|
| RAGY | RAGY | RAGY | RAGY |
| 1 0 0 1 | 1 0 0 1 | 1 0 0 1 | 1 0 0 1 |
| 1 0 0 0 | 1 0 0 0 | 1 0 0 0 | 1 0 0 |
| 1 0 0 1 | 1 0 0 1 | 1 0 0 1 | 1 0 0 1 |
| 1 0 0 0 | 1 0 0 0 | 1 0 0 0 | 1 0 0 0 |
| 1 0 0 1 | 1 0 0 1 | 1 0 0 1 | 1 0 0 1 |
| 1 0 0 0 | 1 0 0 0 | 1 0 0 0 | 1 0 0 0 |
| 1 0 0 1 | 1 0 0 1 | 1 0 0 1 | 1 0 0 1 |
| 1 0 0 0 | 1 0 0 0 | 1 0 0 0 | 1 0 0 0 |
| 1 0 0 1 | 1 0 0 1 | 1 0 0 1 | 1 0 0 1 |
| 1 0 0 0 | 1 0 0 0 | 1 0 0 0 | 1 0 0 0 |
| 1 0 0 1 | 1 0 0 1 | 1 0 0 1 | 1 0 0 1 |
| 1 0 0 0 | 1 0 0 0 | 1 0 0 0 | 1 0 0 0 |
| 1 0 0 1 | 1 0 0 1 | 1 0 0 1 | 1 0 0 1 |
| 1 0 0 0 | 1 0 0 0 | 1 0 0 0 | 1 0 0 0 |

junctions in developing countries.

**REFERENCES**

Adebao AA (2004). Mubi region, a Geographical synthesis, Paraclete and sonspp. 8-9.
Ali D (2007). Design of Tachogenerator for measuring angular velocity of a shaft using frequency to digital conversion techniques. J. League Res. in Niger 8(1): 186-201.
Charles A (1978). Electronic principles and application. Mc Graw-Hill Company, 2nd Edition pp. 117-130.
ECG (2000). Data book by NTE electronics Inc.

**Table 6.** The simplification of the traffic light in abnormal condition

| Signal (Head) | Colored Bulbs | On time (sec) | Off time (sec) |
|---------------|---------------|---------------|----------------|
| Sarki road | Red | 52.50 | 0.00 |
| | Amber | 0.00 | 52.50 |
| | Green | 0.00 | 52.50 |
| | Yellow | 24.50 | 28.00 |
| Stadium road | Red | 52.50 | 0.00 |
| | Amber | 0.00 | 52.50 |
| | Green | 0.00 | 52.50 |
| | Yellow | 24.50 | 28.00 |
| Sahuda road | Red | 52.50 | 0.00 |
| | Amber | 0.00 | 52.50 |
| | Green | 0.00 | 52.50 |
| | Yellow | 24.50 | 28.00 |
| Masalachi road | Red | 52.50 | 0.00 |
| | Amber | 0.00 | 52.00 |
| | Green | 0.00 | 52.00 |
| | Yellow | 24.50 | 28.00 |

Frank JE (2004). Control systems. http//www.power magindia.com/econtrol.html pp. 20-23.
Hughes E (1998). Electronical Technology. McGraw, 7th Edition pp. 101-106.
Jones L (1993). Basic Electronics for Tomorrow's world. Cambridge University Press pp. 78-82.
Loveday G (1984). Essential Electronics. Pitman pp. 15-19.
Nolan W (2004). Variable power supply. http//www.sound.westhost.com.project 44.html pp. 10-12.
Morley EH (1994). Principle of electricity. Long man Group Ltd. pp. 58-62.
Onohaebi O (2006). Design and construction of traffic light controller. J. Electrical Electron 10(1): 22-29.
Paul H (1995). The Art if electronics: Press Syndicate of University of Cambridge pp. 422-448.
Ronald TJ, Neal W (2001). Digital system, principles and application. Pearson Education Pte Ltd, 8th Edition pp. 203-213.
Ralph JS, Richard CD (1992). Circuit devices, and system John and sons. Inc. pp. 147-158.
Road safety (2005). Daily records book.
Theraja BL, Theraja AKA (1997). electronic technology, S. Chandi and Co. Ltd. 2nd Edition pp. 323-336.
Tony VR (2001). Timer/Oscillatortutorials. http://www.uoguelph.cal/antoon/garget/555/555.html pp. 1-12.
Website (2007). http://www.powersupplysolartitle.html 1 - 4

# 17

# Efficacious approach for satellite image classification

**Manish Sharma[1]\*, Rashmi Gupta[2], Deepak Kumar[1] and Rajiv Kapoor[2]**

[1]Electronics and Communication Engineering Department, Maharaja Agrasen Institute of Technology, Sector -22, Rohini, Delhi -110086, India.
[2]Delhi Technological University (Formerly Delhi College of Engineering) Bawana Road, Delhi 110042, India.

**The main idea behind any image classification process is to obtain highest accuracy possible. Minimum distance and parallelepiped method yielded acceptable results for image classification but they are bounded by their inherent limitations. On the other hand, fuzzy based systems are fast and provide good accuracy. In fuzzy, accuracy depends upon the type of membership function used, and how the membership functions in the output of FIS are arranged. In this paper Mamdani fuzzy inference system is used to classify image and how the arrangements and the type of fuzzy membership functions employed in the classification, affected the results obtained, are shown.**

**Key words:** Image classification, fuzzy logic, type of membership function, positioning of membership functions.

## INTRODUCTION

Satellite image processing plays a vital role for research and developments in "remote sensing", GIS, "agriculture monitoring", disaster management and many other fields of study. However, processing these satellite images requires a large amount of computation time due to its complex and lengthy processing criteria. The most common barrier in an image derived from an imaging device is its imperfection. The acquired image can be inconsistent, incomplete, uncertain or a completely a-miss. All these seems to be the main barrier in real time decision making but to switch the job faster, fuzzy (Li et al., 2005; Chao and Cheng, 1998; Bezdek et al., 2005) has proved to be an efficient solution. Since, the main problem lies in providing a better and reliable technique which can provide high performance for digital image analysis (even in situations with uncertainty in Gray level, texture, contours, edges detection, relationship between two segments of an image and all other noisy input conditions), with maximum efficiency and minimum manpower utilization. Fuzzy is one such technology that can implement this with ease and in much less time by

classifying the image with a procedure, which automatically categorizes all the pixels of image into land cover classes and other possible themes. The general architecture of a fuzzy logic system (FLS) which consists of four important components: fuzzifier, rules, inference engine, and defuzzifier are shown in Figure 1.

The fuzzifier transforms the crisp set values to fuzzy sets by applying fuzzification function. The rules and inference engine are the main component of fuzzy logic system which simulates the human reasoning process by making fuzzy inference on the inputs with IF THEN rules (Yiming, 1994). If we consider the satellite images, input data is not in the form of true colour image but for demonstration purpose, a three band {R (red) G (green) B (blue)} true colour image was taken (Figures 2 and 3). Each pixel has a particular colour, colour being described by the amount of red, green and blue components in it.

In this paper, the Mamdani min-operation implication method has been implemented (Jang, 1993). Defuzzifier converts the resultant fuzzy set back to a crisp value set which is the system output. Generally, rules are constructed and the output membership functions are arranged in random order without considering the effect of their position on the output, which leads to decline in accuracy of classification. However, if the arrangements

---

\*Corresponding author. E-mail: manishsharma.mait@yahoo.com.

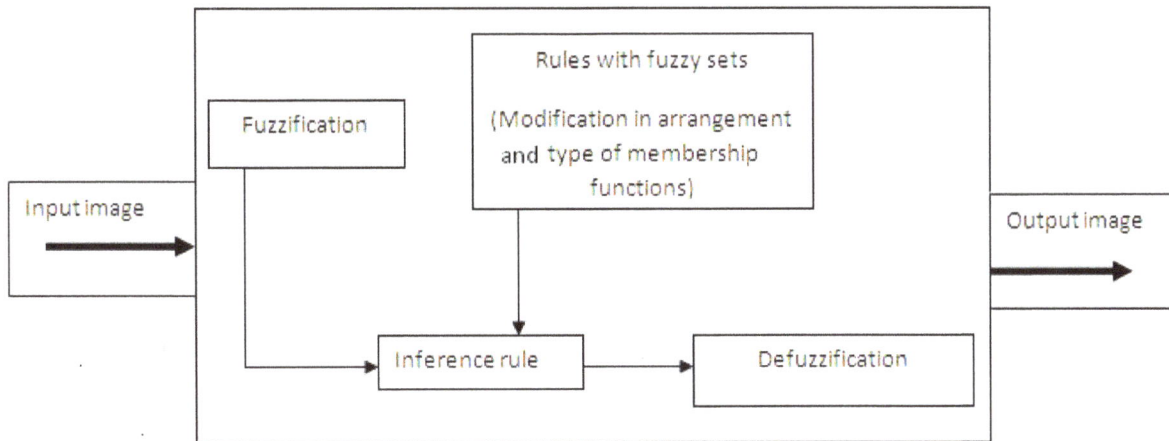

**Figure 1.** Architecture of fuzzy inference system.

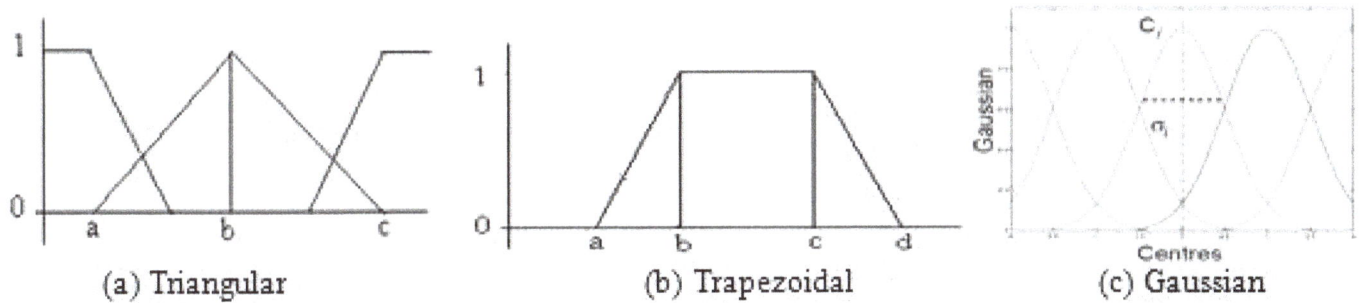

(a) Triangular                     (b) Trapezoidal                     (c) Gaussian

**Figure 2.** Membership functions.

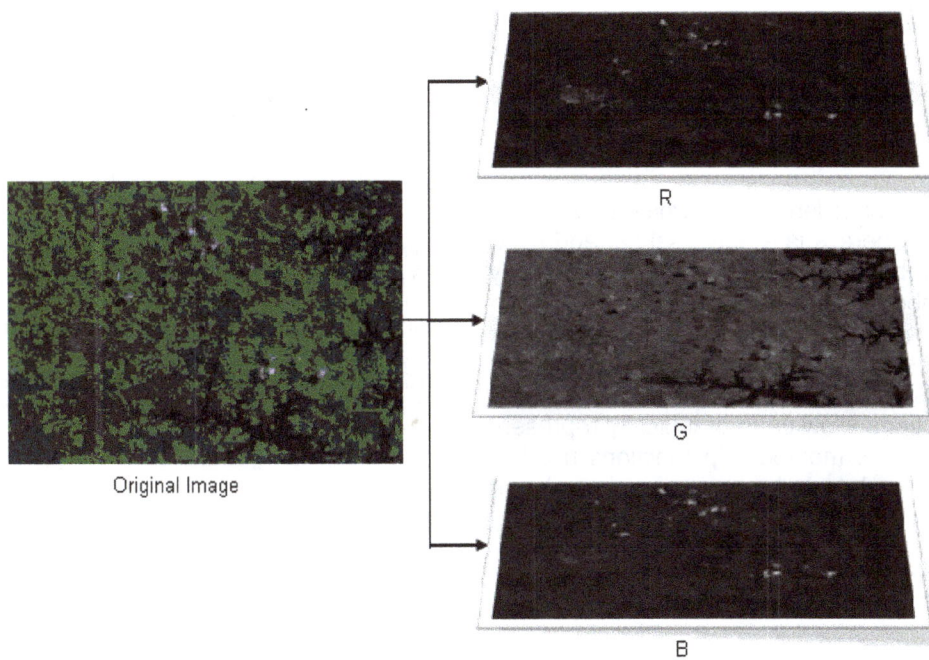

**Figure 3.** Image split into three bands red (R), green (G) and blue (B) bands.

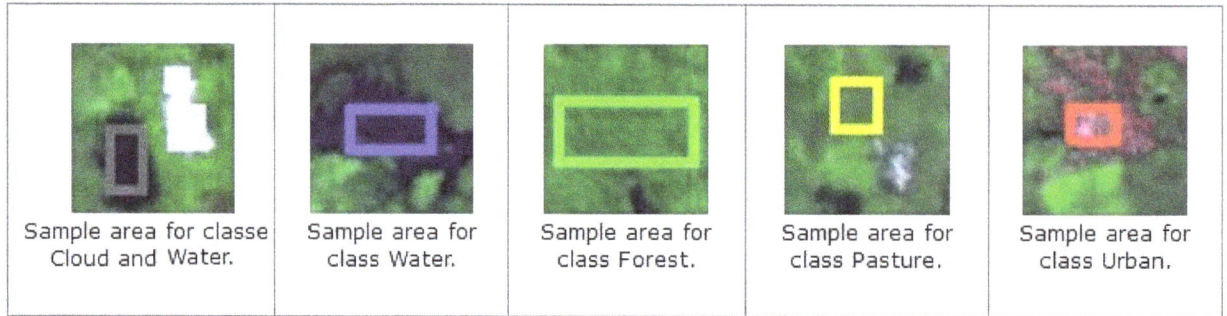

Figure 4. Sample area for various land classes.

of membership functions in output are carefully selected, it leads to a tremendous rise in accuracy. A better arrangement will be, to put membership functions in the output, adjacent to each other if their input membership functions are close or overlapping. Which had been done and rise in accuracy with the same rules is shown in Table 4. The concept was implemented on multi spectral data lines with the spectral pattern (set of radiance measurement obtained in the various wavelength bands for each pixel) used as the numerical basis for categorization using the notion of the normalized fuzzy matrices. This can be implemented with Mamdani-type or Sugeno-type fuzzy inference techniques (Mamdani being implemented in the paper). The method has been implemented by incorporating the suggested fuzzy logic-based representations with assumptions that the fuzziness of all the optimization formulation parameters are true and only spectral and radiometric characteristic of image pixels being considered without using any geometrical and topological relation between the pixels. Finally, changes were made in the arrangement and type of membership function to analyse the variable effects of these changes on the output. The results obtained clearly demonstrate the consistency and robustness of the developed approach.

A fuzzy set is a set of ordered pairs which is given by A= ((x, $\mu_A(x)$): xεX), where X is a universal set and $\mu_A(x)$ is the grade of membership of the object x in A (usually $0 \le \mu_A(x) \le 1$). A membership function $\mu_A(x)$ is characterized by $\mu_A$: X → (0, 1) where X is the universe of discourse, x is a real number describing an object or its attribute and each element of X is mapped to a value between 0 and 1. A membership functions allow us to graphically represent a fuzzy set. The various membership functions used in our classification method d can be represented as shown in Figure 4; where $c_i$ and $\sigma_i^2$ are the centre and width of the $i^{th}$ fuzzy set $A^i$ respectively.

### CLASSIFICATION METHOD

The intent of the classification process was to categorize all pixels

in a digital image into one of several land cover classes or themes. This categorized data can then be used to produce thematic maps of the land cover present in an image. Normally, multispectral data are used to perform the classification and indeed, the spectral pattern present within the data for each pixel was used as the numerical basis for categorization. With the help of already known (mapped) sample area the range of values for input membership functions of FIS can be determined which was used in constructing rules. After pre-processing of the image like removing noise and contouring the area under investigation, FIS (fuzzy inference system) with the names of each input variable (red (r), green (g), and blue (b)) and those of output variable (q) was created using rules. Mamdani's fuzzy inference method is the most commonly used fuzzy methodology and it expects the output membership functions to be fuzzy sets. After the aggregation process, there is a fuzzy set for each output variable that needs defuzification. Sugeno-type system can be used to model any inference system in which the output membership functions are either linear or constant. Here Mamdani type inference system was used. Figure 5 shows a Mamdani fuzzy inference system. It shows a simple diagram with the names of the input red (r), green (g) and blue (b). In each of the input we defined 5 membership functions (mf) because we wanted to classify the image into 5 different land classes (mf1 (water body), mf2 (clouds), mf3 (forest), mf4 (pasture), mf5 (urban)). Here we use the Gaussian/trapezoid/triangular curve for each membership function to study the effects on result. Mfl represents membership function for water body in red input variable. Again we define mf1, mf2, mf3, mf4 and mf5 in each of the other two bands for land classes. The range here lies from 0-255 for each membership function as true colour image was used. The range will vary according to image obtained from respective satellite. Based on the descriptions of the input (red, green and blue) and output variable (5 for each land class) the rules were constructed in the rule editor. Rules are defined as: IF (red is mf1) and (green is mf2) and (blue is mf3) then class (output) is mf4 (here mf1,mf2,mf3,mf4 are used asan example). The inputs were connected with AND function. By using IF-THEN rules and changing the order and type of various membership functions, we obtained different result having different accuracy.

### RESULTS

The Mamdani fuzzy logic system was applied to classify the image into 5 land classes and the accuracy was determined. The type and position of output membership functions were changed to analyse changes in the result. First the input membership function of water body was

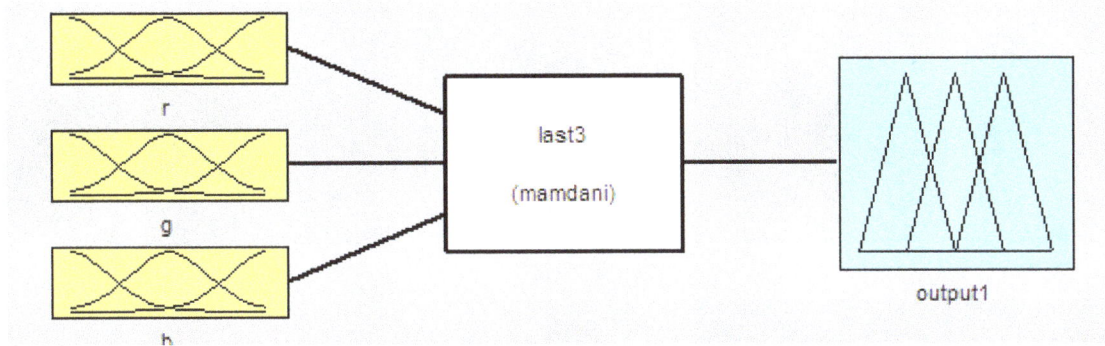

**Figure 5.** Mamdani FIS.

**Table 1.** Input range for RGB component for various classes.

| Class | Max –min value for input band red | Max -min value for input band green | Max -min value for input band blue |
|---|---|---|---|
| Sea/water | 11 to 0 | 12 to 8 | 45 to 30 |
| Cloud | 190 to 140 | 160 to 25 | 245 to 205 |
| Forest | 30 to 10 | 75 to 50 | 50 to 35 |
| Pasture | 50 to 30 | 110 to 90 | 65 to 40 |
| Urban | 120 to 75 | 90 to 60 | 98 to 70 |

changed while keeping all the other input membership function same and the result were studied.

Total number of pixels in image 781 × 671 = 524052

Table 1 shows the loss of pixels that is misclassification when membership functions were changed. The loss (misclassification) is very high in case of trapezoidal membership function.Losses were measured taking number of pixels for waterbody in case of output for gaussmf case as the base.

From the output obtained as shown in Figure 6, it was clear that in other cases original water body pixels were wrongly classified as other land classes. Although membership functions like triangular and trapezoidal gave sharper edges but the loss of pixel that is classification error was clearly visible in case of trapezoidal and triangular membership functions. Since only input membership function for waterbody was changed, the effect on the other classes was minimal. On modifying the arrangement of membership functions and keeping the rules same, different results were obtained. In first attempt of classification, the forest and the pasture land classes, which had input membership functions having values which were very close to each other, in one or the two bands, membership function in the output for these land classes were not placed adjacent to one another.

Figure 7 shows rule editor of first case in this arrangement mf4 which represent urban land class was

placed between forest (mf3) and pasture (mf5) land class which have very similar input values in one or two bands.

Figure 8 shows the output for this (first) arrangement. The misclassification in the case of urban land class was clearly visible. Many of the pixels which were pasture or forest were misclassified under urban land class.

Table 2 shows the colour taxonomy, the yellow colour was used to depict pasture similarly green for forest, blue for water body, white for clouds and red for urban.
In the second case two closely related membership functions were placed adjacent to each other and the unrelated membership function was not placed in-between them.

Figure 9 shows forest (mf3) and pasture (mf4) land class placed adjacent to each other and the urban (mf5 here) membership function, whose input range was not overlapping in any of the bands of input mf of forest and pasture, was not placed in-between them.

Figure 10 shows the result of the second arrangement. The improvement in classification was clearly visible. Original pasture and forest class pixels were not misclassified as urban. The change in arrangement did not affect the output for two other land classes- clouds and waterbody. Idea for accuracy assessment methods of classification results comes from the selecting random sample with known classes and then let methods 'say' what these samples are. With 100 random selected samples, Table 3 shows the comparison of two arrangements.

100 samples from the output of two arrangements were

(a)Original image

(b)Classified output with triangular membership function

(c)Classified output with trapezoidal membership function

(d)Classified output with gaussian2 membership function

(e)Classified output with Gaussian membership function

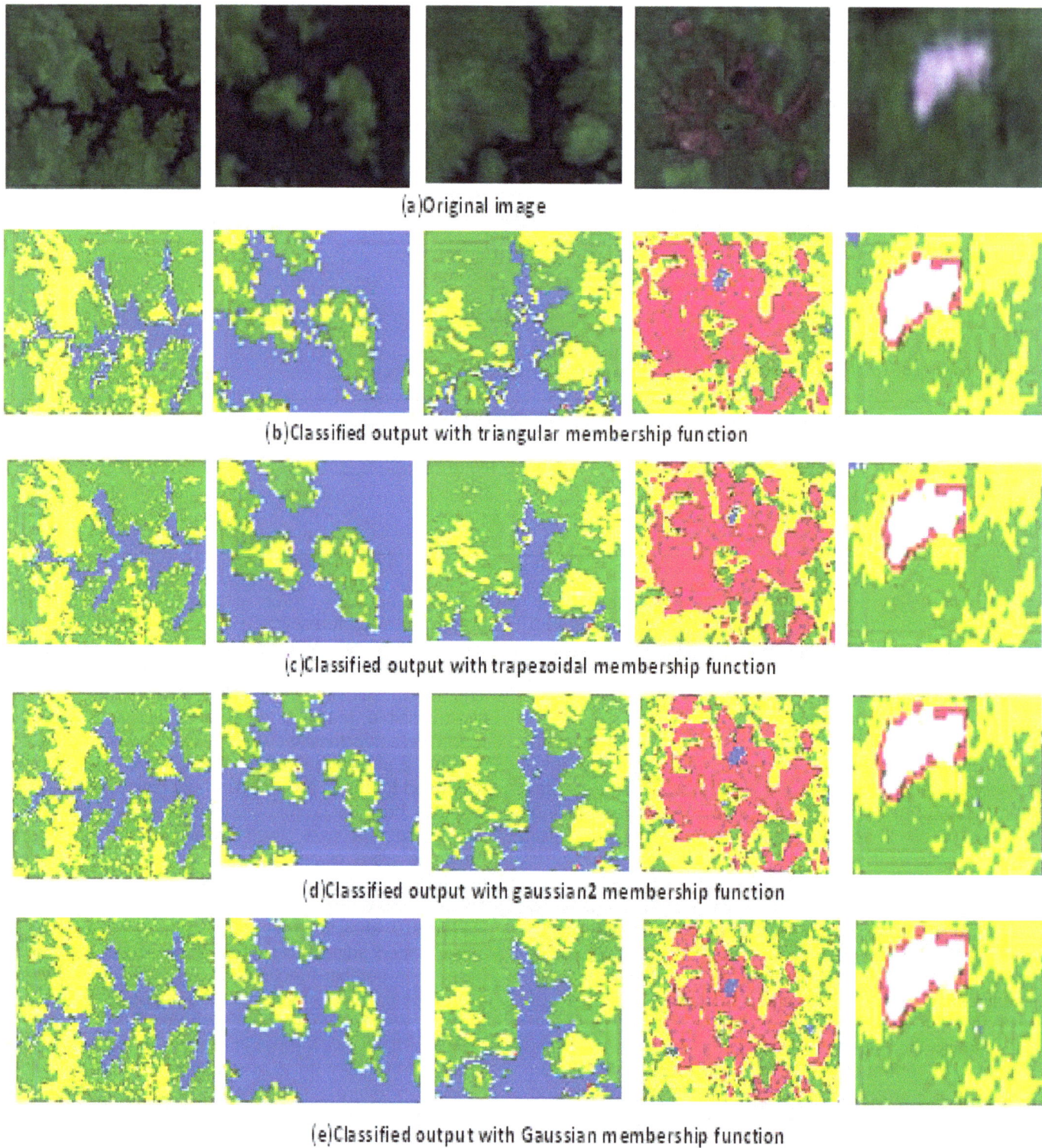

**Figure 6.** Classified results obtained using various membership functions.

taken and they were verified with the original image. The accuracy obtained in the first arrangement was 43% whereas the same for second arrangement was 87%.

**Conclusion**

The positioning of membership functions have a close

1. If (r is mf1) and (g is mf1) and (b is mf1) then (output1 is mf1) (1)
2. If (r is mf4) and (g is mf4) and (b is mf6) then (output1 is mf2) (1)
3. If (r is mf2) and (g is mf3) then (output1 is mf3) (1)
4. If (r is mf3) and (g is mf3) and (b is mf3) then (output1 is mf4) (1)
5. If (r is mf5) and (g is mf6) and (b is mf2) then (output1 is mf5) (1)

If                      and                      and
  r is                    g is                    b is

| mf1 | mf1 | mf1 |
| mf2 | mf2 | mf2 |
| mf3 | mf3 | mf3 |
| mf4 | mf4 | mf4 |
| mf5 | mf5 | mf5 |
| mf6 | mf6 | mf6 |

☐ not            ☐ not            ☐ not

**Figure 7.** Urban mf placed in between forest (rules).

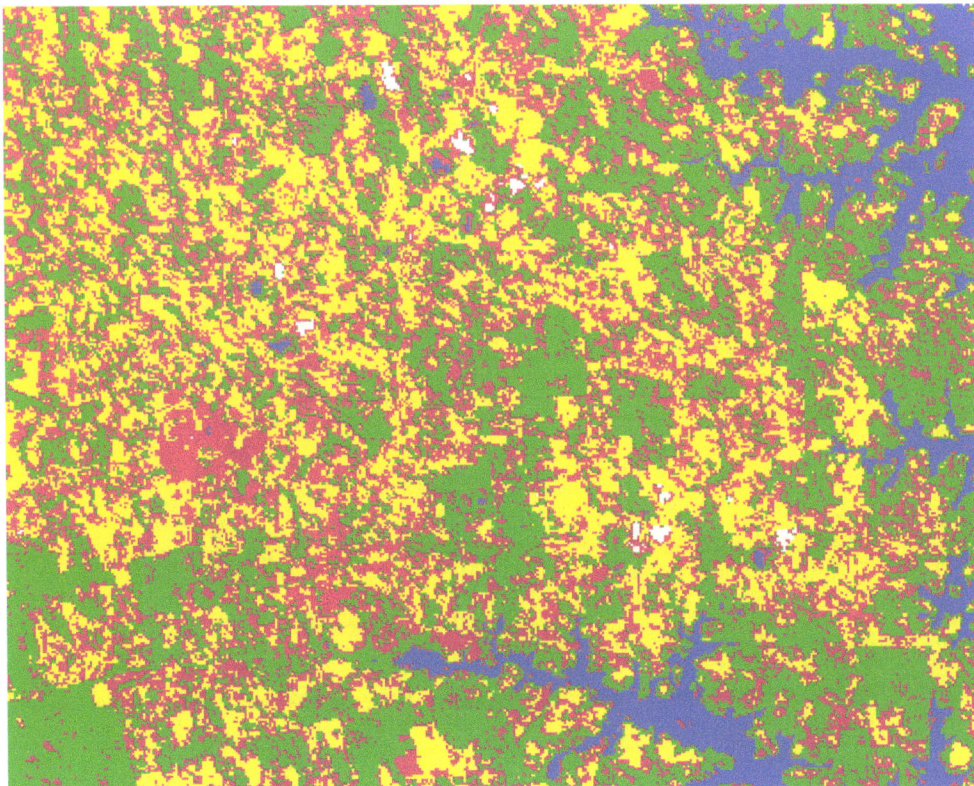

**Figure 8.** Output of 1st arrangement.

**Table 2.** Results of classification with different membership functions.

| Membership function for waterbody | Number of pixel in classified as waterbody | % of total pixel as waterbody | Loss % as compared with gaussmf |
|---|---|---|---|
| Gaussmf | 49095 | 9.36 | - |
| Trapmf | 35687 | 6.81 | 27.31 |
| Gauss2mf | 38098 | 7.27 | 22.39 |
| Trimf | 39827 | 7.60 | 18.87 |

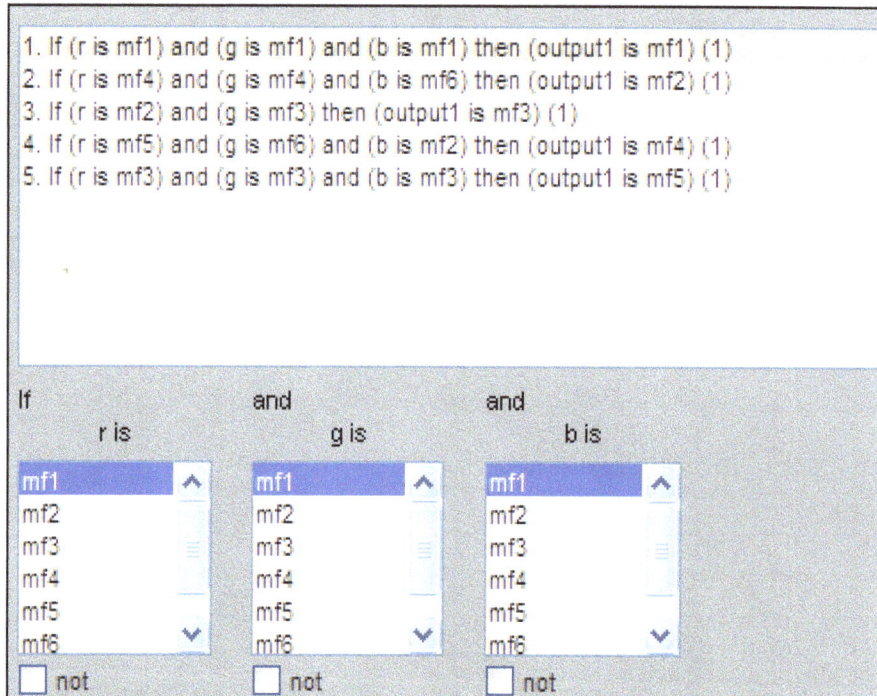

1. If (r is mf1) and (g is mf1) and (b is mf1) then (output1 is mf1) (1)
2. If (r is mf4) and (g is mf4) and (b is mf6) then (output1 is mf2) (1)
3. If (r is mf2) and (g is mf3) then (output1 is mf3) (1)
4. If (r is mf5) and (g is mf6) and (b is mf2) then (output1 is mf4) (1)
5. If (r is mf3) and (g is mf3) and (b is mf3) then (output1 is mf5) (1)

**Figure 9.** 2nd arrangement; urban mf placed after pasture (rules).

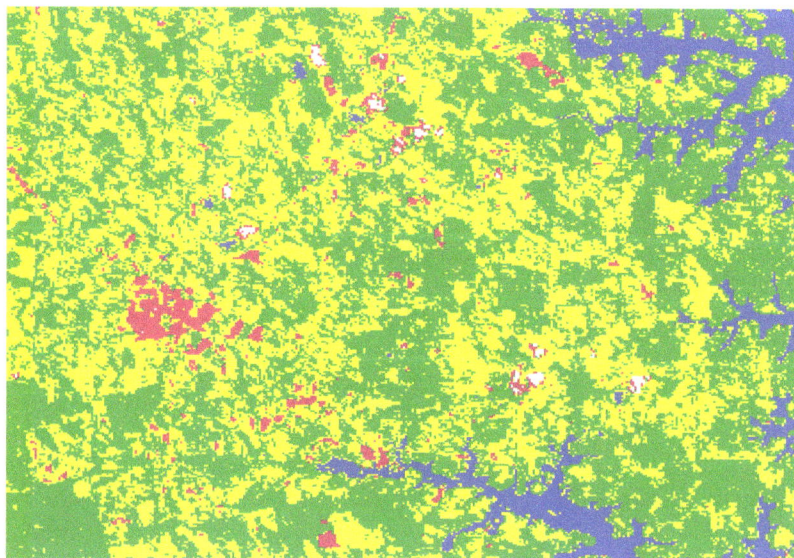

**Figure 10.** Classified output of second arrangement.

**Table 3.** Colour legend for the output image.

| | |
|---|---|
| Yellow | Pasture |
| Green | Forest |
| Blue | Water body |
| White | Clouds |
| Red | Urban |

**Table 4.** Comparison of arrangement of membership function.

| Arrangement of membership functions | Correctly classified sample | Misclassified sample | Accuracy (%) |
|---|---|---|---|
| 1st arrangement | 43 | 57 | 43 |
| 2nd arrangement | 87 | 13 | 87 |

relationship with accuracy of classification if output membership function of classes which are having input membership function overlapping or close to each other triangular lead to misclassification and loss of data of a particular class. Hence, Gaussian membership function appeared to be the best choice.

## REFERENCES

Arakawa K (2000). Fuzzy Rule-Based Image Processing with Optimization Fuzzy Techniques in Image Processing. Studies in Fuzziness and Soft Computing, Springer Verlag.

Bales CLK (2001). Modelling Feature extraction for erosion sensitivity using digital image processing, Master Thesis, Department of Geography, University of New Mexico.

Bezdek JC, Keller J, Krisnapuram R, Pal NR (1999). "Fuzzy Models and algorithms for pattern recognition and image processing." The Handbooks of Fuzzy Sets Series, Series Editors: D. Dubois and H. Prade, Kluwer Academic Publishers, Boston/London/Dordrecht.

Chao CJ, Cheng FP (1998). Fuzzy pattern recognition model for diagnosing cracks in RC structures. J. Comput. Civ. Eng., 12(2): 111-119.

Foody GM (1992). A fuzzy sets approach to the representation of vegetation continua from remotely sensed data: An example from lowland heath. Photogram. Eng. Rem. Sens., 58(2): 443-451.

Gonzalez RC, Woods RE, Eddins SL (2004). Digital image processing using MATLAB, Prentice Hall.

Ishibuchi H, Nojima Y (2008), Pattern Classification with Linguistic Rules, Fuzzy Sets and their Extensions: Representation, Aggregation and Models, 220: 1077-1095, Springer, Berlin.

Jang JSR (1993). ANFIS: Adaptive-Network-Based Fuzzy Inference System. IEEE Trans. Syst. Man. Cybernet., 23(3): 665-685.

Li YW, Chen SY, Nie XT (2005). Fuzzy Pattern Recognition Approach to Construction Contractor Selection. Fuzzy Optim. Decis. Making, 4(2): 103-118.

Wang Y, Jamshidi MO (2004). Fuzzy Logic Applied in Remote Sensing Image Classification. IEEE Int. Conf. Syst. Man Cybernet., 7: 6378 – 6382.

Yiming Y (1994). Rules Based Fuzzy Logic Inference. IEEE Int. Conf. Syst. Man Cybernet., 1: 465-470.

# 18

# Implementation of DSP algorithms on reconfigurable embedded platform

## J. S. Parab*, R. S. Gad and G. M. Naik

Electronics Section, Department of Physics, Goa University, Goa, India.

**Field programmable PLD's are becoming a standard in hardware technologies, as application demands have out placed the conventional processor's ability to deliver. The right combination of price, performance, ease-of use, along with significant power savings, can be achieved by using a Field Programmable Gate Array (FPGA). The versatility of these devices with the EDA tools such as Quartus, ISE, and Mentor Graphics integrated with MATLAB-Simulink gives upper hand for designer in any complex Digital signal processing (DSP) designs. Altera and Xilinx has DSP generator and System generator as target specific tools which support various IP cores libraries for such designs. We have designed an embedded system with Xilinx Spartan III based FPGA for real time audio processing. Various audio effects such as Echo, Reverberation, Fading, Flanging etc. can be demonstrated for real time performance. The designed system has 12-bit ADC tuned for the base-band signal unto 500 KHz. Also, the system can be configured as a DSP processor with IP cores like FFT, convolution etc. The reconstruction of analog signal is achieved with the help of 12-bit DAC converter module. Designed board has many applications in the field of biomedical, consumer, industrial and military. The same audio effects were tested on Altera CYCLONE II based DSP development kit.**

**Key words:** DSP, ASP, ECHO, embedded system, Spartan -III, FPGA, altera.

## INTRODUCTION

Signals play an important role in our daily life. Signals that we encounter frequently are speech, music, picture and video signals (Richard, 2004). Often, signal is spatio-temporal in nature. Speech and music signals represent air pressure as functions of time at a point in space. Most signals we encounters are generated naturally, however, signal can be generated synthetically or by computer. The objective of signal processing is to extract the information carried by the signal and this extraction of information depends on the type and nature of signal carried by it. There are four methods of digital sound synthesis such as wavetable, spectral, non-linear and synthesis by physical modeling. Wavetable synthesis produces recorded or synthesized musical events stored in the digital memory and played back on demand. Spectral synthesis produce sound from frequency domain

model basically by superposition of basis functions with time-varying amplitudes. Non-linear synthesis is a frequency modulation technique of time dependent phase term in the sinusoidal basis function. Physical modeling models the sound production methods of the vibrating physical structures by partial differential equations (Alles, 1980). We have implemented wavetable analysis in our design. The highest sampling frequency reported presently is around 1GHz. Such high frequencies are not usually used in practice since the achievable resolution of the A/D converter given by the word length decreases with an increase in the speed of the converters. For example, the reported resolution of an A/D converter operating at 1GHz is 6 bits (Poulton et al., 1987). On the other hand in most applications, the required resolution of an A/D converter is from 12 - 16 bits. Consequently, the sampling frequency of at most 10 MHz is presently a practical upper limit. Upper limit is becoming larger and larger with advances in technology. We have designed system having 12-bit ADC, having 1MHz sampling frequency.

---

*Corresponding author. E-mail. jsparab@unigoa.ac.in.

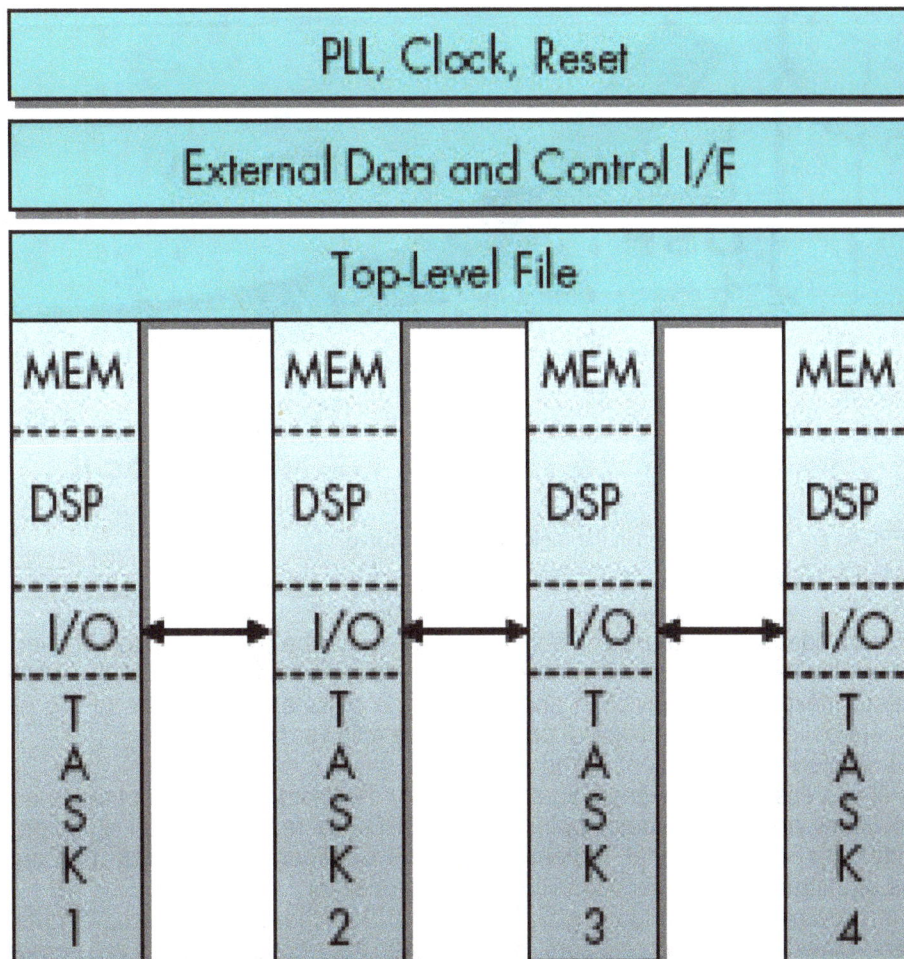

**Figure 1a.** FPGA block diagram.

## PROCESSORS PLATFORMS: FPGA AND DSP

Two types of programmable platforms are used in the Embedded and VLSI design application domain that is DSP and FPGAs.

DSP are a specialized form of microprocessor, while FPGAs are a form of highly configurable hardware. DSP processors have conventionally moved to higher levels of performance through a combination techniques such as increasing clock cycle speeds, increasing the number of operations performed per clock cycle, adding optimized hardware co-processing functionality (such as a Viterbi decoder), implementing more complex instruction sets, minimizing sequential loop cycle counts, adding high performance memory resources, implementing modifications including deeper pipelines and superscalar architectural elements. However, ultimately, each of these design enhancements seeks to increase the parallel processing capability of an inherently serial process.

The following factors make FPGAs promising, particularly for high performance computing applications:

(i) The potential for thousand-fold parallelism (ii) The embedding of control logic (iii) Presence of on-board memory in FPGA also has significant performance benefits. For one, having memory on-chip means that the processor logic's memory access bandwidth is not constrained by the number of I/O pins on device (iv) FPGA with greater capacity can occupy the same board footprint as an older device, allowing performance upgrades without board changes. Advantages of FPGAs for high performance computing are discussed in depth in one of the latest references (Kamat et al., 2009).

Diagram in the Figure 1a shows the how FPGA resources assigned can be tailored to the task requirements, which can be broken up along logical partitions. This makes a well-defined interface between tasks and largely eliminates unexpected interaction between tasks, because each task runs continuously, much less memory is required than in the digital signal processor, which must buffer the data and process in batches. As FPGAs distribute memory throughout the device, each task is permanently allocated the dedicated

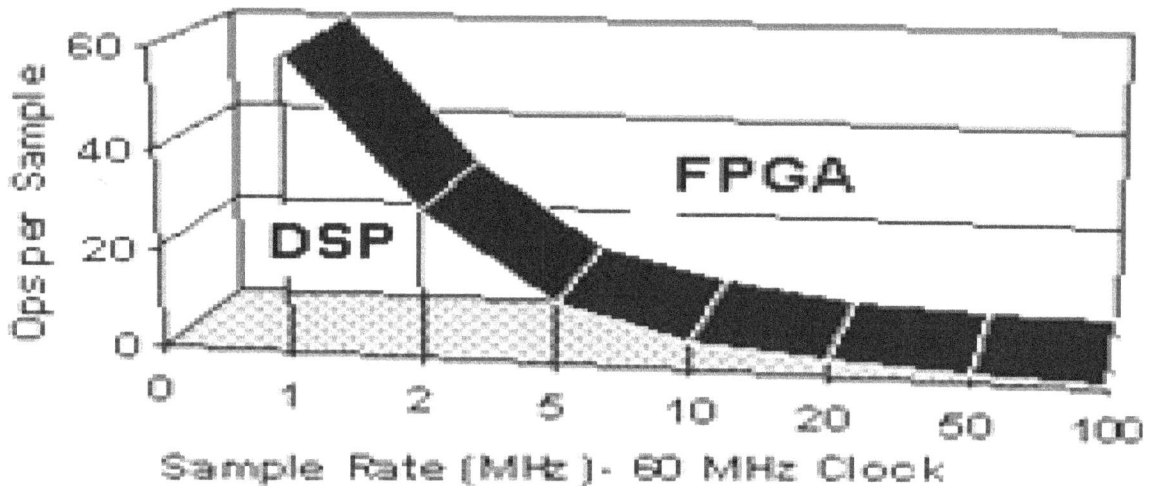

**Figure 1b.** FPGAs are a better solution in the region above the curve.

memory it needs. This provides a high degree of isolation between tasks and results in modification of one task being unlikely to cause unexpected behavior in another task. This, in turn, allows developers to easily isolate and fix bugs in a logical and predictable fashion. In the past, the use of digital signal processors was nearly ubiquitous, but with the needs of many applications outstripping the processing capabilities of digital signal processors (measured in millions of instructions per second (MIPS)), the use of FPGAs is growing rapidly (Figure 1b). The comparison between digital signal processors and FPGAs focuses on MIPS comparison, which, while certainly important, is not the only advantage of an FPGA. Equally important, and often overlooked, is the FPGA's inherent advantage in product reliability and maintainability.

Higher performance implementations of specific DSP algorithms are increasingly available through implementation within FPGAs (http://www.andraka.com /dsp.htm.). Ongoing architectural enhancements, advancement in development tool, speed increases and cost reductions are making FPGA implementation attractive for an increasing range of DSP-dependent applications. FPGA technology advances have increased clock speeds and available logic resources and beyond the range required to implement many DSP algorithms effectively at an attractive price. FPGA implementation provides the added benefits of reducing costs along with design flexibility and future design modification options.

## EMBEDDED SYSTEM IMPLEMENTATION FOR DSP USING XILINX SYSGEN

Objectives were to design a low cost, high speed and reliable DSP system which includes a digital signal processing with real time I/O control. FPGA is used to process the digitized data sampled by the ADC and then output the same to the DAC module.

The aims and objectives of the design are summarized as follows:

(1) To design a general purpose embedded system using FPGA for real time digital signal processing.
(2) Configuring the FPGA as a customized digital signal processor.
(3) Port the soft processor IP core like NIOS II interface with the above design as a co-processor.

The diagram in Figure 2 shows the detailed implementation of the hardware using Spartan III FPGA. A digital signal is applied as an input to the operational amplifier LM324 which acts as a buffer and is then given to the 12-bit ADC AD7891 module. This ADC converts the analog signal into the digital domain. The digitized data is then fed to the FPGA XC3S200 Spartan-III for further processing. The FPGA is programmed using Xilinx ISE webpack over JTAG interface. The FPGA is programmed through IDE Xilinx ISE Webpack 6.3i. The processed data from FPGA is then converted back to analog form using a 12-bit DAC AD7541. The reconstructed analog signal is then given to LM324 for amplification.

The design hardware has several handshaking signals as shown in the Figure 3 for external on board ADC and DAC interface. The data from the ADC after conversion is routed over 12-bit data bus known as 'db'. The processed data over a user's logic is output to DAC over 'dac_out' (12-bit). The DAC output is reconstructed to analog form for representation to external world. The logic in the FPGA can be manually written using HDL for any computation in digital signal domain. With increasing complexities of systems and varied levels of expertise of embedded designers, there is an increasing need for tools that provide a higher level of abstraction, empowering the

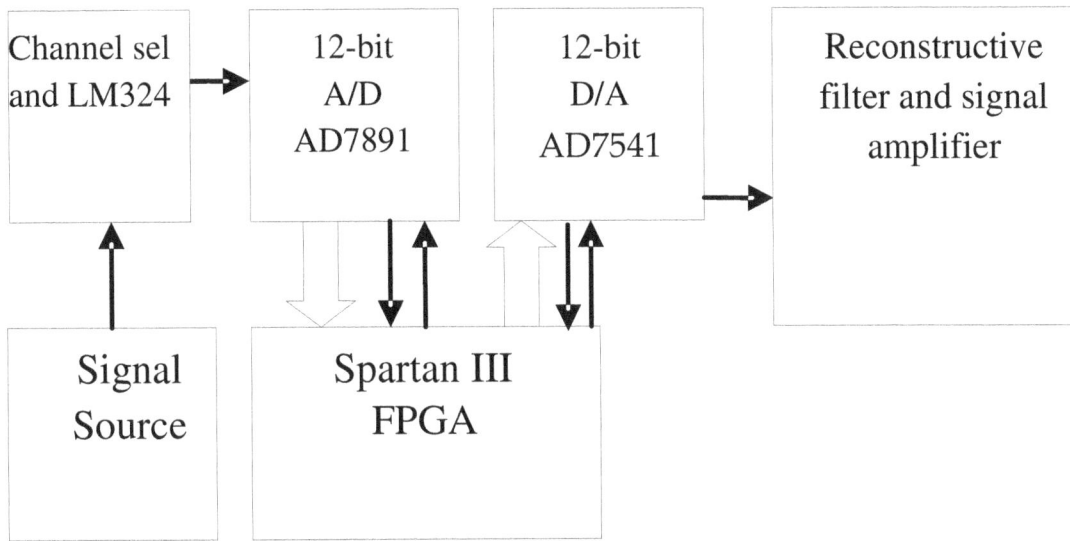

**Figure 2.** Block diagram of embedded system.

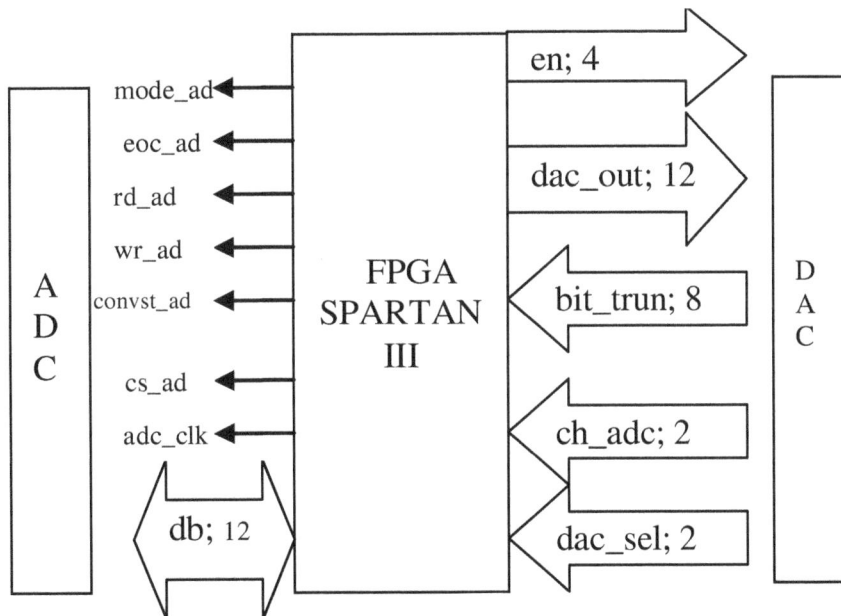

**Figure 3.** Higher level schematic.

empowering the domain experts to use DSPs in building embedded systems rather than spending precious time at the prototyping stage. Graphical programming paradigms have continually evolved to address this problem. There are number of tools that allow a mixture of visual and textual programming such as Signal (Benveniste and Guernic, 1990), Lustre (Halbwachs et al., 1991) and Silage (Hilfinger, 1985).

Completely graphical programming environments such as Ptolemy (http://ptolemy.eecs.berkeley.edu/.) from University of California, Berkeley and National Instruments LabVIEW have evolved to encompass different application domains.

System generator (Xilinx) (http://xilinx.com/products /design_resources/dsp_central/grouping/index.htm)     or DSP generator (ALTERA) (http://altera.com/technology /dsp/dsp-index.jsp.) provides libraries with Simulink customized for respective vendor's targets which gene- rates the HDL code for variety of DSP functionality. The proper EDA interface of the automatic code generator is shown in Figure 4.

Here, the required logic code of user is generated by

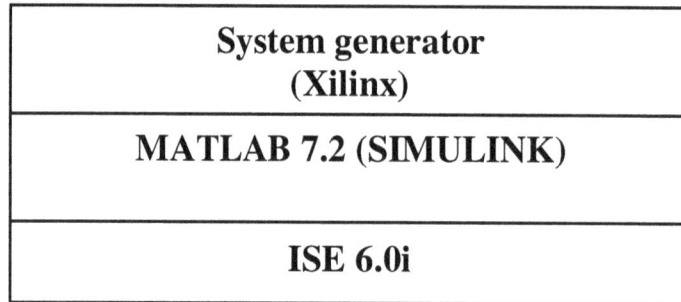

**Figure 4.** Xilinx EDA interface for DSP application.

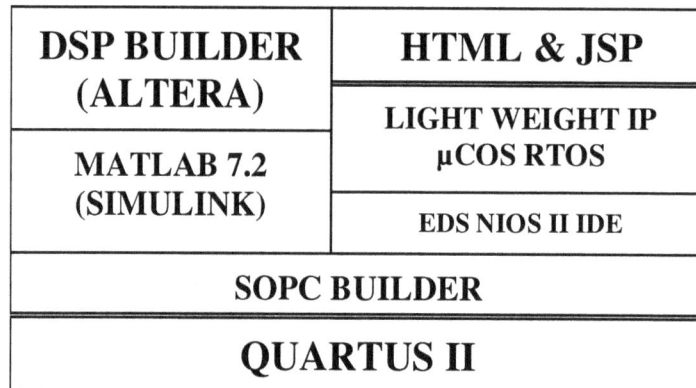

**Figure 5.** ALTERA EDA interface.

the system generator. The generated system generator compiles the intermediate code in form of HDL and the same is available as component of HDL. The component is then mapped with the entity of HDL code of the ADC and DAC interface as shown in figure 3. The 'bit' file after compilation is programmed over FPGA for required performance (http://www.edu.org/comp.lang.vhdl/). The VHDL architecture of the above described mapping required major four processes which will monitor the overall handshake that is conversion of data over ADC, reading ADC, writing ADC data to DAC, selection of DAC channel.

## EMBEDDED SYSTEM IMPLEMENTATION FOR DSP USING ALTERA DSP GENERATOR

The Altera EDA interface of the automatic code generator for DSP application is shown in Figure 5. ALTERA DSP generator is more users friendly, as the input and output port ADC/ DAC are available in the block form compared to XILINX HDL code. The audio signal is given to 12- bit ADC ADS55001, digitized data is then passed through the different audio synthesis blocks like Echo, Revberation, Flaging etc. This effect are selected based on Logic at DIP switches. The synthesized output is then

passed through 14-bit D/A converter DAC904. The analog output of a D/A converter is connected to speakers to hear desired audio effect.

This parallelism can be extended to complex form, to other effect like chorus as shown in the Figure 6.

## AUDIO SYNTHESIS CASE STUDY

The system has been demonstrated to process the audio signal. Various effects like echo, chorus, reverberation, flanging, fading and equalization can be implemented. Echo effect (If the delay is more than 50 - 70 ms, it is perceptible to human ear as an echo.) was tested for satisfactory performance and the effect of change in 50 - 70 ms delay over superimposed ensemble of signal was studied.

### Echo effect

Echo is the repetition of a sound by reflection of sound waves from a surface. It arises in communication systems, when signals encounter a mismatch in impedance (Messerschmitt, 1984). The same has been modeled using Simulink Signal generator as shown in Figure 6.

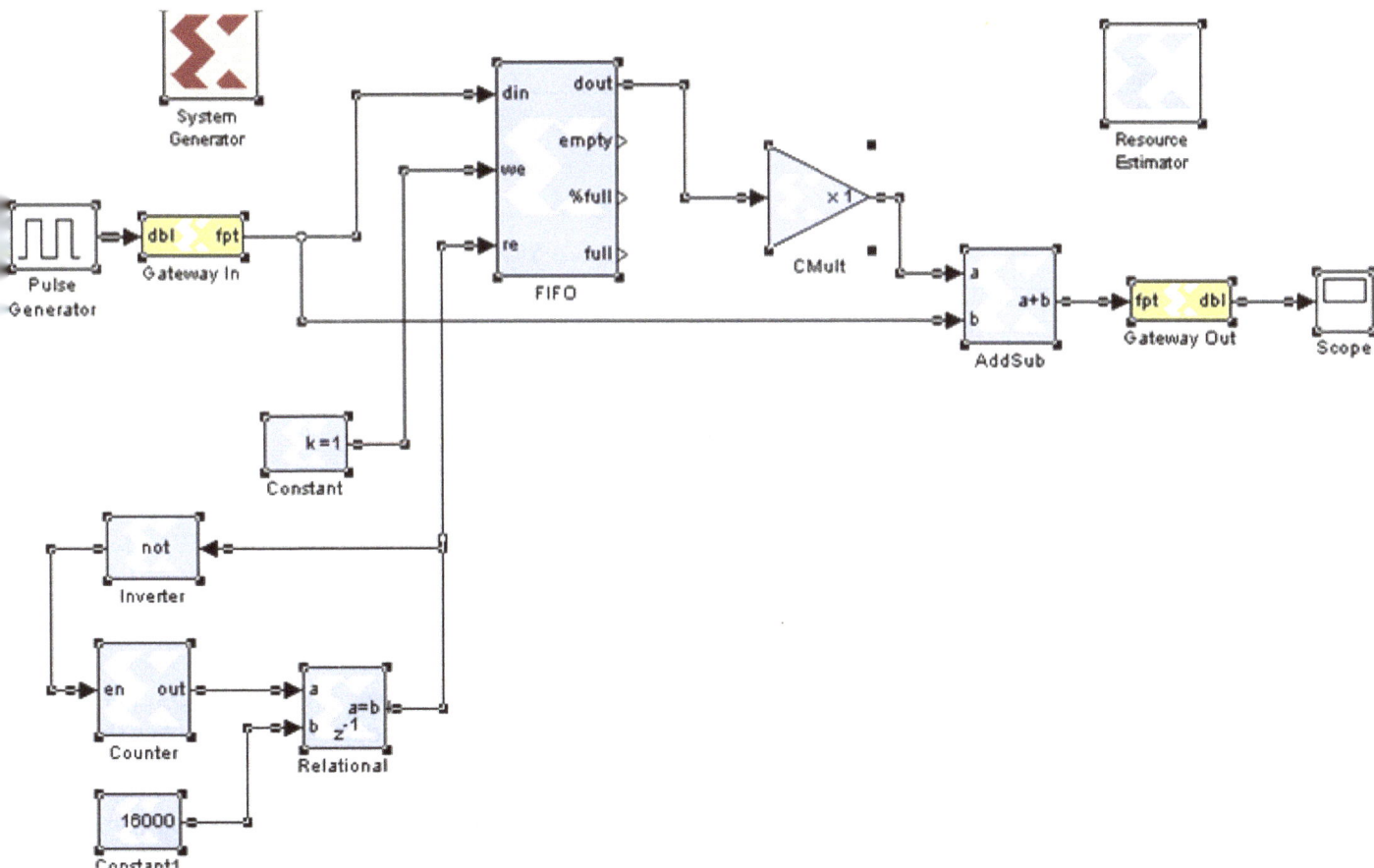

**Figure 6.** Block diagram of the echo in the DSP generator of Xilinx IDE.

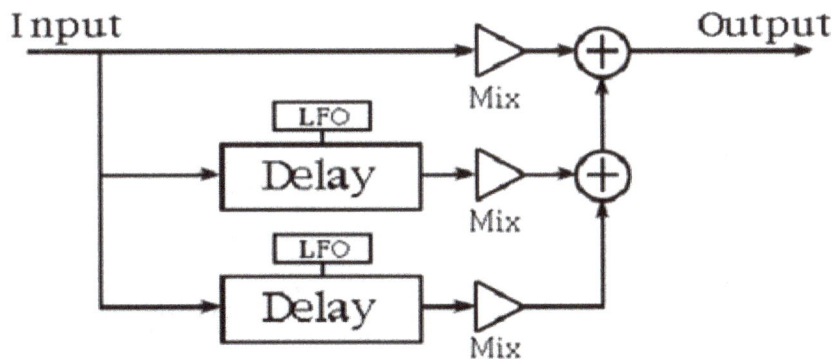

**Figure 7.** Block diagram of the chorus effect.

Figure 6. DELAY, FIFO, ADDER, SUBTRATOR etc are the basic building block for a number of effects in audio signal processing. The running data signal is captured and stored in the FIFO block. Total 2k data samples are captured and stored at any given time. Past signals stored are fed one at a time (after required delay of 50 - 70 ms) to adder block along with running signal. The effects can be understood by varying the 'delay' and 'gain'

of the system described.

**Chorus effect**

More complex example illustrating parallelism is the chorus effect, shown in Figure 7. The chorus effect is often used to alter the sound of an instrument to make it

**Table 1.** Altera cyclone II resources.

| Resource | Available | Used |
|---|---|---|
| Logic elements | 68416 | 2374 (3%) |
| Registers | 1183 | 0 |
| Pins | 422 | 35 (1%) |
| Total memory bits | 1,152,000 | 5,780 ( 1% ) |
| Embedded multiplier 9-bit elements | 300 | 1 |
| PLL | 4 | 0 |

**Table 2.** Xilinx Spartan III resources.

| Resource | Available | Used |
|---|---|---|
| System gate | 200 k | 25 K |
| Logic cells | 4320 | 3135 |
| Dedicated multipliers | 12 | 4 |
| User I/O | 173 | 35 |
| Block RAM | 216K | 30K |

sound as if multiple instruments are playing. If the instrument where a human voice, then this effect would tend to make the single voice sound like a choir. We perceive the multiple voices or instruments, since there is always imprecise synchronization and slight pitch variation when multiple voices or instruments are playing at the same time. These are the principal characteristics of a chorus effect.

Similar way can be used in implementing the parallel computing for complex block-like design (shown in Figure 7) for 32 or more channels. Such implementation can have potential application for the Audio analysis and synthesis Tomography, ECG, EEG etc.

## RESOURCES USED BY ALTERA AND XILINX PLATFORM TO IMPLEMENT AUDIO SYNTHESIS ALGORITHMS

The compared resources used by Altera and Xilinx platform to implement audio synthesis algorithms are given in Table 1 and 2 respectively. After analyzing the compilation report of both platforms, it was found that Altera platform is more versatile and their designs are more optimized than the Xilinx.

## DISCUSSION AND CONCLUSION

The system designed can be upgraded to incorporate Soft core processor like NIOS II (Altera), Micoblaze, Picoblaze (Xilinx) for improving the flexibility of the system. The real time operating system (RTOS) component can also be incorporate to involve the multitasking of the process for real time performance. We

have tested audio effects on both Altera and Xilinx platform. We have concluded that implementing the audio synthesis algorithms like Echo, Reverberation, and Flanging etc. on Altera Platform is easier, user friendly and also provides design flexibility and less compilation and development time than implementing the same on Xilinx Platform.

### REFERENCES

Alles HG (1980). Music synthesis using real time digital techniques. Proc. IEEE, 68: 436-499,

Benveniste A, Le Guernic (1990). P Hybrid Dynamical Systems Theory and the SIGNAL Language," IEEE Tr. Automatic Control, 35 (5): 525-546.

Halbwachs N, Caspi P, Raymond P, Pilaud D (1991). The Synchronous Data Flow Programming Language LUS-TRE,"Proc. IEEE, 79(9):1305-1319.

Hilfinger P (1985). A High-Level Language and Silicon Compiler for Digital Signal Processing", Proceedings of the Custom Integrated Circuits Conference, IEEE Computer Society Press, Los Alamitos, CA, pp. 213-216 .

http://altera.com/technology/dsp/dsp-index.jsp.

http://www.andraka.com/dsp.htm.

http://www.edu.org/comp.lang.vhdl/.

http://xilinx.com/products/design_resources/dsp_central/grouping/index. html

Messerschmitt DG (1984).Echo cancellation in speech and data transmission' IEEE, J. on selected areas in communication, SAC-2:283-297, March [6]

Poulton K, Corcoran JJ, Horna T (1987).  A 1-GHz 6-bit ADC system', IEEE J. Solid-State Circuits, SC-22 : 962-970.

Rajanish KK, Santosh SA, Vinod SG (2009). Unleash the System On Chip using FPGAs and Handel C", Springer, , XXIV, Hardcover. 176p.

Richard GL (2004). Understanding Digital Signal Processing', Prentice Hall, ISBN 0131089897. pp.121-125.

The Ptolemy Project http://ptolemy.eecs.berkeley.edu/.

# 19

# Joint estimation of cochannel signals and direction finding

Khairy Elbarbary[1], Hossam Eldin Abou-Bakr Badr[2,3] and Tarek Bahroun[3]

[1]Electrical Engineering Department, Modern Academy In Maadi, Maadi, Cairo, Egypt.
[2]Egyptian Armed Forces.
[3]Electronic Warfare Department, Military Technical College, Cairo, Egypt.

The need for fast adaptive algorithms for signal separation in dense environments is essential aspect in modern communications systems. For example, enhancement of mobile systems performance to allow different traffics, high quality and minimum delay is achievable in condition that the channel impairments and cohannel interference is reduced. In this paper, the performance of two types of signal separation and interference cancellation algorithms is compared. The paper presents quantitative measures of the two methods and suggests further enhancement of their performance.

**Key words:** Constant modulus algorithms (CMA), iterative least square enumerator (ILSE), signal canceller and multiple signal classification (MUSIC).

## INTRODUCTION

In cellular radio systems, spectral crowding and cochannel interference are becoming increasingly important issues as the number of subscribers grows. Cochannel interference results from frequency reusage, whereby multiple cells operate on the same carrier frequency (Lee, 1989). Depending on geographic considerations and environments conditions, cochannel interference can be the dominant channel impairment. It would be desirable to incorporate "smart" directional antennas into the cellular system to reducing the effects of cochannel interference and in turn allow greater frequency reuse. These antennas should be capable of simultaneously estimating the angles of arrival (AOA's) of several cochannel sources, as well as demodulating the signals themselves (referred to as signal copy).

In recent years, there has been much interest in blind cochannel signal copy algorithms for antenna arrays. For example, a class of blind adaptive algorithms was developed in Tao and Nicholas (2000) that extracts and

separates multiple signals-of-interest on the basis of their differing spectral self coherence refers to the property of a communication signal whereby it is correlated with a frequency-shifted version of itself. Another approach is the two-step procedure described in Ottersten et al. (1989) and Xu et al. (1992) that incorporates a high-resolution direction-finding algorithm followed by a maximum-likelihood scheme to estimate the sources. A signal subspace method, such as the MUSIC (multiple signal classification) algorithm (Schmidt, 1986), is employed to estimate the AOA's. More recently, a decision-feedback approach was presented in Swindlehurst et al. (1995) for the demodulation of digital signals. Symbol decisions based on preliminary signal estimates are used to regenerate the signal waveforms from which improved estimates are derived.

The CM array is an adaptive beamformer designed to blindly recover a cochannel signal (Gooch and Lundell, 1986). It has a conventional weight-and-sum beamformer

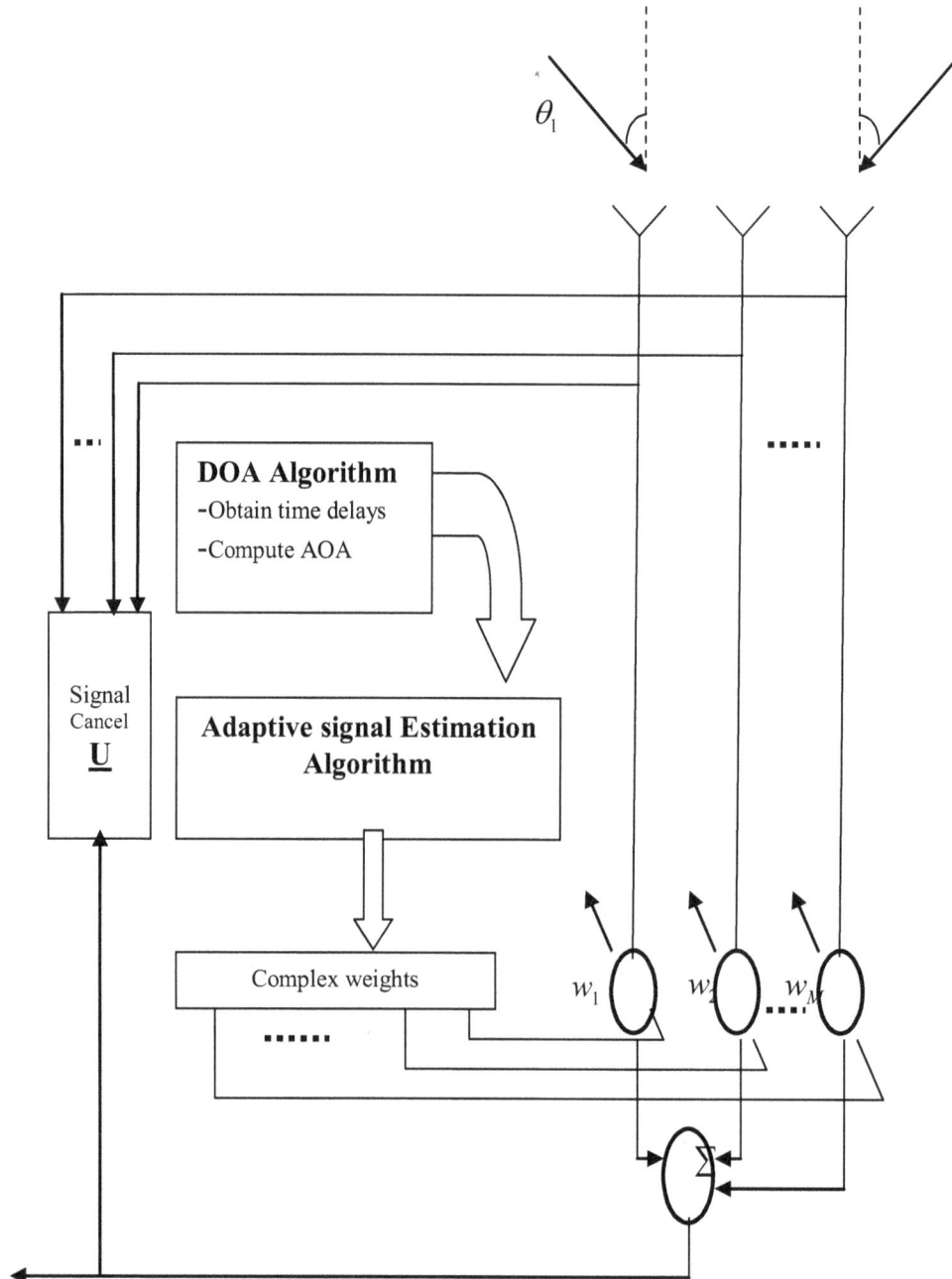

**Figure 1.** One stage CM array followed by signal canceller.

configuration (Widrow and Stearns, 1985) and its weights are adapted by the constant modulus algorithm (CMA) (Treichler and Agee, 1983). The CM array has fast convergence properties and low computational complexity. Moreover, its signal copy performance is insensitive to array imperfections. The multistage CM array consists of a casade of individual CM array stages (Sansrimahachai and Constantindes, 2005; Garth, 2001; Sansrimahachai and Constantinides, 2003). An adaptive signal canceller is included in each stage to remove a captured source from the input before subsequent processing by the follow-on stages.

In this paper, we present a comparison between the two stages algorithm, which utilizes AOA estimator followed by maximum likelihood (ML) multi dimension decision criterion and the multi stages CM array.

## SYSTEM CONFIGURATION AND SIGNAL MODEL

A block diagram of the system is shown in Figure 1. Assume that the antenna elements are uniformly spaced and omni directional so that the array input signals may be expressed as

$$x_m(t) = \sum_{l-1}^{L} s_l(t) e^{-j(m-1)\phi_l} + n_m(t) \quad m=1,\dots,N \quad (1)$$

Where $\{s_l(t)\}$ are the $l^{th}$ (baseband) sources and $\{n_m(t)\}$ are additive white Gaussian noise processes. Because the sources are narrowband, $\phi_l = 2\pi(d/\lambda)\sin(\theta_l)$ where $d$ is the interelement spacing, $\lambda$ is the wavelength of the sources, and $\{\theta_l\}$ are their angles of arrival. By collecting the signals into vectors and assuming that the $\{x_m(t)\}$ are sampled, then Equation (1) may be expressed as:

$$x(k) = As(k) + n(k) \quad (2)$$

Where

$$x(k) \Box [x_1(k),\dots,x_N(k)]^T, s(k) \Box [s_1(k),\dots,s_L(k)]^T, n(k) \Box [n_1(k),\dots,n_N(k)]^T$$

$$A \Box \begin{pmatrix} 1 & \cdots & 1 \\ e^{-j\phi_1} & \cdots & e^{-j\phi_L} \\ \vdots & \vdots & \vdots \\ e^{-j(N-1)\phi_1} & \cdots & e^{-j(N-1)\phi_L} \end{pmatrix} \quad (3)$$

The columns $\{a_i\}$ of the steering matrix $A$ are known as direction vectors because they indicate the response of the array to a narrowband signal emanating from a particular direction. Note that although one is often interested in a uniform linear array as specified by Equation (3), the signal copy performance of the proposed array is independent of the array configuration (Gooch and Lundell, 1986). Our analysis applies to a more general matrix $A$. It is assumed that $L \le N$ for the proposed array, unlike most direction-finding algorithms (e.g., the MUSIC algorithm) where $L < N$ must be chosen.

The correlation matrix of the array output data, is defined as

$$R_x \Box E[x(k)x^H(k)].$$

It is assumed that the incident signals is independent of the additive white Gaussian noise thus $R_x$ is given by

$$R_x = AR_s A^H + R_n \quad (4)$$

Where $R_s \Box E[s(k)s^H(k)]$ and $R_n \Box E[n(k)n^H(k)]$. Assume that the signals and the noise at each array element are mutually uncorrelated, thus, $R_s$ and $R_n$ can be represented by the diagonal matrices $\sum_s$ and $\sum_n$, respectively. Furthermore, assume that the sensor noise powers are identical so that $\sum_n = \sigma_n^2 I$ and (4) becomes

$$R_x = A \sum_s A^H + \sigma_n^2 I \quad (5)$$

The $i^{th}$ diagonal component of $\sum_s$ is $\sigma_{si}^2 = E[|s_i(k)|^2]$, corresponding to the power of the $i^{th}$ source. It is well known that the rank of $A \sum_s A^H$ is equal to the number of sources with different AOA's (L) so that $N - L$ eigen values of $R_x$ are equal to $\sigma_n^2$.

## CM ARRAY AND ADAPTIVE SIGNAL CANCELLER

The CM array estimates one component, $s_i(k)$, of $s(k)$ from $x(k)$ in an on-line adaptive manner without directly estimating $R_x$. It also provides a correction for the estimate of the source direction vector $a_i$ and, thus, the angle of arrival $\theta_i$. Observe in Figure 1 that, the input vector $x(k)$ is processed by a weight-and sum beamformer, yielding the output

$$y(k) = w^H(k)x(k) \quad (6)$$

Where $w(k) \Box [w_1(k),\dots,w_N(k)]^T$ are the adaptive weights adjusted by the constant modulus algorithm such that

$$w(k+1) = w(k) + 2\mu_{cma} x(k)\varepsilon_c^*(k) \quad (7)$$

with

$$\varepsilon_c(k) = y(k)/|y(k)| - y(k) \quad (8)$$

The step size $\mu_{cma} > 0$ controls the convergence rate of (7), and the superscript * denotes complex conjugate.

This update is identical to that of the complex least-mean-square (LMS) algorithm (Widrow and Stearns, 1985), except that the desired signal is replaced by $y(k)/|y(k)|$.

It has been shown for constant modulus signals that the capture behavior of the CM array depends on the initial weight vector $w(o)$ and the relative signal powers at the array output. Specifically, for $L = N = 2$ sources (and a different version of CMA), it was demonstrated that the CM array will lock onto the source with the greatest power at the output of the array while nulling the other source (Gooch and Lundell, 1986). Since the array output primarily contains the captured source, a signal canceller may be used to remove $s_i(k)$ from $x(k)$, generating a modified input vector that can be processed by a follow-on CM array stage in a multistage system (Sansrimahachai and Constantindes, 2005; Shynk et al., 1996). Figure 1 show the signal canceller processes; the array output via $u(k) \square [u_1(k),...,u_N(k)]^T$ result is subtracted from the array input to yield an error vector

$$e(k) = x(k) - u(k)y(k) \qquad (9)$$

The canceller weights may be updated by gradient-descent algorithm using

$$u(k+1) = u(k) + 2\mu_{1ms}y^*(k)e(k) \qquad (10)$$

This recursion implements a set of $N$ independent signal-weight LMS algorithm updates. It is straightforward to show that for convergence in the mean, the step size is bounded by $0 < \mu_{1ms} < 1/\sigma_y^2$ where $\sigma_y^2 = E\left[|y(k)|^2\right]$ is the variance of the CM array output (this variance is actually time-varying because the CM array weights are continually updated by Gooch and Lundell (1986). Thus, the convergence properties of the canceller weights depend on those of the CM array, whereas the CM array weights are independent of the adaptive canceller. All canceller weights converge with the same time constant (because of the single input $y(k)$) given approximately by

$$\tau \approx 1/\left(2\mu_{1ms}\sigma_y^2\right) \qquad (11)$$

## DIRECTION OF ARRIVAL ESTIMATION AND SIGNALS SEPARATION ALGORITHM

A core problem in the area of blind signal separation/

equalization is the following. Consider $L$ independent sources, transmitting binary symbols {+1;−1} at equal rates in a wireless scenario. The signals are received by a central antenna array, consisting of $N$ elements antenna array. Assuming synchronized sources, equal transmission delays, negligible delay spread, and sampling at the bit rate, each antenna receives a linear combination of the transmitted symbol sequences and a weighted combination of the antennas output is obtained. The blind CM array depends on restoring the constant amplitude property for capturing one of the $L$ sources without determination of the order of this source within the combined received signal. Usually in mobile system the direction of arrival and the power of the received signal are variable parameters. This variation of individual sources limits the capturing capability of the CM array. A promising method to overcome this problem is the utilization of two stage system based on AOA estimator allowed by iterative least square enumerator (ILSE) for simultaneous detection of all the $L$ incident signals. The subspace methods such as MUSIC algorithm provides an accurate AOA estimator which are utilized as an initial estimate of the array weights. The simultaneous signals detection is then carried out by examining the most likelihood vertex of $2^L$ vertices of the hybrid cube represents the all possible signals.

## MUSIC Algorithm

The multiple signal classification (MUSIC) method is a profit from the eigen structure properties of the array correlation matrix to obtain very-high-resolution estimates with lower computational complexity when compared to ML estimation schemes. The basic idea of the MUSIC method is to separate signal from noise by the orthogonal property of their spaces through eigen-decomposition of the correlation matrix of the received signal.

Let us analyze the properties of the spatial correlation matrix $R_x$ described in Equation (4). It is clear that if the number of array sensors is large than the number of signal sources (that is, $N > L$), when $R_s$ is positive definite (that is, the signals $s_i(t)$ are not fully correlated), the matrix $R_x - \sigma^2 I$ will have rank L and a null space of dimension N–L. Then matrix $R_x$ will have L eigen values greater than $\sigma^2$ and N–L eigen values equal to $\sigma^2$; these eigen values may be sorted from largest to smallest such that

$$\lambda_1 > \lambda_2 > ... > \lambda_L > \lambda_{L+1} = \lambda_{L+2} = ... \lambda_N = \sigma^2$$

The eigenvectors $\{e_1, e_2, ..., e_L\}$ are corresponding to the largest eigen values span the $L$-dimensional signal subspace. These eigen vectors can be grouped in the columns of matrix $E_s$.

The eigen vectors $\{e_{L+1}, e_{L+2}, \ldots, e_L\}$ corresponding to the smallest eigen values span the (*N-L*)-dimensional noise subspace. These signal eigen vectors can be grouped in the columns of matrix $E_N$.

Then it is clear that, as the columns of matrix A are orthogonal to the eigen vectors that span the noise subspace.

For an exactly known $R_x$, the desired angles $\theta_i, i = 1, 2, \ldots, L$ can be found by evaluating the MUSIC spatial spectrum defined as

$$P_{MUSIC}(\theta) = \frac{1}{a(\theta)^H E_N E_N^H a(\theta)} \qquad (12)$$

Ideally, $P_{MUSIC}(\theta)$ will peak to infinity each time a true $\theta_i, i = 1, 2, \ldots, L$ angle is tested. The MUSIC method works only when the rank of matrix $R_x - \sigma^2 I$ is equal to L, that is, when the signals are uncorrelated.

### Iterative least square with enumeration algorithm

In the blind signal separation scenario, both A and S are unknown, and the objective is, given X, to find the factorization X = AS such that S belongs to the binary alphabet. Alternatively, we try to find a weight matrix W of full row rank L such that $S = W^* X$. Uniqueness of this factorization is important, and was established in Lee (1989) if A is full rank and the columns of S exhaust all $2^L$ distinct (up to a sign) possibilities, then this is sufficient for the factorization to be unique up to trivial permutations and scaling by ±1 of the rows of S and columns of A. Hence, once any such factorization of X is found, S contains the binary signals that were originally transmitted, or their negative, but not some ghost signal. This scenario by itself is perhaps naive, but it is the core problem in more realistic blind (FIR-MIMO) scenarios (Tao and Nicholas, 2000), where long delay multi path is allowed, and sources are not synchronized and are modulated by arbitrary pulse shape functions.

One of the first papers to consider this problem appeared in full in Lee (1989). In that paper, arbitrary finite alphabets are considered although only BPSK was tested extensively. A fixed-point iteration algorithm is proposed, it is called ILSE which is based on clever enumeration of candidate matrices S. Clearly, ILSE is a conditional maximum likelihood estimator. The ILSE algorithm utilizes an accurate initialization of the array weights by the aid of the MUSIC Algorithm. The iterative

solution of this LS optimization problem is given by the following steps

$$\min_S \|X - AS\|_F^2 = \min_{S(1)} \|X(1) - AS(1)\|_F^2 + \ldots + \min_{S(N)} \|X(N) - AS(N)\|_F^2$$

$ILSE$,
1. $Given\ A_0, k = 0$
2. $k = k + 1$
   - $Minimize\ (1)\ for\ S_k\ (by\ enumeration)$
   - $A_k = XS_k^*(S_k S_k^*)^{-1}$
3. $Continue\ until\ (A_k - A_{k-1}) = 0$

The basic idea behind ILS solutions of is simple, that each time, compute an LS update for one of the unknown matrices conditioned on a previously obtained estimate for the other matrix, proceeds to update the other matrix, and repeat until convergence of the LS cost function is reached.

### COMPUTER SIMULATIONS

Computer simulations are performed using MATLAB to verify the performance of the MUSIC, the ILSE and the CMA Algorithms. Assume that three signals arrive from faraway signal sources from directions $\theta = 10^o, 30^o, 50^o$. The three signals have equal power and they have the same signal to noise ratio. The array is a linear array consists of 4 elements separated by half the wavelength. Only 256 snapshots are taken into consideration to estimate the correlation matrix of the received signal. The performance of the MUSIC method is evaluated where, the incident signals are assumed to be uncorrelated. The output of the MUSIC method under these conditions is presented in Figure 2. It is clear that the MUSIC method has successfully determined the correct DOA of the incident signals. These results are applied as initial estimate of the array weights vector in iterative manner for both the CM array with interference canceller as well as the ILSE where both of them were able to estimate the signals and their corresponding probability of error are plotted in Figures 3 and 4, respectively. One can see although the ILSE is less complex compared with the CM array, followed by signal canceller, it provides a similar behavior for the considered environments. Moreover the ILSE is capable to tolerate the DOA estimation error; this is indicated in Figure 5, where the ILSE was provided with 5% error in the DOA estimate.

### Conclusions

The Multi stage CM array with signal canceller provides an acceptable performance for successive signal separation and interference cancellation in stationary environments. The CMA is completely blind algorithm and it depends on restoring the constant envelope

MUSIC

snr = 10  dB   # Ele=4   # Snapshots = 256

AOA  = [ 10 30 50 ]

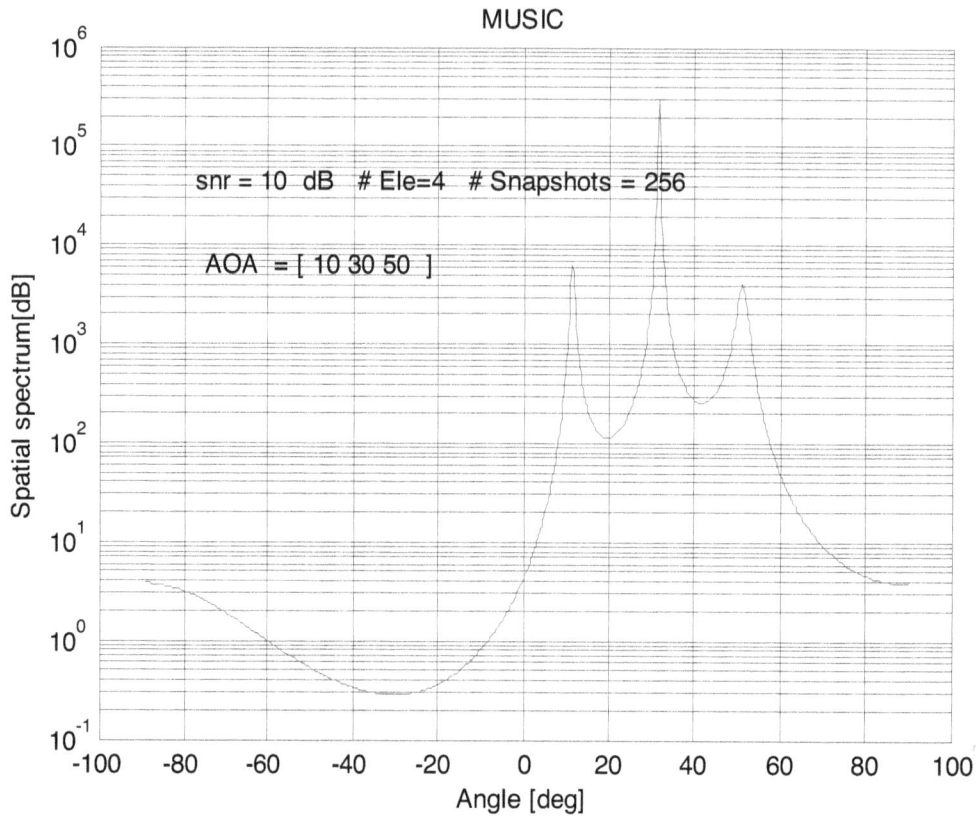

**Figure 2.** AOA estimation based on MUSIC spatial spectrum estimation.

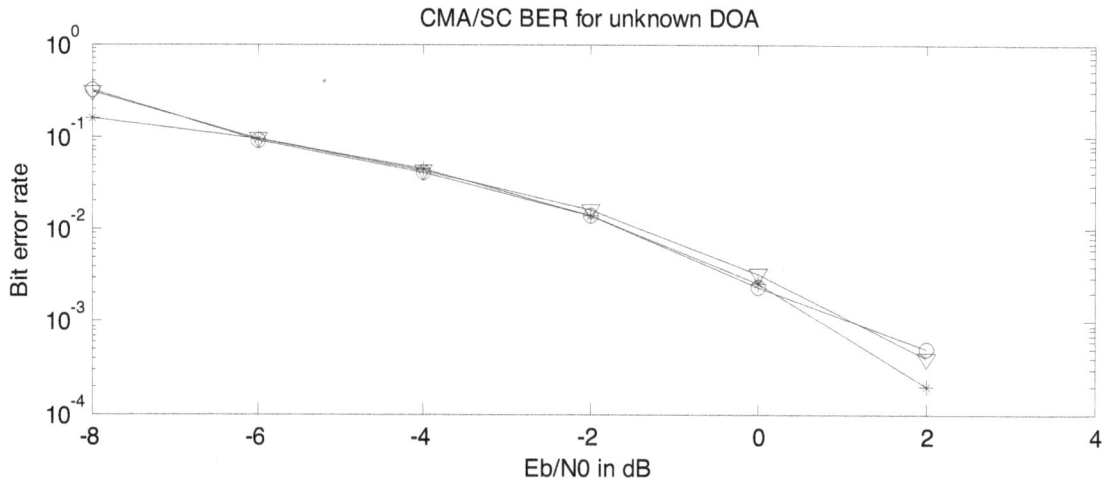

CMA/SC BER for unknown DOA

**Figure 3.** Bit error rate performance of 3 stages CM array with signal Canceller

property of capturing the incident signals in successive manner with the aid of LS signal canceller. The CM array performance decays in rapid changing environments such as cellular channels where the direction and power relations between signals are changed rapidly. The ILSE is a promising algorithm for joint DOA estimation and

signal copy. Convergence of the ILSE cost is guaranteed because each (conditional LS) update may either improve or maintain, but cannot worsen the fit. The final output is generally dependent on the initialization. For that reason, an initial weight vector based on MUSIC algorithm enhances the performance of the algorithm, An

ILSE enumeration BER for known DOA

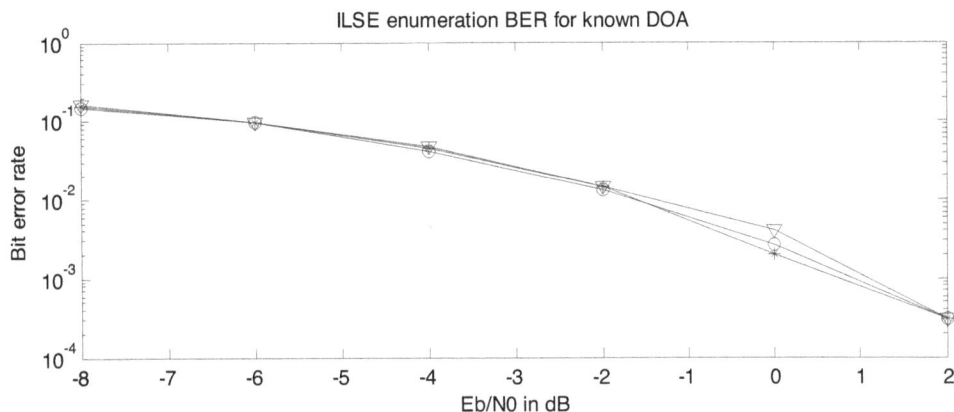

**Figure 4.** Bit error rate for 3 detected users by ILSE algorithm with perfect AOA estimate.

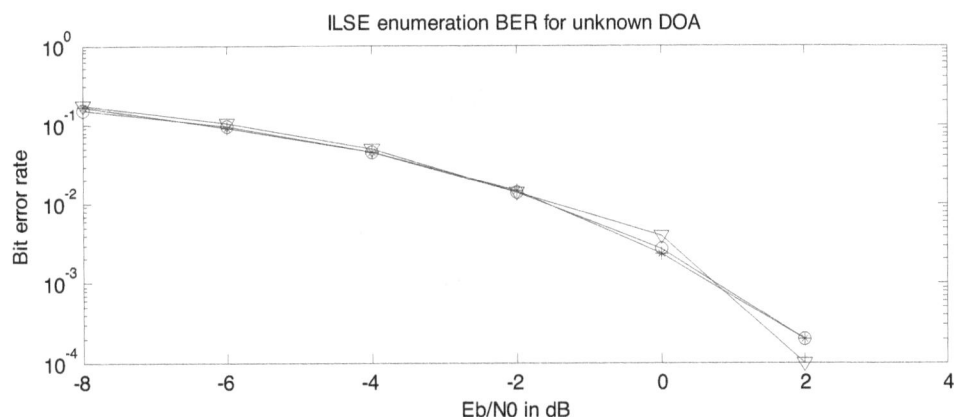

ILSE enumeration BER for unknown DOA

**Figure 5** Bit error rate for 3 detected users by ILSE algorithm with 5% error in AOA estimate.

interesting point for further research is to find out a fast AOA estimator, which needs a few array snapshot to provide an accurate AOA estimate to be utilized as an initial weight vector for the signal separation algorithm. Of course such fast initialization will enhances the possibility of real time processing in the fast changing and dense environment encountered in cellular communications.

## REFERENCES

Garth L (2001). "A dynamic convergence analysis of blind equalization of real signals," IEEE Trans. Commun. 49(4):624-634.

Gooch RP, Lundell JD (1986). "The CM array: An adaptive beamformer for constant modulus signals." Proc. IEEE Int. Conf. Acoust. Speech, Sig. Process. Tokyo, Japan. pp. 2523-2526.

Lee WCY (1989). Mobile Cellular Telecommunications Systems. New York: McGraw-Hill.

Li T, and Sidiropoulos ND (2000). "Blind digital signal separation using successive interference cancellation iterative least squares". IEEE Trans. Sig. Process. 48(11):3146-3152.

Ottersten B, Roy R, Kailath T (1989). "Signal waveform estimation in sensor array processing." Proceedings of 23rd Asilomar Conference Signals Syst. Comput., Pacific Grove, CA. pp. 787-791.

Sansrimahachai P, Ward DB, Constantindes AG (2005). "Blind sources separation of instantaneous MIMO systems based on the least-squares constant modulus algorithm." IEE Proc. Vis. Image Sig. Process. 152(5):616-622.

Sansrimahachai P, Ward DB, Constantinides AG (2003). "Multiple-input multiple-output least-squares constant modulus algorithms." Proc. GLOBECOM. 4:2084-2088.

Schmidt RO (1986). "Multiple emitter location and signal parameter estimation." Proc. RADC Spectrum Estimation Workshop, Griffiss Air Force Base, NY, 1979, pp. 243-258, 1979; repr. IEEE Trans. Antennas Propag. AP-34:276-280.

Shynk JJ, Keerthi AV, Mathur A (1996). "Steady-state analysis of the multistage constant modulus array." IEEE Trans. Sig. Process. 44:948-962.

Swindlehurst A, Daas S, Yang J (1995). "Analysis of a decision directed beamformer." IEEE Trans. Sig. Process. 43:2920-2927.

Treichler JR, Agee BG (1983). "A new approach to multipath correction of constant modulus signals". IEEE Trans. Acoust., Speech Sig. Process. ASSP-31:459-472.

Widrow B, Stearns SD (1985). Adaptive Signal Processing. Englewood Cliffs, NJ: Prentice-Hall,.

Xu G, Cho Y, Paulraj A, Kailath T (1992). "Maximum likelihood detection of co-channel communications signals via exploitation of spatial diversity." Proc. 26th Asilomar Conference on Signals System Computing, Pacific Grove, CA, Oct., pp.1142-1146.

# 20

# Design of a double clad optical fiber with particular consideration of leakage losses

Chakresh Kumar, Girish Narah and Aroop Sharma

Department of Electronics and Communication Engineering, Tezpur University, Napaam-784028, Sonitpur, Assam, India.

In this paper we present a double clad optical fiber that consists of core, inner cladding and outer cladding. The refractive index of the core and the outer cladding are the same and the value of refractive index of the core is greater than the refractive index of inner cladding. The cutoff number $V_c$ is calculated and plotted with respect to the ratio of the radius of inner cladding and the radius of the core. Finally the leakage losses are calculated considering both the bending effect and the non-bending effect. And a comparison is made between the double clad optical fiber and a single clad optical fiber.

**Key words:** Bessel's function, cutoff value (Vc), cut off wavelength, loss co-efficient (2α).

## INTRODUCTION

In the leaky waveguides, the low refractive index surrounding region has a finite thickness comparable to the penetration depth of the guided field and beyond this distance the medium has a refractive index equal to or greater than that of the guiding region. In such a case, the waves do not undergo total internal reflection and thus the reflection coefficient is less than unity. Such a phenomenon is known as frustrated total internal reflection (FTIR). Hence, in the waveguides, there are no perfectly guided modes. On the other hand, such waveguides have leaky modes that are characterized by a finite loss coefficient. The losses associated with these modes are calling the leakage loss. One of the characteristics of leakage loss is that large differential leakage loss between the fundamental and higher order modes is responsible for single mode operation required in LMA fibers (Ajeet et al., 2008, 2010).

### EXPERIMENTAL DESIGN

#### Propagation of ray in a leaky structure

In the design of the double clad optical fiber we considered that the optical fiber consists of one core and two cladding that is, the inner and the outer cladding. The refractive index of the core and the outer cladding is the same.

Figure 1 shows the variation of refractive index with radius of the core, inner cladding and the outer cladding. $n_1$ is the refractive index of the core and $n_3$ is the refractive index of the cladding (value of both of these are the same). $n_2$ is the refractive index of the inner cladding. The value of $n_3$ is kept higher than that of $n_2$ to make the structure leaky. In the designed fiber, we have taken the radius of the core, $x_1 = 5$ µm, the radius of the inner cladding, $x_2 = 30$ µm, radius of the outer cladding, $x_3 = 50$ µm, refractive index of the core, $n_1 = 1.5$, refractive index of the inner cladding, $n_2 = 1.4$, refractive index of outer cladding, $n_3 = 1.5$. We have taken these values for analysis throughout this paper. Guided modes are those modes that are mainly confined to the film and hence their field should decay in the cover (Ajoy and Thyagarajan, 2011). Thus,

$$\beta^2 > k_0^2 n_2^2$$

In the leaky modes field is oscillatory in nature. Thus in leaky modes

$$\beta^2 < k_0^2 n_2^2$$

Where β is the propagation constant.

In Figure 2, the variation of electric field with fiber radius has been analyzed for each of the layer, that is, core, inner cladding and the outer cladding. Light propagates in core and inner cladding

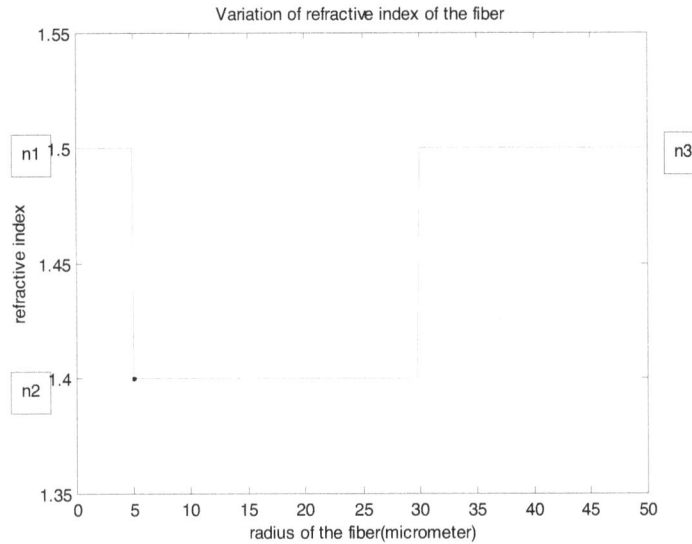

**Figure 1.** Variation of refractive index with the radius.

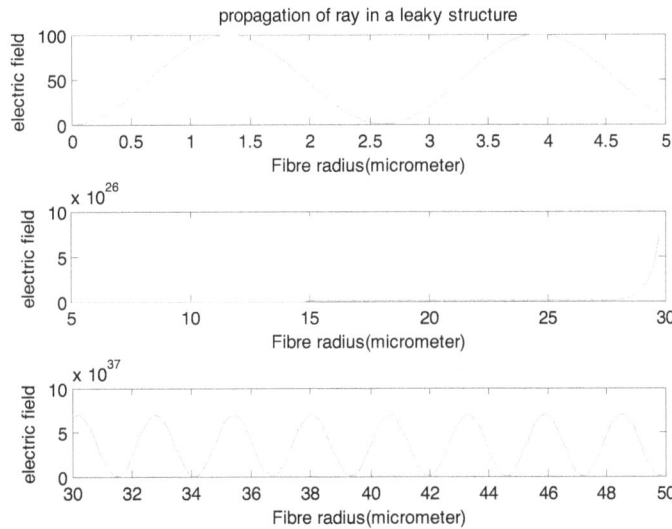

**Figure 2.** Variation of electric field with the radius.

in the guided mode, which implies that light will propagate in an exponential manner in these two regions. As soon as the light enters in the outer cladding, due to its leaky behavior, the light undergoes oscillatory motion. The overall propagation of the ray inside the fiber is shown in the Figure 3.

**Cut-off characteristics**

If β/k> $n_3$, then there are no leakage losses. However, if the propagation constant is smaller, so that β/k < $n_3$, then the mode is said to be "cut off" because power radiates through the outer cladding. Notice that if Δ = 0, as in the Figure 1, then β/k < $n_3$ and there are leakage losses at all wavelengths, where $\Delta = \frac{(n_1 - n_3)}{n_3}$,

$$\Delta' = \frac{(n_2 - n_3)}{n_3}.$$

The propagation characteristics of double-clad light guides are determined from the eigenvalue equation for the weakly guiding approximation (Leonard et al., 1982; Maxim et al., 2005).

$$V_c^2 = \left(2\pi \frac{x_1 n_3}{\lambda_c}\right)^2 (2\Delta) \tag{1}$$

Where $V_c$ is the cut off number and $\lambda_c$ is the cut off wavelength. Figure 4 summarizes the cutoff behavior for the fundamental mode of the double-clad fiber. The solid curves show the cutoff value $V_c$ as a function of the cladding radius ratio b/a for several values of the refractive index parameters H = - Δ/Δ'. A truly guided mode exists only in the region above a given curve while the area below

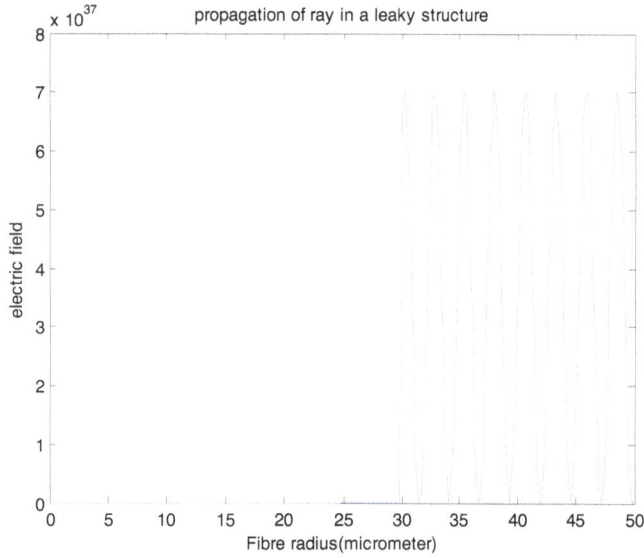

**Figure 3.** Overall propagation of ray inside the double clad optical fiber.

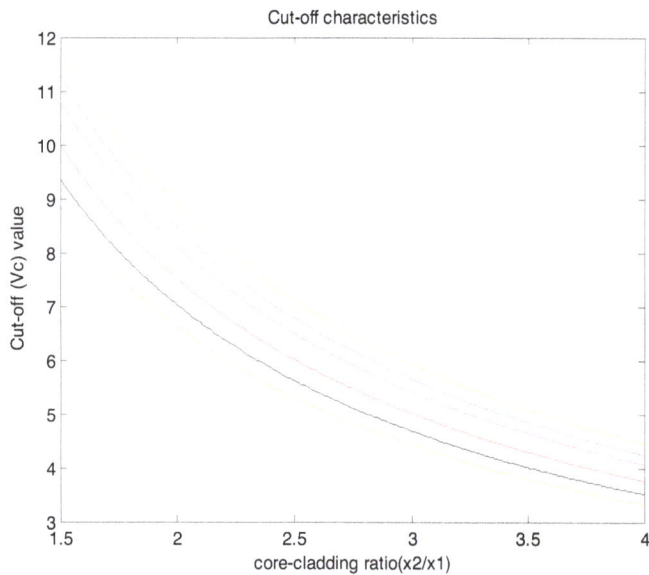

**Figure 4.** Graph of $V_c$ Vs $x_2/x_1$ for different values of $H = -\frac{\Delta'}{\Delta}$.

the curves indicates the region where the $HE_{11}$ mode is cutoff because of leakage losses. No cutoff occurs if $V_c = 0$.

## RESULTS AND DISCUSSION

### Leakage loss calculation

The radiation losses associated with double-clad fiber with wide depressed cladding can be derived in terms of $V_2$, $\gamma$, $\sigma$ and $\kappa$ (Leonard et al., 1982).

Consider a step-index profile with either $x_2 \to \infty$ or $\Delta' = 0$. The electromagnetic field solution computed for the step-index fiber is then used as a zero-order approximation for calculating radiation losses of the double-clad fiber. This is done by introducing a reflected wave at the index step $r = x_2$ and a transmitted wave in the outer cladding at $r = x_2$. The corresponding wave amplitudes are found by requiring that the boundary conditions (Leonard et al., 1982) should be satisfied at $r = x_2$ with the zero-order field solution being regarded as an "incident" wave on the index step at $r = x_2$. The power loss coefficient $2\alpha$ can be computed from the power that is radiated radially per unit length of fiber, divided by the power carried by the guided mode along the fiber axis.

Using the above procedure, the equation is derived for the power loss coefficient, $2\alpha$

$$2\alpha = \frac{2\pi \kappa^2 \gamma \sigma e^{-2\gamma x_2}}{\beta n_3^2 k^2 |\Delta'| V_2^2 K_1^2 (\gamma x_1)} \qquad (2)$$

Where

$$V_2 = k x_1 n_3 \left(2(\Delta - \Delta')\right)^{\frac{1}{2}}$$
$$= [(\kappa x_1)^2 + (\gamma x_1)^2]^{\frac{1}{2}} \qquad (3)$$

$$\kappa = [n_3^2 (1 + \Delta)^2 k^2 - \beta^2]^{\frac{1}{2}} \qquad (4)$$

$$\gamma = [\beta^2 - n_3^2 (1 + \Delta')^2 k^2]^{\frac{1}{2}} \qquad (5)$$

$$\sigma = (n_3^2 k^2 - \beta^2)^{\frac{1}{2}} \qquad (6)$$

$$\Delta = \frac{(n_1 - n_3)}{n_3} \qquad (7)$$

$$\Delta' = \frac{(n_2 - n_3)}{n_3} \qquad (8)$$

In Figure 5, radiative leakage losses are plotted as a function of wavelength with the cladding-to-core ratio of $x_2/x_1$ as the variable parameter. These losses are never zero because the $HE_{11}$ fundamental mode is cutoff at all wavelengths. Care must be taken to choose the ratio of $x_2/x_1$ large enough to ensure low leakage losses within the wavelength range of interest. For example- $x_2/x_1 = 6$ is required to keep losses below 0.2 dB/km for wavelengths shorter than 1.6 μm. Throughout this paper we have assumed that the double-clad fiber has a piecewise constant refractive index distribution. Instead of solutions of the straight fiber, we use simplified WKB-type solutions (Leonard et al., 1982) of the curved structure in the derivation (2). In this way, Equation (9) for the loss of the curved double-clad fiber has been derived.

$$2\alpha(R) = \frac{4x_2 \kappa^2}{\beta V_2^2 K_1^2 (\gamma x_1)} \int_0^\pi \frac{\sigma(x_2) \gamma^2(x_2) e^{-2u}}{u[\gamma^2(x_2) + \sigma^2(x_2)]} d\phi \qquad (9)$$

Where $K_1$ is the Bessel's constant

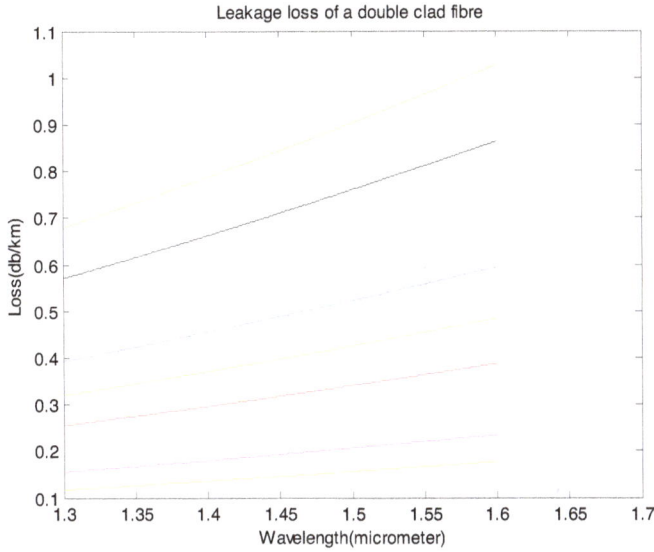

**Figure 5.** Variation of leakage loss Vs wavelength.

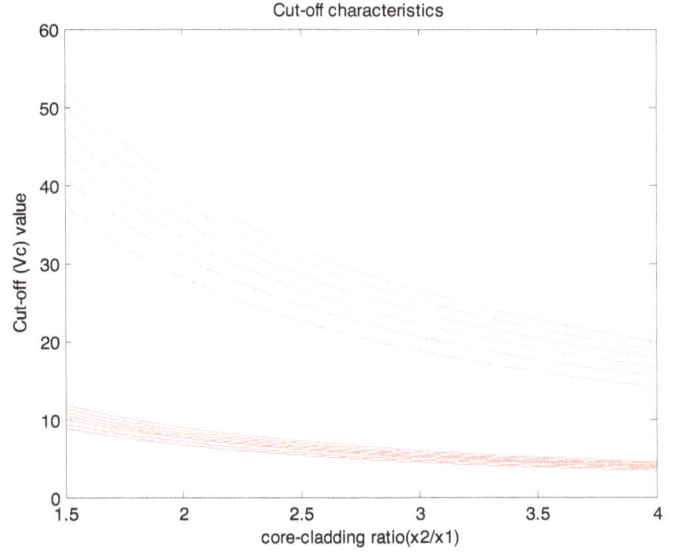

**Figure 7.** Comparison of cut off number Vs the core cladding ratio of the two fibers.

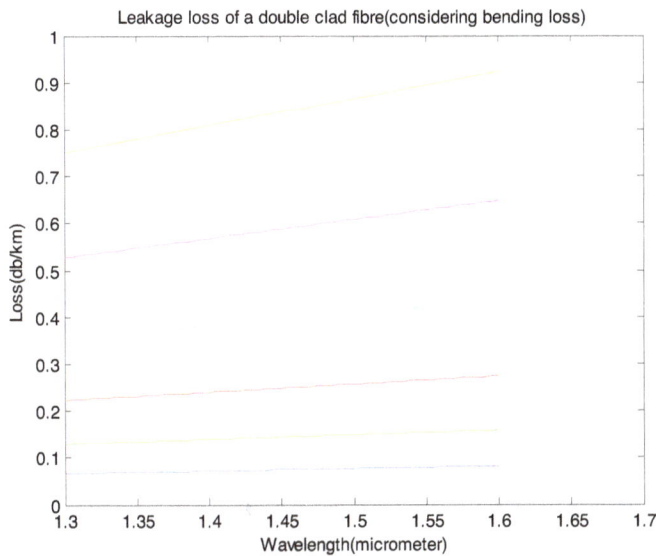

**Figure 6.** Variation of leakage loss considering the bending effect vs wavelength with cladding-to-core ratio $x_2/x_1$ as the variable parameter.

$$\bar{\gamma}^2(x_2) = \beta^2 - n_3^2(1+\Delta')^2\left(1+2\frac{x_2}{R}\cos\phi\right)k^2 \qquad (10)$$

$$\bar{\sigma}^2(x_2) = n_3^2\left(1+\frac{2x_2}{R}\cos\phi\right)k^2 - \beta^2 \qquad (11)$$

$$u = \frac{R|\gamma^3+\gamma^3(x_2)|}{3\,n_3^2(1+\Delta')^2k^2\cos\phi} \text{ for } \phi \neq \frac{\pi}{2}$$

$$= \gamma x_2 \text{ for } \phi = \frac{\pi}{2} \qquad (12)$$

$$V_2 = kx_1n_3\left(2(\Delta-\Delta')\right)^{\frac{1}{2}}$$

In Figure 6, we observed that the predicted losses for a straight fiber could be significantly increased due to bending effects induced by cabling the fiber. As $(x_2/x_1)$ increases, the bending loss also increases. For $x_2/x_1$ around (3 - 5) the increase in the bending loss is quite low as compared to the increase in the bending loss as the value of $(x_2/x_1)$ goes beyond 5.

## Comparisons

The design fiber with double clad has been compared with ordinary fiber at different aspects. In Figure 7, cut off number of designs fiber is denoted by the red curves and the cutoff number of the single clad fiber is denoted by blue curves. The graph is plotted by varying the value of $H = -\frac{\Delta'}{\Delta}$ for each fiber. The cutoff number ($V_c$) of the single clad fiber is significantly higher than that of the double clad fiber. Due to the low voice number, the double clad fiber is sensitive to bending loss and absorption loss at the cladding interface, and due to the high Vc number in single clad fiber the scattering losses in the core or at the core–cladding interface increases (Snyder and Love, 1983).

In Figure 8, the leakage losses of the designed fiber are drawn by the red lines and the leakage losses of the same fiber under bending condition are drawn with dotted lines. As clearly mentioned in the graph, for a fixed ratio of the radius of the core and cladding $(x_2/x_1)$ the leakage losses under bending effect is more than the normal leakage losses when the value of the wavelength ($\lambda$) is (1.3 - 1.45) μm and beyond this wavelength the leakage losses are more than the bending losses. As the ratio of the radius of the core and cladding $(x_2/x_1)$ increases, the

Leakage loss comparison of a double clad fibre for bending and non-bending condition

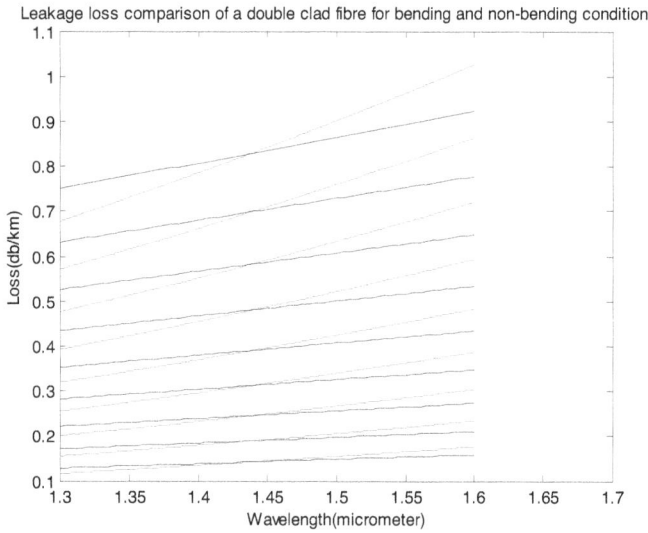

**Figure 8.** Comparison between the leakage losses for bending and non bending conditions in the double clad optical fiber.

Leakage loss comparison of the double clad fibre and the single clad fiber

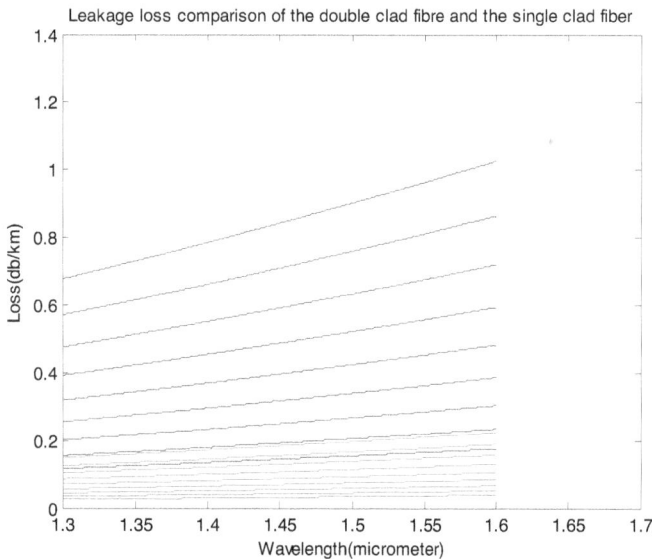

**Figure 9.** Comparison to the loss of leakage losses Vs the wavelength of double clad and single clad optical fiber.

slope of loss curve increases by increasing the wavelength, that is, when the ratio of $x_2/x_1$ is around (5 - 6), the difference in leakage loss at $\lambda$=1.3 µm and at $\lambda$=1.6 µm is quite less. But when the ratio of $x_2/x_1$ is around (6.5 - 9) then the difference of the leakage loss at $\lambda$=1.3 µm and at $\lambda$=1.6 µm is comparably high.

In Figure 9, comparison of the leakage losses of the designed fiber and a single clad fiber is clearly shown. The losses in the double clad fiber are drawn by the dotted lines and the leakage losses of the single clad fiber are drawn with red lines. As the ratio of $(x_2/x_1)$ keep on increasing the leakage losses in the double clad fiber

Bending loss comparison of the double clad fibre and the single clad fiber

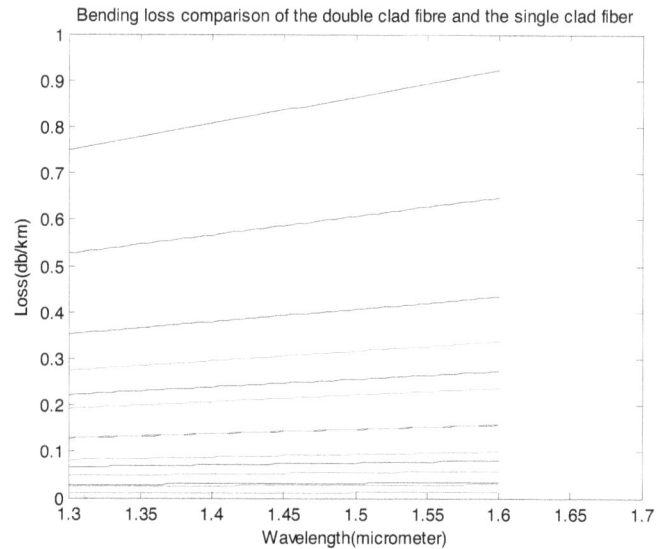

**Figure 10.** Comparison of the bending losses of the single clad and the double clad optical fiber Vs the wavelength.

**Table 1.** Table for cut-off $V_c$ for different values of wavelength.

| Wavelength ($\lambda_c$) in µm | Cut-off value ($V_c$) |
|---|---|
| 1.2 | 14.33 |
| 1.25 | 13.76 |
| 1.3 | 13.23 |
| 1.35 | 12.74 |
| 1.4 | 12.29 |
| 1.45 | 11.86 |
| 1.5 | 11.47 |

tends to increase more as compared to the increase in the leakage losses of the single clad fiber. It occurs because of the introduction of the leaky layer in the double clad fiber, that is, the outer cladding.

In Figure 10, the bending losses of the designed fiber are drawn by the dotted lines and the bending losses of the single clad fiber are drawn with red lines. By increasing the ratio of $(x_2/x_1)$, bending losses of double clad fiber become significantly higher than that of the single clad fiber.

## Conclusion

In this paper we have proposed a design of double clad optical fiber and analyzed its characteristics at different conditions with the single clad optical fiber. It is observed that the cutoff conditions in terms of normalized curves which show the $HE_{11}$ mode cutoff $V_c$ number plotted as a function of $x_2/x_1$ by constantly varying $H = -\frac{\Delta}{\Delta}$ . From Table 1 the cut off wavelength of the double clad optical fiber is found to be 1.5 µm, since at this value of $\lambda_c$, the value of $V_c$ is the smallest. The $V_c$ number of the design fiber is less than the single clad fiber.

The leakage losses under the bending effect are more than the normal non bending effect for values of λ= (1.3 - 1.45) μm as shown in Figure 8. As the ratio of $x_2/x_1$ is around (5 - 6), the difference in leakage loss at λ=1.3 μm and at λ=1.6 μm is quite less. But when the ratio of $x_2/x_1$ is around (6.5 - 9) the difference of the leakage loss at λ=1.3 μm and at λ=1.6 μm is comparably high. These leakage losses and bending losses in double clad fiber increases with an increase in the ratio of $x_2/x_1$ as compared to the single clad fiber.

Overall we conclude that the designed fiber has less scattering loss and more sensitive to micro-bend loss as compared to the single clad fiber. The combined leakage losses and the bending losses of the double clad fiber are more than that of the single clad fiber and these losses are significantly higher when the values of $x_2/x_1$ are more than 5 and a value of wavelength is more than 1.45 μm.

**REFERENCES**

Leonard GC, Dietrich M, Wanda LM (1982). "Radiating eaky-mode losses in single mode light guides with depressed-index cladding". IEEE MTT-30:10.

Ajoy G, Thyagarajan K (2011). "Introduction to Fiber Optics".

Ajeet K, Vipul R, Youngjoo C, Won-Taek H (2010). "Dual shape large core single mode fiber for high power applications." Research Institute for Solar and Sustainable Energies, Gwangju Institute of Science and Technology, Korea.

Ajeet K, Vipul R, Charu K, Bernard D (2008). "Co-axial dual-core resonat leaky fiber for optical amplifier" J. Optics.

Maxim VE, Andrey GR, Alexander BM (2005). "Guided and Leaky Modes of Complex waveguide Structures." J. Light wave Technol. 23:8.

Snyder AW, Love JD (1993). Optical Waveguide Theory. Chapman and Hall, London.

# 21

# Runge kutta to precise the detection of lesion for magnetic resonance imaging (MRI) image

## K. El kourd and S. El kourd

Electronic institute, Biskra University, Algeria.

**In this paper, we translate non linear model to linear one and used numerical analysis: "Runge kutta4 (RK4)" is mathematical solutions to study the approximation's solutions of ordinary differential equations, then pass to the statistical study for multiple regression to applied anova technique and with it, we extract the place of the lesion on medical image MRI by two ways: Distribution of gaussien curve (hypothesis test of ho) and directly on the pathologic image. The logicial applied here is Matlab.**

**Key words:** Runge kutta, linear regression, anova.

## INTRODUCTION

Analysis of variance became widely known after being included in Fisher's 1925 book, Statistical Methods for Research Workers (David, 1986). In statistics, analysis of variance (ANOVA) is a collection of statistical models, and their associated procedures, in which the observed variance in a particular variable is partitioned into components attributable to different sources of variation. In its simplest form, ANOVA provides a statistical test of whether or not the means of several groups are all equal, and therefore generalizes *t*-test to more than two groups. Doing multiple two-sample t-tests would result in an increased chance of committing an error. For this reason, ANOVAs are useful in comparing two, three, or more means (David, 1986).

### Background

#### Runge kutta

In numerical analysis, the Runge–Kutta (RK) methods [German pronunciation: (ʀʊŋəˈkʊta)] are an important family of implicit and explicit iterative methods for the approximation of solutions of ordinary differential equations. These techniques were developed around 1900 by the German mathematicians (Atkinson, 1989; Ascher et al., 1998).

## DESCRIPTION

The position of the spaceship can be described as $(x(t),y(t))$ for any time point t. Similarly, the velocity is $(x'(t),y'(t))$ so the spaceship moves in two-dimensional space. Its acceleration is described by the following system of differential equations (Atkinson, 1989): $x''=f(x,y,x',y')$ and $y''=g(x,y,x',y')$, given the initial values: xo, y, uo, vo. $u=x'$, $w=y'$.

In the RK4 method, you calculate four intermediate approximations, k1, k2, k3 and k4; the final approximation will be given by a weighted average of these intermediate approximations.

$$u_{i+1}=u_i+(1/6)*(k1+2*k2+2*k3+k4) \tag{1}$$

After translating the non linear model (exponential equation) to linear one, the linear regression and the fitted regression line was calculated and anova test was done under the following:

**Regression:** Description about the relationship between two variables where one is dependent and the other is independent (Armstrong, 2012; York, 1966).

**1. Fitted regression line:** The true regression line corresponding to equation (2) is usually never known. However, the regression line can be estimated by estimating the coefficients $\beta_1$ and $\beta_o$ for an observed data set (Armstrong, 2012; York, 1966):

$$E(Y) = \beta_0 + \beta_1 x \tag{2}$$

The actual values of y (which are observed as yield from the chemical process from time to time and are random in nature), are assumed to be the sum of the mean value, E(Y), and a random error term: The actual values of y (which are observed as yield from the chemical process from time to time and are random in nature), are assumed to be the sum of the mean value, E(Y), and a random error term, Equation (3):

$$Y = E(Y) + \epsilon = \beta_0 + \beta_1 x + \epsilon \tag{3}$$

The estimates, $\tilde{\beta}_1$ and $\tilde{\beta}_o$, are calculated using least squares. The estimated regression line, obtained using the values of, $\tilde{\beta}_1$ and $\tilde{\beta}_o$, is called the *fitted* line. The least square estimates, $\tilde{\beta}_1$ and $\tilde{\beta}_o$, are obtained using the following equation:

$$\hat{\beta}_1 = \frac{\sum_{i=1}^{n} y_i x_i - \frac{\left(\sum_{i=1}^{n} y_i\right)\left(\sum_{i=1}^{n} x_i\right)}{n}}{\sum_{i=1}^{n}(x_i - \bar{x})^2} \tag{4}$$

$$\hat{\beta}_0 = \bar{y} - \hat{\beta}_1 \bar{x} \tag{5}$$

Where $\bar{y}$ is the mean of all the observed values and $\bar{x}$ is the mean of all values of the predictor variable at which the observations were taken. $\bar{y}$ is calculated using $\bar{y} = (1/n)\sum_{i=1}^{n} y_i$ and $\bar{x}$ is calculated using:

$$\bar{x} = (1/n)\sum_{i=1}^{n} x_i$$

Once $\tilde{\beta}_1$ and $\tilde{\beta}_o$ are known, the fitted regression line can be written as

$$\hat{y} = \hat{\beta}_0 + \hat{\beta}_1 x \tag{6}$$

Where $\tilde{y}$ is the *fitted* or *estimated* value based on the fitted regression model. It is an estimate of the mean value, E(Y). The fitted value, $\hat{y}_i$, for a given value of the predictor variable, $x_i$, may be different from the corresponding observed value, $y_i$. The difference between the two values is called the *residual* (Armstrong, 2012; York, 1966):

$$e_i = y_i - \hat{y}_i \tag{7}$$

To calculate the Statistic $F_o$, it must pass by the six titles (Uts and Hekerd, 2004; Dudok, 2010; Plonsky, 2007; El Kourd and El kourd, 2013; Gear, 1971; Jon and Predrag, 2006; Henson and Penny, 2005; http://www.weibull.com/DOEWeb/introduction.ht;12/6/2012, 15:45; Research Methods I, ANOVA and Multiple Regression; Viviane Kostrubiec. Les comparaisons multiples: entre mythe et réalité).

**Total sum of squares (SST):** On simple linear regression that the total sum of squares, $SS_T$, is obtained using the following equation:

$$SS_T = \sum_{i=1}^{n}(y_i - \bar{y})^2 = \sum_{i=1}^{n} y_i^2 - \frac{(\sum_{i=1}^{n} y_i)^2}{n} \tag{8}$$

The total sum of squares in matrix notation is:

$$SS_T = \mathbf{y'y} - (\tfrac{1}{n})\mathbf{y'Jy} = \mathbf{y'}\left[\mathbf{I} - (\tfrac{1}{n})\mathbf{J}\right]\mathbf{y} \tag{9}$$

Where y is the vector of observed values, I is the identity matrix of order n; and J represents an n x n square matrix of ones.

**Model sum of squares (SSR):** Similarly, the model sum of squares or the regression sum of squares, $SS_R$, can be obtained in matrix notation as:

$$SS_R = \sum_{i=1}^{n} \hat{y}_i^2 - \frac{(\sum_{i=1}^{n} y_i)^2}{n}$$
$$= \mathbf{\hat{y}'\hat{y}} - (\tfrac{1}{n})\mathbf{y'Jy}$$
$$= \mathbf{y'}\left[\mathbf{H} - (\tfrac{1}{n})\mathbf{J}\right]\mathbf{y} \tag{10}$$

Where H is the hat matrix and is calculated using:

$$\mathbf{H} = \mathbf{X(X'X)^{-1}X'} \tag{11}$$

**Error sum of squares:** The error sum of squares or the residual sum of squares, $SS_E$, is obtained in the matrix notation from the vector of residuals, e, as:

$$SSE = SS_T - SS_R \tag{12}$$

$$SS_E = \mathbf{e'e}$$
$$= (\mathbf{y} - \mathbf{\hat{y}})'(\mathbf{y} - \mathbf{\hat{y}})$$
$$= \mathbf{y'(I - H)y} \tag{13}$$

**Mean squares (MST):** Mean squares are obtained by dividing the sum of squares with their associated degrees of freedom. The number of degrees of freedom associated with the total sum of squares, $SS_T$, is (n-1) since there are n observations in all, but one degree of freedom is lost in the calculation of the sample mean $\bar{y}$. The total mean square is:

$$MS_T = \frac{SS_T}{n-1} \tag{14}$$

**Regression mean square (MSR):** The number of degrees of freedom associated with the regression sum of squares, $SS_R$ is k.

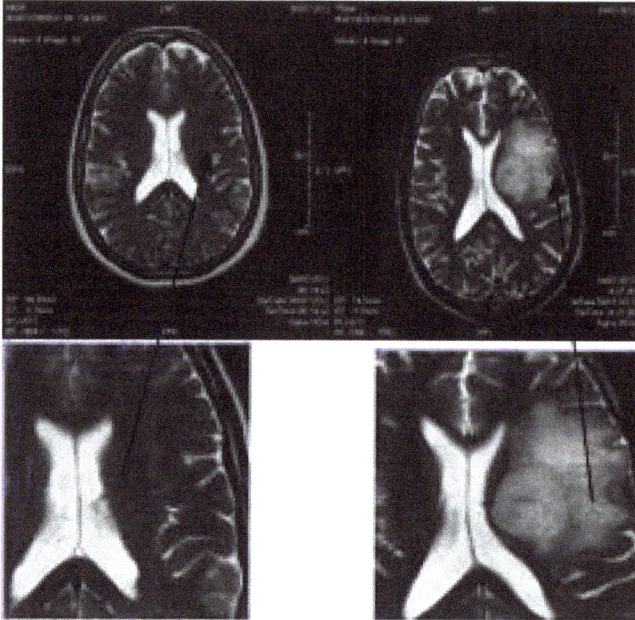

**Figure 1.** Normal image (left) and pathological image (right).

There are (k+1) degrees of freedom associated with a regression model with (k+1) coefficients, $\beta_o, \beta_1, \beta_2 \cdots \beta_k$ However, one degree of freedom is lost because the deviations, $(\tilde{y}_i - \bar{y})$, are subjected to the constraints that they must sum to zero $\sum_{i=1}^{n}(\tilde{y}_i - \bar{y})^2$. The regression mean square is:

$$MS_R = \frac{SS_R}{k}$$

(15)

The number of degrees of freedom associated with the error sum of squares is: n-(k+1), as there are n observations in all, but (k+1) degrees of freedom are lost in obtaining the estimates of $\beta_o, \beta_1, \beta_2 \cdots \beta_k$ to calculate the predicted values, $\tilde{y}_i$. The error mean square is ( Ascher et al., 1998; rmstrong, 2012; York, 1966; Uts and Hekerd, 2004; Dudok, 2010):

$$MS_E = \frac{SS_E}{n - (k + 1)}$$

(16)

The error mean square, MS$_E$, is an estimate of the variance, $\sigma^2$ of the random error terms, e$_i$.

**Mean square error (MS$_E$):** MS$_E$ is estimate variance $(\tilde{\sigma}^2)$ of random error e$_i$.

$$\hat{\sigma}^2 = MS_E$$

(17)

**Calculation of the statistic F$_o$:** Once the mean squares MS$_R$ and MS$_E$ are known, the statistic to test the significance of regression can be calculated as follows (Atkinson, 1989):

$$F_0 = \frac{MS_R}{MS_E}$$

(18)

## EXPRIMENTAL RESULTS

### Algorithm

(1) Read the images
(2) Choose a sample of image
(3) Calculate the error between y and y estimate then plot it for one vector .
(4) Applied analyse mathematic with runge kutta4.
(5) Estimate the linear regression with parameters estimates (image x).
(6) Application the analyses of variance 'ANOVA' .
(7) Compeer the result of Anova with Runge kutta and the technical of Anova with linear model

### Analysis of data (the protocol radiologic)

Our protocol is for a patient aged 55 years; He made MRI scan with injection of contrast medium. The machine used: Type "SIEMENS", and with the field B = 1.5 Tesla. The sequences performed T1 and T2.

The result MRI scan of a tumor appears in the middle of the field on the left side and the third ventricle is to evoke the tumor (Figure 1).

The left bottom is a section of normal image and the right one is the section of image; it has surface (200×200), our analysis was then applied, but before that the lesion or the error between the both images was detected by calculating the error for one vector such as:

$$e = y - \hat{y}$$

(19)

Where: y is the vector, and $\tilde{Y}$ is data estimate (Figure 2). Figure 2 from length (0 and 45) of image present the place of disease (red color). Y is with blue color and $\hat{Y}$ is with green one. From (Equation 18), we extract the disease for all image with calculate Fcal by two ways directly on the pathological image and from Gaussian curve (Figure 3). Figure 3a present the pathological image. Figure 3b detects the lesion directly on the pathological image, which were represented with white color. Figure 3 present the curve of anova technique as anova equation:

F- cal=MS$_R$/MS$_E$

Where: Fcal present with green color its value beside zeros number. MS$_R$: The within-groups variation which present with blue star color. MS$_E$: The between-groups, it is with red color. The last Figure 3d present gauss curve (Gaussian distribution of Anova –test) for $\alpha = 0.01$, and from the table of fisher test we have:

*Degree of liberty: ddl:
$v = n$ -1= 200-1=199.

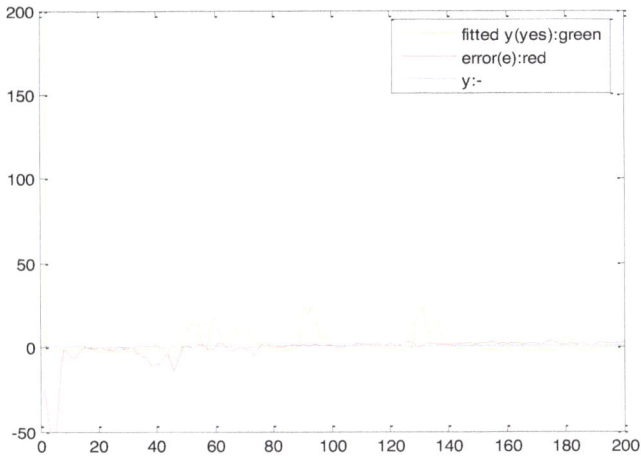

**Figure 2.** Presentation error e, y and $\tilde{Y}$.

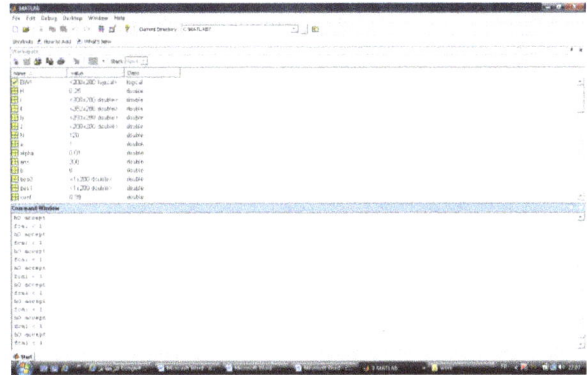

**Figure 4.** Presentation of hypothesis ho or h1 as display in Figure 3d.

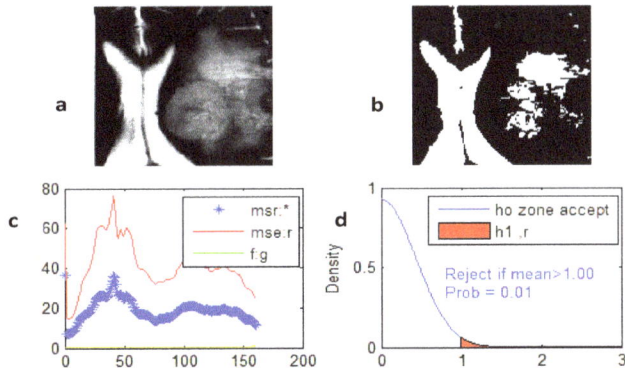

**Figure 3.** Ways of detecting lesion. (a) Pathological image; (b) Detection of lesion withb anova; (c)Computation of anova; (d) Anova $F$-tab.

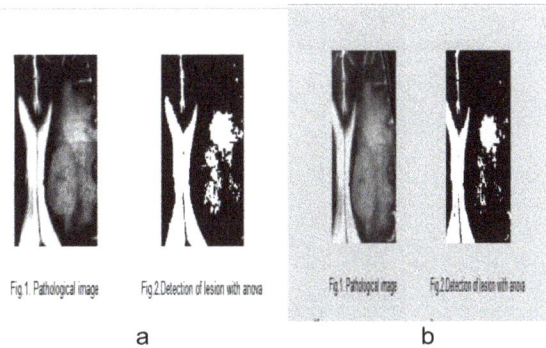

**Figure 5.** Comparaison between Application Anova for non linear model on the pathological image (a) and linearization with anova simple (b).

*F-tab = 1.
*Hypothesis ho will be:

$$\begin{cases} \text{If } F\text{-cal} \leq F\text{-tab} \Rightarrow \text{f-test with (anova)} \rightarrow \text{accept } H_0. \\ \text{If } F\text{-cal} > F\text{-tab} \Rightarrow \text{test } F\text{-test reject } H_0 \text{ (red color)} \end{cases}$$

Figure 4 present the hypothesis accept or reject.

## Comparison between two methods of ANOVA

The result with model non linear presented in Figure 5a is more precise in front of anova with linear model (Figure 5a)

## Conclusion

From these study we can distinct:

1. Numeric analysis with Runge kutta is used to pass from non linear to linear model.
2. Runge kutta used for approximate the solutions of ordinary differential equations; from here come our idea. It can precise results and achieve to the complex places in image, where doctors cannot see it clearly.
3. The statistical study is used to do the comparison between two models: A new and old one in different specialty.
4. In this article we have used a medical image with MRI scan.
5. The result with Anova pass by," non linear model" then do the linearization , is more precise then used directly the linear model of Anova.

We propose for researcher to use always Anova technical, but for non linear regression and for multi-images.

## Conflict of Interest

The authors have not declared any conflict of interest.

## REFERENCES

David GH (1986). On the history of ANOVA in unbalanced, factorial designs: The First 30 Years. Am. Statist. 40(4):265-270.

Atkinson KA (1989). An introduction to numerical analysis (2nd ed.), New York: John Wiley & Sons,

Ascher Uri M, Petzold LR (1998). Computer methods for ordinary differential equations and differential-algebraic equations, Philadelphia: Soc. Indust. Appl. Math.

Armstrong JS (2012). Illusions in regression analysis. Int. J. Forecast. 28(3):689.

York D (1966). Least-squares fitting of a straight line. Canad. J. Phys. 44:1079-1086.

Uts J, Hekerd R (2004). Mind on Statistics. Chapter 16 - Analysis of Variance. Belmont, CA: Brooks/Cole-Thomson Learning, Inc. Dudok de Wit T (2010). Analyse Numérique, Licence de physique – 3ème année, Université d'Orléans –Faculté des Sciences,Université d'Orléans.

Plonsky M (2007). One Way ANOVA. Retrieved from: http://www.uwsp.edu/psych/stat/12/anova-1w.htm.

El Kourd K, El kourd A (2013), The detection of disease by statistic test of analyze of variance. J. IEEE.

Gear CW (1971). Numerical initial value problems in ordinary differential equations (EnglewoodCliffs, NJ: Prentice-Hall).PMCid:PMC1411756 http://www.weibull.com/DOEWeb/introduction.ht;12/6/2012, 15:45.

Research Methods I, ANOVA and Multiple Regression, Viviane Kostrubiec. Les comparaisons multiples: entre mythe et réalité. Laboratoire Adaptations Perceptivo-Motrices et Apprentissage (EA 3191). Université Paul Sabatier–Toulouse III.

Jon R, Predrag P (2006). t-tests, ANOVA and regression. Henson R, Penny W (2005)

.ANOVAsandSPM.InstituteofCognitiveNeuroscience,WellcomeDepartm entofImagingNeuroscience,UniversityCollegeLondon.

# 22

# Reduction of side lobe level in non-uniform circular antenna arrays using the simulated annealing algorithm

**A. Zangene[1] , H. R. Dalili Oskouei[2] and M. Nourhoseini[3]**

[1]Amirkabir University of Technology, Tehran, Iran.
[2]University of Aeronautical Science and Technology (Shahid Sattari), Tehran, Iran
[3]Amirkabir University of Technology, Tehran, Iran.

This paper investigates the reduction of side lobe level in antenna arrays. Reduction of side lobe level in antenna arrays has some limitations including fixed width of the beam. We have modeled the side lobe level reduction as an optimization problem using simulated annealing technique for side lobe level reduction of a specific beam width. The advantage of this method compared with other methods is that it can get out of local minimums and converge to the optimized answer. Efficiency of simulated annealing algorithm in pattern extraction of desired circular antenna, which is used frequently in modern telecommunication and radar systems, is investigated and the results are compared with that of genetic and evolutionary algorithms.

**Key words:** Simulated annealing, antenna array, circular array, non-uniform antenna array, side lobe level.

## INTRODUCTION

Antenna arrays have various applications in wireless and mobile communication systems. In most application antenna should be designed in such a way that it can transmit produced beam in various directions and distances. To fulfill this goal, an array of antennas must be used. Higher transmission power, lower power consumption, radiating beam control, and higher efficiency can be obtained using antenna arrays. Arrays can have different forms such as linear, circular, and planar with different applications, like radar, sonar, imaging, biomedicine and mobile communications (Panduro et al., 2006; Balanis, 1997; Shihab et al., 2008 and Dessouky et al., 2006).

Due to the particular structure of circular antenna, attention toward circular arrays has increased in recent studies, (Panduro et al., 2006); Shihab et al., 2008; Dessouky et al., 2006; Pathak et al., 2009). In circular antenna arrays with non-uniform distribution, elements are placed on a circular ring with non-uniform distances (Figure 1).

This group of antennas has important functionality with different applications such as radio navigation, air and space navigation, sound tracking, etc (Dessouky et al., 2006; Granville et al., 1994; Locatelli et al., 1994; Ingber, 1993; Aydin and Fogarty, 2004). Recently, antenna arrays have been suggested for wireless communications especially as intelligent antennas. In many studies (Panduro et al., 2006; Balanis, 1997; Shihab et al., 2008;

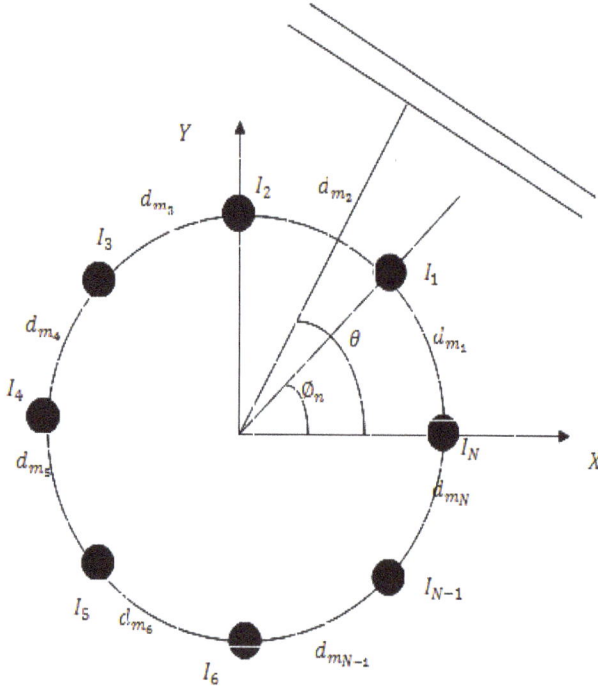

**Figure 1.** Structure of an antenna array with n elements (Panduro et al., 2006).

Dessouky et al., 2006; Pathak et al., 2009) it has been tried to reduce side lobe level as much as possible in non-uniform distribution state. Reduction of side lobe level has ideal influences on telecommunication systems.

Genetic and evolutionary algorithms have been used to reduce side lobe level in paper (Panduro et al., 2006; Shihab et al., 2008). Various parameters such as beam width, and noise sensitivity, and some other factors (antenna gain, radiation pattern, antenna size) should be considered to reduce side lobe level. The objective of this problem is to find the best distances between elements and the stimulation amplitude of each element so that side lobe level is minimal.

Therefore, the problem of finding the best set of distances between elements and their stimulation amplitude can be proposed as an optimization problem. Here, we have used simulated annealing technique to specify the best distances between elements and their stimulation amplitude so that the final produced pattern will have the maximum reduction in side lobe level. Simulated annealing algorithm adds a random aspect to the descent along with the gradient which is controllable by the T (temperature) parameter. This algorithm allows both transitions to the higher energy state or to the lower energy state; therefore using this algorithm, it is possible to get out of local minimums and get to the global optimized answer. In our experiments beam width and the main beam formation angle were fixed to 50 and 0° respectively.

## PROBLEM STATEMENT

Suppose that n isotropic elements with interspaces $d_m$ are placed on the circumference of a circular ring with radius *a* on x-y plane (Figure 1). Assuming that elements are isotropic, it can be concluded that the propagation pattern of this array of antennas can be explained by its array factor.

Array factor for a circular array on x-y plane is stated as below (2):

$$AF(\theta,I,d_m) = \sum_{n=1}^{N} I_n e^{jka(\cos(\theta-\theta_0)-\cos(\theta_0-\phi_n))} \quad (1)$$

In which:

$$ka = \frac{2\pi a}{\lambda} = \sum_{i=1}^{N} d_{m_i} \quad (2)$$

$$\phi_n = \frac{(2\pi\sum_{i=1}^{n} d_{m_i})}{\sum_{i=1}^{N} d_{m_i}} \quad (3)$$

$\theta$ is the angle at which the main beam is generated, $d_m$ (a 1*10 matrix) is the distance between antenna array elements, and $I_m$, which is also a 1*10 matrix, is the stimulation amplitude of each element.

In the $d_m$ array, each component $d_{m_i}$ is the distance of the ith element from the (i+1)th one.

$k = \frac{2\pi a}{\lambda}$ is the constant value of the phase difference between elements, $\theta$ is the intersection angle of the beam with x-y plane, $\lambda$ is the beam wave length, and $\theta_0$ is the angle at which the main beam has the most propagation. As indicated before, finding the best set of places and stimulation amplitude of the elements can be proposed as an optimization problem. Therefore, to solve this problem using the simulated annealing algorithm, an objective function must be defined through which the simulated annealing algorithm can get to the optimized answer. Assuming that $\theta_0$ is the angle at which the maximum propagation occurs and θ varies in the range [-π,π], $\theta_{msl}$ is the angle at which the first side lobe, which is the highest one, is generated, BWFN_desired is the width of the desired beam, which assumed to be equal to the constant value 50, and BWFN ($I$, $d_m$) is the first null beam width, the objective function is stated as follows:

$$f_1 = \frac{\left| AF(\theta_{msl1}, I, d_m) + AF(\theta_{msl2}, I, d_m) + AF(\theta_{msl3}, I, d_m) \right|}{AF(\theta_0, I, d_m)} + \left| BWFN_{desired} - BWFN(I, d_m) \right| \tag{4}$$

Based on the defined objective function, the best set of $I$ and $d_m$ is obtained when $f_1$ is minimal. Reduction of all side lobes is considered at the same time using the defined objective function.

In this problem, it is also assumed that the circumference of the circle on which the elements are located is constant.

## THE PROPOSED ALGORITHM

The main objective of this work is to maximally reduce side lobes for a circular antenna array in which elements are distributed non-normally. There are some limitations to maximally reduce side lobes in a circular antenna array such as constant width of the desired beam, number of elements and the circumference.

Simulated annealing which is used in this paper is a generic probabilistic met heuristic method for obtaining optimized main point for the desired objective function in a large search space which was first presented by Kirkpatrick in 1983 (Pathak et al., 2009). This method is usually used when the search space is discrete. For some specific problems, simulated annealing technique can be more efficient than searching the whole state space. It may be possible to obtain the best answer by searching the whole space state, but it is not possible considering the time needed for the process. Furthermore, most of the time, we get the answer which is close to the best answer in a specific period of time.

The name and idea of this algorithm has been extracted from the annealing technique in metallurgy. In this process, metal is heated to the melting temperature and then is cooled gradually under control. Heating causes the atoms in the crystalline structure of the metal to leave their primary position (primary positions are considered as local minimums) and place randomly in new locations. Then, in the gradual cooling process, the states with lower energy levels with respect to the primary state of the metal have more chances of converging.

In this technique, any point in the search space is considered as a state with energy E. when the system transits from one state to another, the probability of accepting the new state is defined by P ($E_{current}$, $E_{new}$, T) which depends on the current state energy, new state energy, and the parameter T. In this algorithm, if the new state energy is lower than the current state energy, current state to new state transition is done with the probability of 1.

$$\Delta E = E_{new} - E_{current} \tag{5}$$

$$\rho(\Delta E) = \begin{cases} e^{\frac{-\Delta E}{T}} & \Delta E > 0 \\ 1 & \Delta E <= 0 \end{cases} \tag{6}$$

And if the new state energy is equal to or higher than the current state energy, algorithm accepts this state transition with the probability of $e^{\frac{-\Delta E}{T}}$ which is dependent on parameter T. At first, this probability has the maximum value and gradually after running the algorithm when $T \to 0$, it tends toward zero. Transition to higher energy states, provides the possibility of getting out of local minimums for the algorithm (Smith et al., 1998; Koulmas et al., 1994; Kirkpatrick et al., 1983).

It can be shown that for any finite problem, the probability that the simulated annealing algorithm will give an answer close to the total optimized answer, with the assumption of no time limitation, tends to zero (Granville et al., 1994; Locatelli, 2001).

It is also possible to use an adaptive neighborhood in this algorithm; so that the neighborhood radius will accept all the states at the beginning and continuing the algorithm it is reduced gradually until converging to the best answer in the end. Simulated annealing algorithm with adaptive neighborhood radius is applicable when the distance between the optimized answer and the current answer is shorter than the step length (Ingber, 1993). The Flow chart of our proposed algorithm for solving side lobe reduction problem is shown in Figure 2.

Based on this technique, it is possible to search the whole state space normally in the primary stages and to reduce the search space to obtain the best answer during the algorithm process. In reference (Aydin and Fogarty, 2004), a number of advantages and disadvantages of the simulated annealing technique have been proposed. This technique has been used for solving major and practical problems such as flow shop scheduling (Low, 2005; Burke et al., 2003), time tabling (Framinana and Schusterb, 2006; Cerny, 1985), travelling salesman (Lin and Kernighan, 1973; Salcedo-Sanz et al., 2004), communication systems (Paik and Soni, 2007; Locatelli, 2000), continuous optimization, and etc.

## RESULTS AND DISCUSSION

The presented algorithm in the previous part was implemented and the results were studied for the design of a circular antenna array with non-normal distribution. In the experiments, to maximally reduce side lobes, the angle with maximum propagation was assumed to be zero at $\theta_0 = 0$. The experiments were carried out for different number of elements 8, 10, and 12 and the resulted array factor for each one was reported.

Variation distance value and the coefficient, which was used for gradual cooling in the algorithm, was set to 0.01

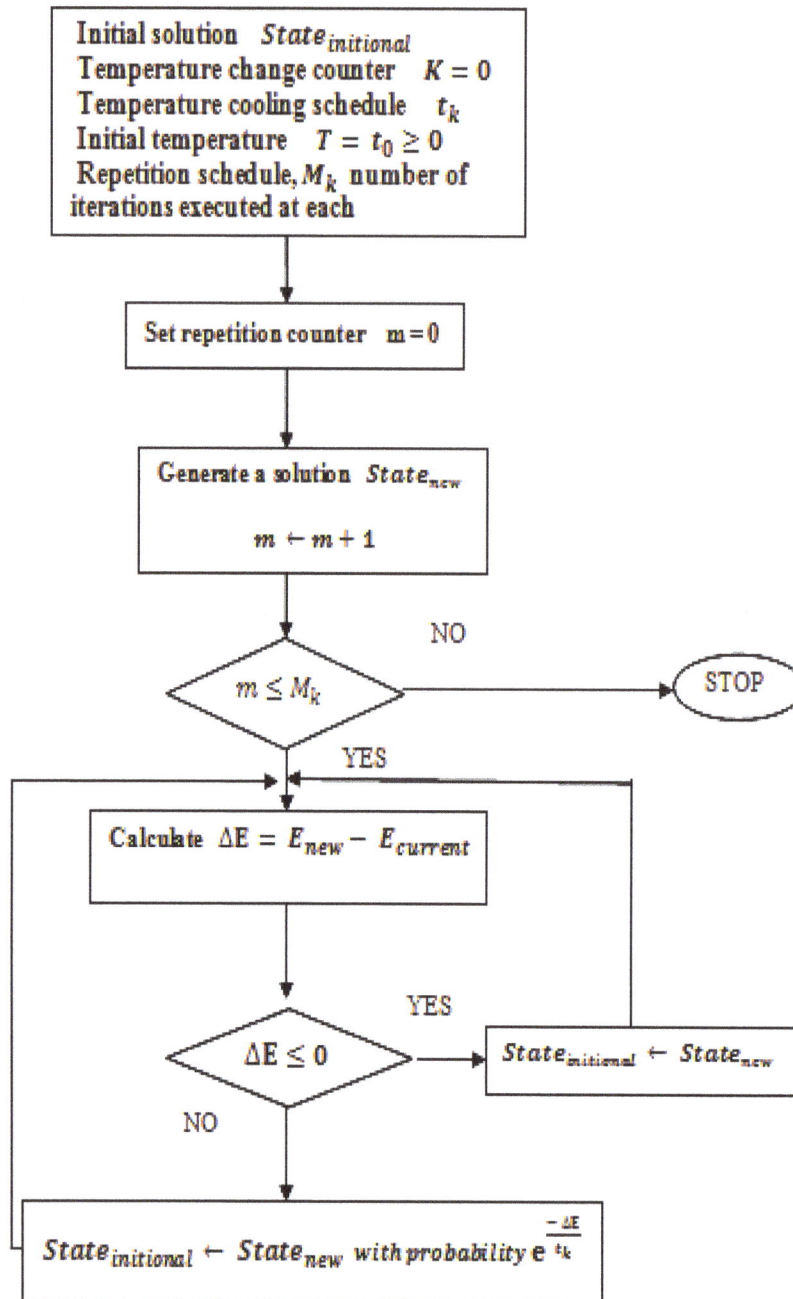

**Figure 1.** Flow chart for the simulated annealing algorithm.

and 0.7 respectively in the implementation of the simulated annealing algorithm. The algorithm continues until 5 successive output values converge to a unit value. Maximum number of repetitions is assumed to be 10000. As seen in the Figures 3 and 4, for 10 elements, the first side lobe level by using the normal distribution, the genetic algorithm and the proposed algorithm is -7.9, -11.1 and -11.9 dB respectively. In conclusion, by using the simulated annealing algorithm, the side lobe level with respect to the main lobe has 0.8dB reduction in

comparison to the genetic algorithm and 4.1 d B reduction in comparison to the algorithm of normal distribution of elements.

According to the results, superiority of this algorithm in comparison to the genetic algorithm can be observed, because the Genetic algorithm may fall in local minimums while the simulated annealing algorithm can converge to an optimized answer by starting from an appropriate primary point and by some repetitions. In Figures 5 and 6, propagation patterns for an antenna array with 12

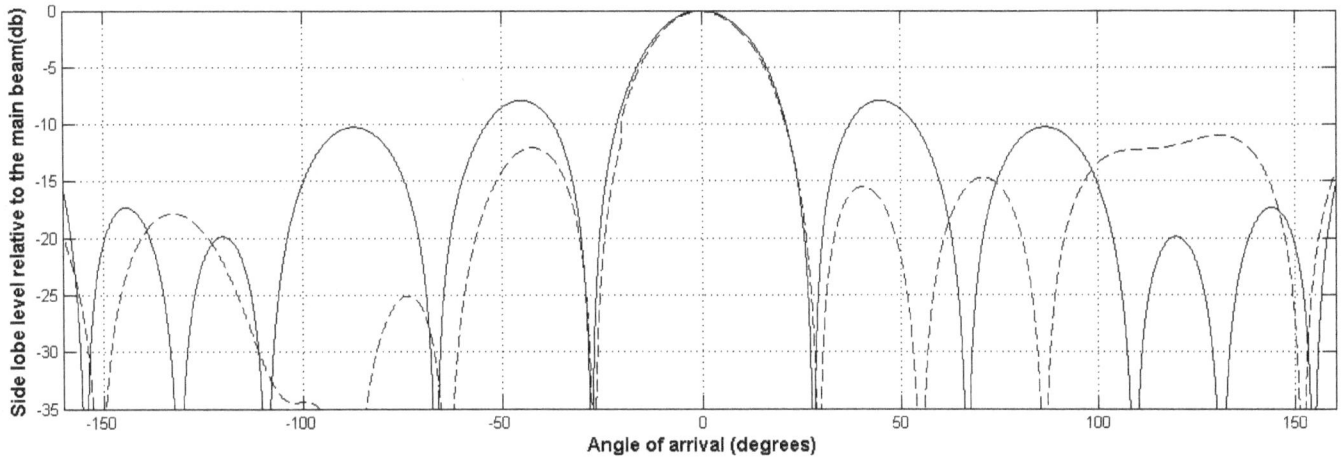

**Figure 3.** Comparison between propagation patterns of normal distribution (——) of elements and simulated annealing algorithms (-----) for an antenna array with 10 elements.

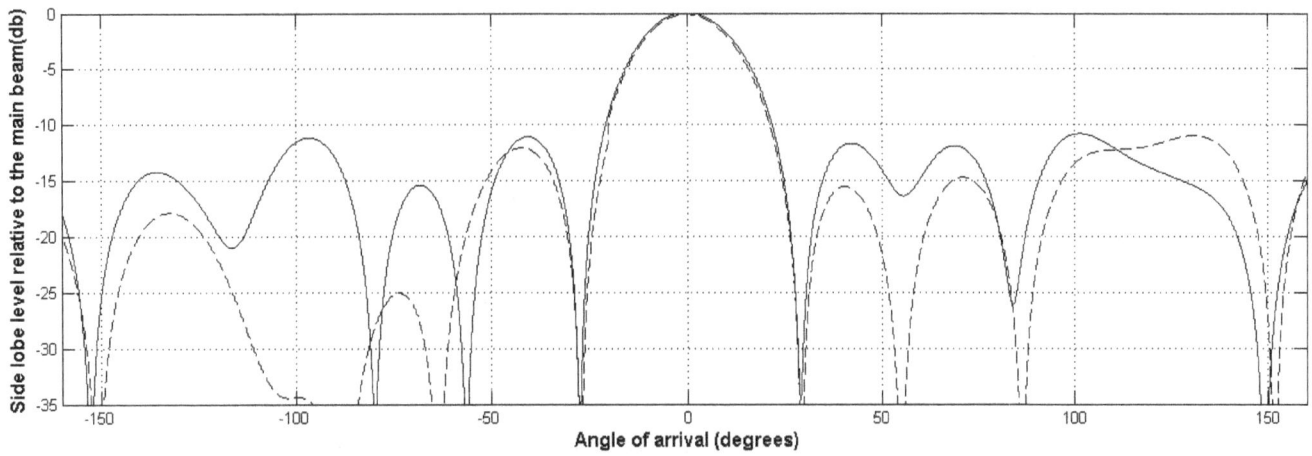

**Figure 4.** Comparison between propagation patterns of genetic ( —— ) and simulated annealing algorithms (- - - - -) for an antenna array with 10 elements.

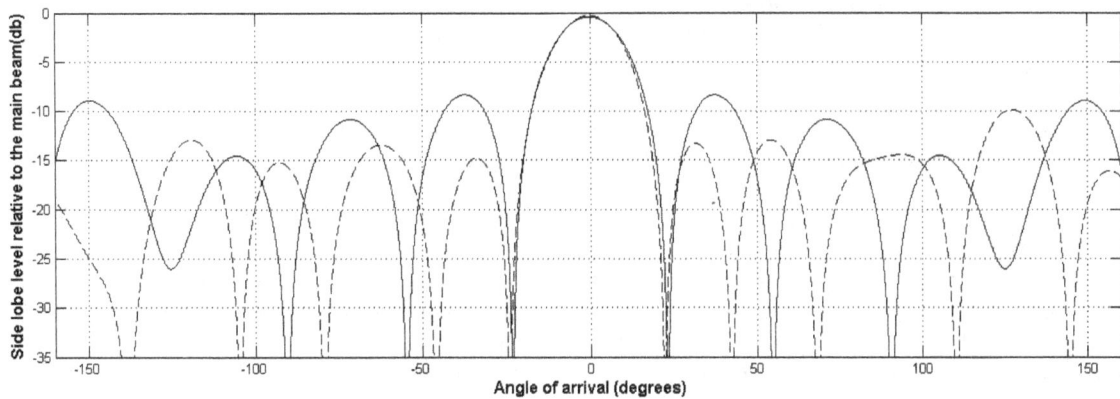

**Figure 5.** Comparison between propagation patterns of normal distribution of elements( —— ) and simulated annealing algorithms (- - - - -) for an antenna array with 12 elements.

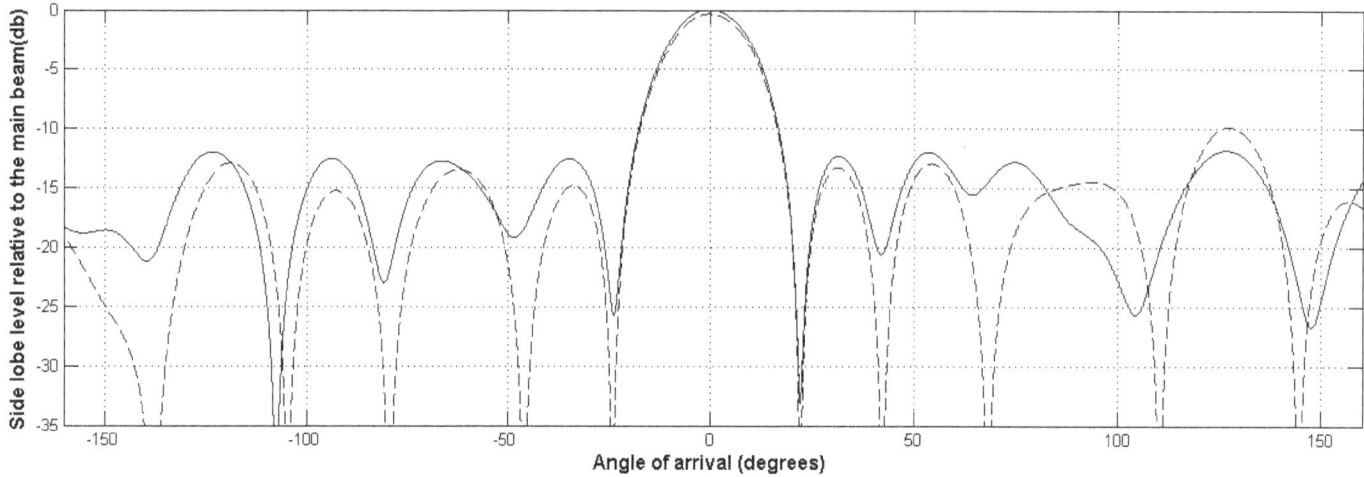

**Figure 6.** Comparison between propagation patterns of Genetic ( —— ) and simulated annealing algorithms (- - - - -) for an antenna array with 12 element

**Table 1.** Examples for the non-normal distribution of antenna array elements on a circular surface for two different numbers of elements.

| N | SLL(dB) | BWFN(deg) | $d_{m_1}, d_{m_2}, d_{m_3}, \cdots, d_{m_N} ; I_1, I_2, I_3, \cdots, I_N$ | Aperture |
|---|---------|-----------|----------------------------------------------------------------------------|----------|
| 8 | -10.5 | 71.5 | $0.05961\lambda$ , $0.314\ \lambda$ , $0.7763\lambda$ , $0.7425\lambda$ , $0.6297\lambda$ , $0.8969\lambda$ , $0.4633\lambda$ , $0.5267\lambda$ , $0.3289$ , $0.2537$ , $0.7849$ , $1$ , $1.0171$ , $0.5183$ , $0.5176$ , $0.4612$ | $4.40\ \lambda$ |
| 10 | -11.9 | 55.3 | $0.3258\lambda$ , $0.4934\ \lambda$ , $0.3505\lambda$ , $1.6573\lambda$ , $0.6213\lambda$ , $0.9948\lambda$ , $0.4968\lambda$ , $0.2431\lambda$ , $0.5878\lambda$ , $0.3148\lambda$ , $0.9845$ , $0.3883$ , $0.4092$ , $0.9674$ , $0.6586$ , $0.5533$ , $0.5834$ , $0.6215$ , $0.4665$ , $0.6148$ | $6.08\ \lambda$ |
| 12 | -15 | 45.6 | $0.5181\lambda$ , $0.3665\lambda$ , $1.4283\lambda$ , $0.8367\lambda$ , $0.3249\lambda$ , $0.5702\lambda$ , $0.4983\lambda$ , $0.8135\lambda$ , $0.8195\lambda$ , $0.4992\lambda$ , $0.3595\lambda$ , $0.7370\lambda$ , $0.1864$ , $0.5116$ , $0.2046$ , $0.6486$ , $0.7512$ , $0.8393$ , $0.5577$ , $0.4095$ , $0.5225$ , $0.4875$ , $0.4833$ , $0.8453$ | $7.07\ \lambda$ |

elements are shown. In these figures, the propagation pattern obtained from the simulated annealing algorithm is compared with the patterns obtained from normal distribution of elements and the genetic algorithm in Figure 5 and 6 respectably.

As seen in the figures, the first side lobe level is -15 dB for the simulated annealing algorithm,-12.8 dB for the genetic, and -7 dB for normal distribution of elements. Therefore we have 1.2 and 5 dB reduction in first side lobe using simulated annealing algorithm in comparison of genetic algorithm and normal distribution of 12 elements (Panduro et al., 2006).

Results obtained from simulated annealing algorithm for distribution of elements on a circular array with their distance $\{d_{m_i}\}$, amplitude $\{I_i\}$ for a set of 8, 10 and 12 elements are presented. In Table 1, Based on the results,

by increasing the number of elements and the circumference of the circle on which elements are placed, side lobe levels reduce.

## Conclusion

In this paper, simulated annealing algorithm has been used for maximally reducing side lobes in circular antenna arrays with non-normal distribution of elements with the assumption of constant beam width. Simulated annealing algorithm can converge to an optimized answer by starting from an appropriate primary point and by some repetitions. This algorithm can provide the possibility of getting out of local minimums and converging to the local optimized answer by adding a probability aspect to the descent along with the gradient.

Based on the obtained results from the experiments, the efficiency of this algorithm in getting out of local minimums (the genetic algorithm may get stuck in local minimums) and getting to a close optimized answer has been demonstrated.

In conclusion, simulated annealing algorithm shows better efficiency and reduction in side lobes in comparison to the other methods.

## Conflict of Interest

The authors have not declared any conflict of interest.

## ACKNOWLEDGEMENT

This research is supported by the Iran Telecommunication Research Center and the authors gratefully acknowledge the institute.

## REFERENCES

Panduro M, Mendez A, Dominguez R, Romero G (2006). Design of non-uniform circular antenna arrays for side lobe reduction using the method of genetic algorithms, Int. J. Electron. Commun. (AEU) 60:713-717.

Balanis CA (1997). Antenna Theory: Analysis and Design, John Wiley & Sons, New York.

Shihab M, Najjar Y, Dib N, Khodier M (2008). Design of non -uniform circular antenna arrays using particle swarm optimization, J. Elect. Eng. 59(4):216-220.

Dessouky M, Sharshar H, Albagory Y (2006). Efficient Side lobe Reduction Technique For Small-Sized Concentric Circular Arrays, Progress In Electromagnetics Res. PIER 65:187-200.

Granville V, Krivanek M, Rasson JP (1994). Simulated annealing, A proof of convergence, IEEE Transactions On Pattern Analysis and Machine Intelligence, 16:652.

Locatelli M (2001).Convergence and first hitting time of simulated annealing algorithms for continuous global optimization. Math. Methods. Oper. Res. 54:171-199.

Ingber L (1993). Simulated annealing: practice versus theory, J. Math. Comput. Model.18(11):29-57.

Aydin ME, Fogarty TC (2004). A distributed evolutionary simulated annealingalgorithm for combinatorial optimisation problems, J. Heuristics 10:269-292.

Low CY (2005). Simulated annealing heuristic forflowshop scheduling problems withunrelated parallel machines, Comput. Operat. Res. 32:2013-2025.

Burke EK, Eckersley A, McCollum B (2003). Using simulated annealing to studybehaviour of various exam timetabling data sets, in: Proceedings of the Fifth Meta heuristics Int. Conference (MIC 2003), Kyoto, Japan, August.

Framinana JM, Schusterb C (2006). An enhanced timetabling procedure for the nowait job shop problem: a complete local search approach, Comput. OperationsRes. 331:1200–1213.

Cerny V (1985). A thermodynamical approach to the travelling salesman problem: anefficient simulation algorithm, J. Optimization Theory Appl. 45:41-51.

Lin S, Kernighan BW (1973). An effective heuristic algorithm for the traveling salesman problem, Operations Res. 21(2):498-516.

Salcedo-Sanz S, Santiago-Mozos R, Bousono-Calzon C (2004). A hybrid hopfieldnetwork-simulated annealing approach for frequency assignment in satellitecommunications systems, IEEE Trans. Syst. Man Cybernet. Part B: Cybernet. 34(2):1108-1116.

PMid:15376856

Paik CH, Soni S (2007). A simulated annealing based solution approach for the two layered location registration and paging areas partitioning problem in cellularmobile networks, Eur. J. Operat. Res. 178:579-594.

Locatelli M (2000). Convergence of a simulated annealing algorithm for continuousglobal optimization, J. Global Optimization 18:219–234.

Smith K, Palaniswami M, Krishnamoorthy M (1998). "Neural techniques for combinatorial optimization with applications," Neural Networks, IEEE Transactions on, 9:1301-1318.

PMid:18255811

Koulmas AC, Antony SR, Jaen R (1994). A survey of simulated annealing applications to operations research problems, Omega, 22:41.

Kirkpatrick S, Gelatt CD, Vecchi MP (1983). Optimization by simulated annealing, Science 220(4598):671-680.

PMid:17813860

Pathak NN, Mahanti GK, Singh SK, Mishra JK, Chakraborty A (2009). "Synthesis of thinned planar circular array antennas using modified particle swarm optimization," Progress In Electromagnetics Res. Lett. 12:87-97.

# 23

# Estimating global solar radiation on horizontal surface from sunshine hours over Port Harcourt, Nigeria

**Mfon David Umoh[1], Sunday O. Udo[2] and Ye-Obong N. Udoakah[3]**

[1]Department of Physics, University of Uyo, Uyo, Akwa Ibom State, Nigeria.
[2]Department of Physics, University of Calabar, Calabar, Cross River State, Nigeria.
[3]Department of Electrical/Electronics Engineering, University of Uyo, Uyo Akwa Ibom State, Nigeria.

A model for estimating sunshine hours from some meteorological parameters was developed. An eleven year (1997 to 2007) period of relative humidity, maximum and minimum temperatures, rainfall and wind speed measured at Port Harcourt, Nigeria (Latitude 4°56'26.2''N) was analyzed. The results of the correlations show that the four variable correlations with the highest value of R gives the best result when considering the error term Root Mean Square Error (RMSE). The developed model can be used in estimating global solar radiation for Port Harcourt and other locations with similar climatic conditions.

**Key words:** Global solar radiation, sunshine hours.

## INTRODUCTION

Development of a solar energy research programme must always start with a study of solar radiation data at the site or region of interest. Long term measurements of solar radiation on a horizontal surface exist for only relatively few meteorological stations (Akpabio, 2002). To this effect, the development of emperical models for the estimation of solar radiation in a developing country such as Nigeria has become imperative. Sunshine hours has been the best alternative way of estimating global solar radiation. This is because sunshine hours is easy to use and is reliable. It is also easily measured and readily available. Several empirical models have been developed to calculate global solar radiation using various parameters (Umoh and Udoh, 2010, Umoh and Akpan, 2011a, b; Khan and Ahmad, 2012; Hussein and Ahmed 2012; Jakhrani et al., 2013; Kaya, 2012; Al-Dulaimy, 2010). The parameter used as input in the calculations include, sunshine duration, mean

temperature, soil temperature, relative humidity, number of rainy days, altitude, latitude, total perceptible water, albedo, atmospheric pressure, cloudiness and evaporation. The main objective of the study is to develop an equation that correlate monthly average daily sunshine hours with certain meteorological parameters for Port Harcourt, Nigeria. Global solar radiation is then computed from this equation.

### METHODOLOGY

In order to obtain the set of equations necessary for this research, measured data of monthly average sunshine hours, relative humidity, difference in maximum and minimum temperatures, rainfall data and wind speed of Port Harcourt corresponding to the period 1997 to 2007 was obtained from the Nigerian Meteorological Agency (NIMET) in Oshodi, Lagos. Port Harcourt is located at Latitude 4°56'26.2''N (Figure 1). Monthly averages (over the eleven

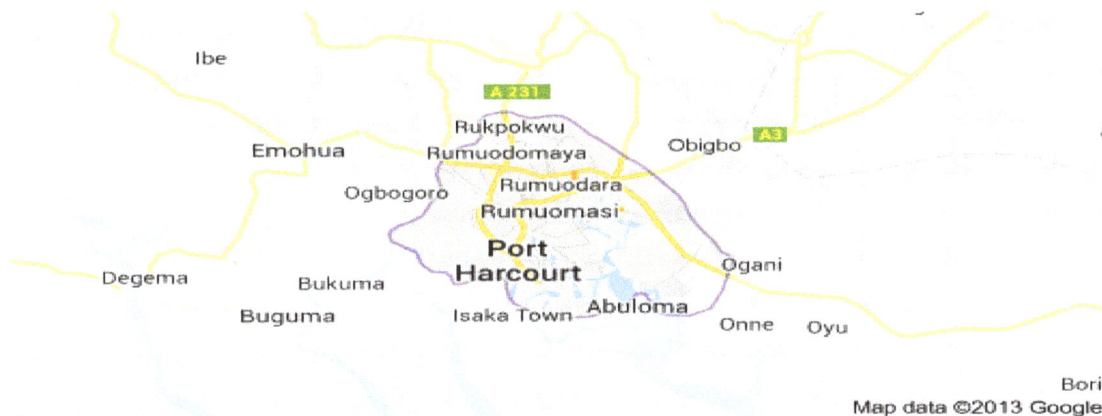

**Figure 1.** Map of Port Harcourt. Source: Google Map.

**Table 1.** Sunshine hours and relevant meteorological data for Port Harcourt.

| Month | January | February | March | April | May | June | July | August | September | October | November | December |
|---|---|---|---|---|---|---|---|---|---|---|---|---|
| S (h) | 4.30 | 4.24 | 3.84 | 4.41 | 4.85 | 3.74 | 2.28 | 2.36 | 3.30 | 4.07 | 5.08 | 5.37 |
| RH (%) | 53.55 | 56.64 | 64.82 | 70.36 | 75.73 | 78.09 | 81.64 | 81.18 | 79.36 | 74.91 | 63.55 | 53.91 |
| T (°C) | 11.60 | 10.96 | 9.81 | 8.95 | 8.04 | 7.32 | 6.51 | 6.44 | 6.90 | 8.04 | 9.46 | 11.26 |
| RF (mm) | 34.00 | 69.00 | 107.00 | 157.00 | 284.00 | 308.00 | 341.00 | 263.00 | 389.00 | 226.00 | 79.00 | 20.00 |
| W (m/s) | 2.95 | 3.18 | 3.40 | 3.36 | 2.19 | 3.18 | 3.09 | 3.43 | 2.97 | 2.36 | 2.23 | 2.15 |

year period) of the data in preparation for correlations are presented in Table 1. Multiple linear regression equation for estimating S with four parameters is as follows:

$$Y = a + bX_1 + CX_2 + dX_3 + eX_4$$

Where a......e, are the regression coefficients and $X_i$ is the correlated parameter. The estimated values were compared to measured values in each regression equation through correlation coefficient R and standard error of estimate σ (Akpabio et al., 2004).

## Correlations

Table 1 presents the various meteorological parameters. These parameters are all linked to sunshine hours in various degrees. In order not to overlook any particular parameter or group of parameters, multiple linear regressions of four parameters (RH, T, RF, and W) were employed to estimate the sunshine hours. Here S is the monthly average daily sunshine hours; RH is the monthly average daily relative humidity in percentage; RF is the monthly average daily rainfall in millimeters, W is the monthly average daily wind speed in m/s.
The various linear regression analyses are as follows

### One variable correlation

This correlation gives the highest value of R as 0.717 for T and lowest value of R as 0.547 for W.

$$S = 0.699 + 0.375T \quad (R = 0.717, \sigma = 0.70775) \tag{1}$$

$$S = 7.422 - 1.162W \quad (R = 0.547, \sigma = 0.85003) \tag{2}$$

### Two variable correlation

This correlation gives the highest value of R as 0.737 for T and W and lowest value of R as 0.689 for RH and RF.

$$S = 3.436 + 0.315T - 0.749W \quad (R = 0.791, \sigma = 0.65459) \tag{3}$$

$$S = 8.109 - 0.058RH - 0.3396RF \quad (R = 0.689, \sigma = 0.77560) \tag{4}$$

### Three variable correlation

This correlation gives the highest value of R as 0.797 for T, RF and W and lowest value of R as 0.737 for RH, T and RF.

$$S = 1.917 + 0.453T + 2.167RF - 0.783W \quad (R = 0.797, \sigma = 0.6854) \tag{5}$$

$$S = -13.674 + 0.123RH + 1.050T - 0.334RF \quad (R = 0.737, \sigma = 0.76735) \tag{6}$$

### Four variable correlation

$$S = -27.306 - 0.272RH + 1.806T - 0.281RH - 1.114W \quad (R = 0.863, \sigma = 0.6258) \tag{7}$$

## RESULTS AND DISCUSSION

MPE gives information on long term performance of the examined regression equation; a positive MPE value

**Table 2.** Comparison of estimated and measured sunshine Hours data for Port Harcourt.

| Month | S | Equation 1 | Equation 2 | Equation 3 | Equation 4 | Equation 5 | Equation 6 | Equation 7 |
|---|---|---|---|---|---|---|---|---|
| January | 4.30 | 5.05 | 3.99 | 4.88 | 4.99 | 4.94 | 5.08 | 3.83 |
| February | 4.24 | 4.81 | 3.73 | 4.51 | 4.80 | 4.54 | 4.78 | 4.22 |
| March | 3.84 | 4.38 | 3.47 | 3.98 | 4.31 | 3.93 | 4.56 | 4.09 |
| April | 4.41 | 4.06 | 3.52 | 3.74 | 3.97 | 3.68 | 4.33 | 4.07 |
| May | 4.85 | 3.71 | 3.72 | 3.58 | 3.60 | 3.68 | 3.99 | 4.03 |
| June | 3.74 | 3.44 | 3.73 | 3.36 | 3.46 | 3.41 | 3.51 | 3.37 |
| July | 2.28 | 3.14 | 3.83 | 3.17 | 3.20 | 3.19 | 3.09 | 2.96 |
| August | 2.36 | 3.11 | 3.44 | 2.90 | 3.30 | 2.72 | 2.99 | 2.35 |
| September | 3.30 | 3.29 | 3.97 | 3.38 | 3.35 | 3.56 | 3.20 | 3.16 |
| October | 4.07 | 3.71 | 4.68 | 4.20 | 3.67 | 4.20 | 3.91 | 4.75 |
| November | 5.08 | 4.25 | 4.83 | 4.75 | 4.39 | 4.63 | 4.05 | 4.43 |
| December | 5.37 | 4.92 | 4.92 | 5.37 | 4.97 | 5.38 | 4.77 | 5.18 |

**Table 3.** Error calculations.

| Equations | R | MBE | RMSE | MPE |
|---|---|---|---|---|
| $S = 0.699 + 0.375T$ | 0.717 | 0.0525 | 0.6464 | 4.7617 |
| $S = 7.422 - 1.162W$ | 0.547 | -0.00083 | 0.7755 | 5.4808 |
| $S = 3.436 + 0.315T - 0.749W$ | 0.791 | -0.00167 | 0.5665 | 2.7483 |
| $S = 8.109 - 0.058RH - 0.3396RF$ | 0.689 | 0.0175 | 0.6892 | 4.2380 |
| $S = 1.917 + 0.453T + 2.167RF - 0.783W$ | 0.797 | 0.00167 | 0.5601 | 2.6975 |
| $S = -13.674 + 0.123RH + 1.050T - 0.334RF$ | 0.737 | 0.0350 | 0.6272 | 3.9792 |
| $S = -27.306 - 0.272RH + 1.806T - 0.281RH - 1.114W$ | 0.863 | -0.1107 | 0.4665 | -1.1560 |

provides the average amount of over estimation in the calculated values while a negative MPE gives under estimation (Akpabio and Etuk, 2002). On the whole, a low MPE is desirable. The test on RMSE conveys information on the short term performance of the different equations since it enables a term –by – term comparison of the actual variations between the estimated and measured values. For more accurate estimation, lower values of RMSE should be obtained (Akpabio and Etuk, 2002). $R^2$ denotes the multiple coefficient of determination, which is a measure of how well the multiple regression equation fits the sample data. A perfect fit would result in $R^2 = 1$. A very good fit results in a value near 1. A very poor fit results in a value of $R^2$ close to 0. However, the $R^2$ has serious flaws, which is because, as more variables are included $R^2$ increases. This is not supposed to be so. Consequently, it is better to use the adjusted $R^2$ when comparing different multiple regression equations because it adjusts the $R^2$ value based on the number of variables and the sample size (Triola, 1998).

Equations (1), (3), (5) and (7) have the highest value of correlation coefficient while Equations (2), (4) and (6) have the lowest values of R. However, the applicability of the proposed correlations is tested by estimating the sunshine duration values for Port Harcourt location used in the analysis. Estimated values of sunshine duration for Port Harcourt along with the measured data are shown in Table 2. Inspection of the table shows that the models estimate sunshine hours fairly accurately.

A study of Table 3 indicates that based on the RMSE, Equation (7) produces the best correlation while Equation (2) gives the worst with larger value of RMSE. For MBE the result shows that Equation (2) is the best while Equation (7) is the worst. With respect to MPE, Equation (7) offers the best correlation while Equation (2) gives the worst.

Hence for Port Harcourt,

$$S = -27.306 - 0.272RH + 1.806T - 0.281RH - 1.114W$$

The value of $R^2 = 0.745$ in the equation indicates that 74.5% of the variation in sunshine hours can be explained by the relative humidity, temperature, rainfall and wind speed. Hence the adjusted $R^2$ value is 0.600. This shows that 60% of the variation in sunshine hours can be explained by the relative humidity, temperature, rainfall and wind speed.

Figure 2 shows plots of Equation (7) with the least value of RMSE together with the monthly average daily sunshine hours measured for eleven years. Equation (7)

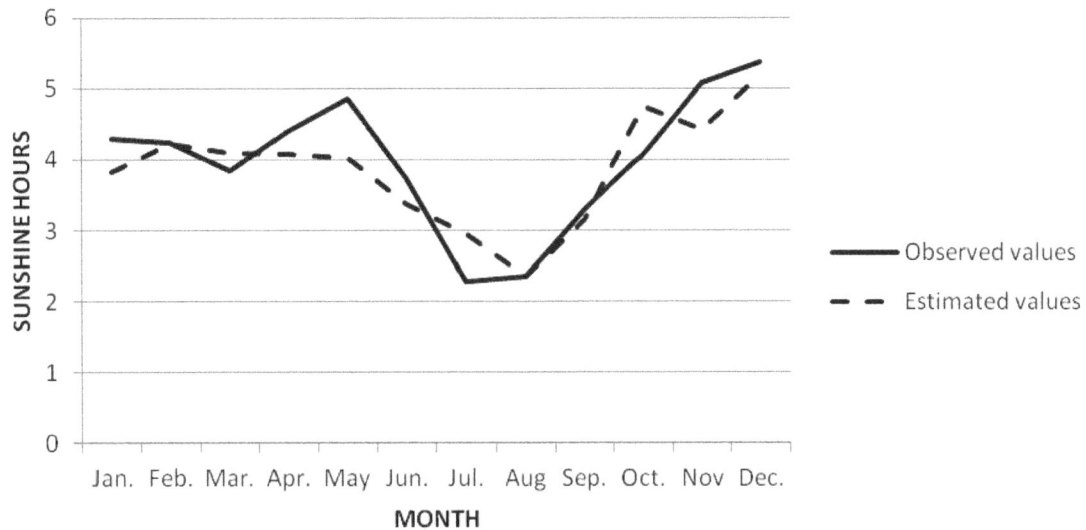

**Figure 2.** Comparison of measured and estimated data of monthly average daily sunshine hours for Port Harcourt, Nigeria.

shows almost exact fit to the sunshine hours data. Based on Equation 7, the values of global solar radiation (H) were computed and presented in Table 2.

## COMPUTATION OF GLOBAL SOLAR RADIATION

The average global solar radiation is such that certain constants are known, thus yielding a value measured in kilo joules per meter squared daily.

$H = H_o \{a + b[S/S_o]\}$ (Nwokoye, 2006).

Solar radiation cannot be easily measured at every place. All over the world, attempts are only being made to have solar radiation data computed based on measured meteorological data. The average daily global radiation on horizontal surface H, for a location is now possible if the sunshine hours S are measured and known (Nwokoye, 2006).

$H_o = 24/\pi * I_{sc} *[1 + 0.033\text{Cos} (360/365) *dn] * [(\omega\text{Sin}\varphi\text{Sin}\delta)+(\text{Cos}\varphi\text{Cos}\delta\text{Sin}\omega)]$

and

$W = \text{Cos}^{-1}(-\text{tan}\delta\text{tan}\varphi)$

$S_o = (2/15) \omega$ (Nwokoye, 2006).

## Conclusion

Multiple regressions was employed in this study to develop several correlation equations used to describe the dependence of sunshine hours on other meteorological data for Port Harcourt, Nigeria. The result shows that the four variable correlations which is the equation with the highest R give the best result when considering the error term (RMSE). Hence the multiple regression equation can be employed for the purpose of estimating sunshine hours for Port Harcourt and for locations that have the same climate and latitude as Port Harcourt. The equation with the least value of RMSE is

$S = -27.306 - 0.272RH + 1.806T - 0.281RH - 1.114W$

Based on Table 4, the greatest amount of global solar radiation was received in October (15.79 MJ/m$^2$) and the least amount of global solar radiation was received in August (11.21 MJ/m$^2$).

## Definition of terms

**RH =** Relative humidity
**T =** Difference in maximum and minimum temperature
**RF =** Rainfall
**W =** Wind speed
**MBE =** Mean bias error
**MPE =** Mean percentage error
**RMSE =** Root mean square error
**H =** Global solar radiation
**H$_O$ =** Daily extraterrestrial radiation
**H/H$_O$ =** Clearness index
**S =** Maximum sunshine duration or day length
**S$_O$ =** Daily sunshine duration
**S/S$_O$ =** Measure of cloud cover
**I$_{sc}$ =** Solar constant
**dn =** Day number

**Table 4.** Value of global solar radiation for Port Harcourt, Nigeria.

| Month | H (MJ/m$^2$) | H$_o$ (MJ/m$^2$) | S (h) | S$_o$ (h) | S/S$_o$ | H/H$_o$ |
|---|---|---|---|---|---|---|
| January | 12.99 | 34.56 | 3.83 | 11.75 | 0.33 | 0.38 |
| February | 14.47 | 36.44 | 4.22 | 11.85 | 0.36 | 0.40 |
| March | 14.85 | 38.06 | 4.09 | 11.97 | 0.34 | 0.39 |
| April | 14.80 | 38.63 | 4.07 | 12.11 | 0.34 | 0.38 |
| May | 14.62 | 38.17 | 4.03 | 12.23 | 0.33 | 0.38 |
| June | 13.10 | 37.71 | 3.37 | 12.28 | 0.27 | 0.35 |
| July | 12.36 | 37.90 | 2.96 | 12.26 | 0.24 | 0.33 |
| August | 11.21 | 38.56 | 2.35 | 12.16 | 0.19 | 0.29 |
| September | 13.08 | 38.44 | 3.16 | 12.03 | 0.26 | 0.34 |
| October | 15.79 | 37.10 | 4.75 | 11.89 | 0.40 | 0.43 |
| November | 14.44 | 35.10 | 4.43 | 11.77 | 0.38 | 0.41 |
| December | 15.45 | 34.02 | 5.18 | 11.72 | 0.44 | 0.45 |

$\omega$ = Sunset hour angle
$\delta$ = declination
$\varphi$ = Latitude of research location.

## ACKNOWLEDGEMENT

The authors are grateful to the Nigeria Meteorological Agency (NIMET), Oshodi, Lagos for providing all the necessary data.

## Conflict of Interests

The author(s) have not declared any conflict of interests.

### REFERENCES

Al-Dulaimy F, Al-Shahery G (2010). Estimation of global solar radiation on horizontal surfaces over Haditha, Samara, and Beji, Iraq. The Pacific J. Sci. Technol. 11(1):73-82.

Akpabio LE, Etuk SE (2002). Relationship between global solar radiation and sunshine duration for Onne, Nigeria. Turk J Phys 27:161-167.

Akpabio LE, Udo SO, Etuk SE (2004). Empirical correlations of global solar radiation with meteorological data for Onne, Nigeria. Turk. J. Phys. 28:205-212.

Hussein T, Ahmed T (2012). Estimation of hourly global solar radiation in Egypt using mathematical model. Int. J. Latest Trends Agr. Food Sci. 2(2):74-82.

Jakhrani A, Samo S, Rigit A, Kamboh S (2013). Selection of models for calculation of incident solar radiation on tilted surfaces. World Appl. Sci. J. 22(9):1334-1343.

Kaya M (2012). Estimation of global solar radiation on horizontal surface in Erzincan, Turkey. Int. J. Phys. Sci. 7(33):5273-5280.

Khan M, Ahmad M (2012). Estimation of global solar radiation using clear sky radiation in Yemen. J. Eng. Sci. Technol. Rev. 5(2):12-19.

Nwokoye AOC (2006). Solar energy technology, other alternative energy resources and environmental science. Rex Charles and Patrick Limited, Booksmith House, Anambra.

Triola MF (1998). Elementary Statistics; Adison Wesley Longman Inc, USA.

Umoh M, Akpan U (2011a). Horizontal global solar radiation based on sunshine hours over Enugu, Nigeria. Canadian J. Pure Appl. Sci. 5(2):1553-1557.

Umoh M, Akpan U (2011b). Estimating global solar radiation from sunshine hours for Uyo, Nigeria. Canadian J. Pure Appl. Sci. 5(1):1433-1437.

Umoh M, Udoh S (2010). Estimation of global solar radiation from sunshine hours for Warri, Nigeria, Nigeria. Global J. Environ. Sci. 9(1-2):51-56.

# Permissions

All chapters in this book were first published in JEEER, by Academic Journals; hereby published with permission under the Creative Commons Attribution License or equivalent. Every chapter published in this book has been scrutinized by our experts. Their significance has been extensively debated. The topics covered herein carry significant findings which will fuel the growth of the discipline. They may even be implemented as practical applications or may be referred to as a beginning point for another development.

The contributors of this book come from diverse backgrounds, making this book a truly international effort. This book will bring forth new frontiers with its revolutionizing research information and detailed analysis of the nascent developments around the world.

We would like to thank all the contributing authors for lending their expertise to make the book truly unique. They have played a crucial role in the development of this book. Without their invaluable contributions this book wouldn't have been possible. They have made vital efforts to compile up to date information on the varied aspects of this subject to make this book a valuable addition to the collection of many professionals and students.

This book was conceptualized with the vision of imparting up-to-date information and advanced data in this field. To ensure the same, a matchless editorial board was set up. Every individual on the board went through rigorous rounds of assessment to prove their worth. After which they invested a large part of their time researching and compiling the most relevant data for our readers.

The editorial board has been involved in producing this book since its inception. They have spent rigorous hours researching and exploring the diverse topics which have resulted in the successful publishing of this book. They have passed on their knowledge of decades through this book. To expedite this challenging task, the publisher supported the team at every step. A small team of assistant editors was also appointed to further simplify the editing procedure and attain best results for the readers.

Apart from the editorial board, the designing team has also invested a significant amount of their time in understanding the subject and creating the most relevant covers. They scrutinized every image to scout for the most suitable representation of the subject and create an appropriate cover for the book.

The publishing team has been an ardent support to the editorial, designing and production team. Their endless efforts to recruit the best for this project, has resulted in the accomplishment of this book. They are a veteran in the field of academics and their pool of knowledge is as vast as their experience in printing. Their expertise and guidance has proved useful at every step. Their uncompromising quality standards have made this book an exceptional effort. Their encouragement from time to time has been an inspiration for everyone.

The publisher and the editorial board hope that this book will prove to be a valuable piece of knowledge for researchers, students, practitioners and scholars across the globe.

# List of Contributors

**Ravi Pratap Singh Kushwah**
Department of Electronics, Madhav Institute of Technology and Science, Gwalior -474 005, India

**P. K. Singhal**
Department of Electronics, Madhav Institute of Technology and Science, Gwalior -474 005, India

**Sameh Oueslati**
Department of Physics, Laboratory of Signal Processing, Faculty of Sciences, University Tunis, El Manar 1060, Tunis
Department: Image and Information Processing, Higher National School of Telecommunication of Bretagne, Technopole of Brest Iroise, 29285 Brest – France

**Adnane Cherif**
Department of Physics, Laboratory of Signal Processing, Faculty of Sciences, University Tunis, El Manar 1060, Tunis

**Bassel Solaiman**
Department: Image and Information Processing, Higher National School of Telecommunication of Bretagne, Technopole of Brest Iroise, 29285 Brest – France

**P. Ramachandran**
Department of Electrical Engineering, Dr. MGR University, Chennai, India

**R. Senthil**
University College of Engineering, Villupuram Anna University, Chennai, India

**Abdoulaye M'bemba Camara**
Department of Electrical and Electronics Engineering, University of Fukui, Fukui 910-8507, Japan

**Yosuke Sakai**
Department of Electrical and Electronics Engineering, University of Fukui, Fukui 910-8507, Japan

**Hidehiko Sugimoto**
Department of Electrical and Electronics Engineering, University of Fukui, Fukui 910-8507, Japan

**Mohammad Golkhah**
Department of Electrical Engineering University of Manitoba, Winnipeg, Canada

**Sahar Saffar Shamshirgar**
Islamic Azad University of Sciences and Researches, Tehran, Iran

**Mohammad Ali Vahidi**
K. N. Toosi University of Technology, Tehran, Iran

**Lütfü SARIBULUT**
Department of Electrical-Electronics Engineering, Adana Science and Technology University, Adana/TURKEY

**Ahmet TEKE**
Department of Electrical-Electronics Engineering, Çukurova University, Adana/TURKEY

**Mohammad BARGHI LATRAN**
Department of Electrical-Electronics Engineering, Çukurova University, Adana/TURKEY

**Mehmet TÜMAY**
Department of Electrical-Electronics Engineering, Çukurova University, Adana/TURKEY

**Hayder A. Ahmed**
Physics Department, College of Science, Basrah University, Basrah, Iraq

**Arafat J. Jalil**
Physics Department, College of Science, Basrah University, Basrah, Iraq

**T. Ananthapadmanabha**
Department of Electrical and Electronics Engineering, The National Institute of Engineering, Mysore-08, Karnataka, India

**A. D. Kulkarni**
Department of Electrical and Electronics Engineering, The National Institute of Engineering, Mysore-08, Karnataka, India

**Benjamin A. Shimray**
Department of Electrical and Electronics Engineering, The National Institute of Engineering, Mysore-08, Karnataka, India

**R. Radha**
Department of Electrical and Electronics Engineering, The National Institute of Engineering, Mysore-08, Karnataka, India

**Manoj Kumar Pujar**
Manoj Kumar Pujar Bangalore Transmission Zone, KPTCL Karnataka India

**Jalal J. Hamad Ameen**
School of Electrical and Electronics Engineering, University Sains Malaysia (USM), Malaysia

**Widad Binti Ismail**
School of Electrical and Electronics Engineering, University Sains Malaysia (USM), Malaysia

**Vinodkumar Jacob**
Department of Electronics and Communication Engineering, MACE, Kothamangalam, 686666, Kerala, Ph. 91 98461 21223, India

**M. Bhasi**
School of Management Studies, Cochin University of Science and Technology, Kochi, Kerala, India

**R. Gopikakumari**
School of Engineering, Cochin University of Science and Technology, Kochi, Kerala, India

**Yinfang Xu**
National Key Laboratory of Antennas and Microwave Technology, Xidian University, Xi'an 710071, China

**Yongjun Xie**
National Key Laboratory of Antennas and Microwave Technology, Xidian University, Xi'an 710071, China

**Zhenya Lei**
National Key Laboratory of Antennas and Microwave Technology, Xidian University, Xi'an 710071, China

**Chao Deng**
National Key Laboratory of Antennas and Microwave Technology, Xidian University, Xi'an 710071, China

**Ahmed Nabih Zaki Rashed**
Department Electronics and Electrical Communication Engineering, Faculty of Electronic Engineering, Menouf 32951, Menoufia University, Egypt

**Lini Mathew**
Electrical Engineering Department, National Institute of Technical Teachers Training and Research Sector-26, Chandigarh, India

**Vivek Kumar Pandey**
Electrical Engineering Department, Bharat Institute of Technology By-Pass Road, Partapur, Meerut, U.P., India

**T. Ananthapadmanabha**
The National Institute of Engineering, Mysore, India

**R. Prakash**
HMS IT, Tumkur, India

**Manoj Kumar Pujar**
KPTCL, Bangalore, India

**Anjani Gangadhara**
The National Institute of Engineering, Mysore, India

**M. Gangadhara**
NIE, Mysore, India

**Tilak Thakur**
Department of Electrical Engineering, Punjab Engineering College, Deemed University, Chandigarh, India

**Jaswanti Dhiman**
Department of Electrical Engineering, Chandigarh College of Engineering and Technology (CCET), Chandigarh, India

**D. A. Shalangwa**
Department of Physics Adamawa State University, Mubi. Nigeria

**Manish Sharma**
Electronics and Communication Engineering Department, Maharaja Agrasen Institute of Technology, Sector -22, Rohini, Delhi -110086, India

**Rashmi Gupta**
Delhi Technological University (Formerly Delhi College of Engineering) Bawana Road, Delhi 110042, India

**Deepak Kumar**
Electronics and Communication Engineering Department, Maharaja Agrasen Institute of Technology, Sector -22, Rohini, Delhi -110086, India

**Rajiv Kapoor**
Delhi Technological University (Formerly Delhi College of Engineering) Bawana Road, Delhi 110042, India

**J. S. Parab**
Electronics Section, Department of Physics, Goa University, Goa, India

**R. S. Gad**
Electronics Section, Department of Physics, Goa University, Goa, India

**G. M. Naik**
Electronics Section, Department of Physics, Goa University, Goa, India

**Khairy Elbarbary**
Electrical Engineering Department, Modern Academy In Maadi, Maadi, Cairo, Egypt

**Hossam Eldin Abou-Bakr Badr**
Egyptian Armed Forces Electronic Warfare Department, Military Technical College, Cairo, Egypt

**Tarek Bahroun**
Electronic Warfare Department, Military Technical College, Cairo, Egypt

**Chakresh Kumar**
Department of Electronics and Communication Engineering, Tezpur University, Napaam-784028, Sonitpur, Assam, India

**Girish Narah**
Department of Electronics and Communication Engineering, Tezpur University, Napaam-784028, Sonitpur, Assam, India

**Aroop Sharma**
Department of Electronics and Communication Engineering, Tezpur University, Napaam-784028, Sonitpur, Assam, India

**K. El kourd**
Electronic institute, Biskra University, Algeria

**S. El kourd**
Electronic institute, Biskra University, Algeria

**A. Zangene**
Amirkabir University of Technology, Tehran, Iran

**H. R. Dalili Oskouei**
University of Aeronautical Science and Technology (Shahid Sattari), Tehran, Iran

**M. Nourhoseini**
Amirkabir University of Technology, Tehran, Iran

**Mfon David Umoh**
Department of Physics, University of Uyo, Uyo, Akwa Ibom State, Nigeria

**Sunday O. Udo**
Department of Physics, University of Calabar, Calabar, Cross River State, Nigeria

**Ye-Obong N. Udoakah**
Department of Electrical/Electronics Engineering, University of Uyo, Uyo Akwa Ibom State, Nigeria